AF581853

Analysis, Design and Fabrication of Micromixers, Volume II

Analysis, Design and Fabrication of Micromixers, Volume II

Editor

Kwang-Yong Kim

MDPI • Basel • Beijing • Wuhan • Barcelona • Belgrade • Manchester • Tokyo • Cluj • Tianjin

Editor
Kwang-Yong Kim
Department of Mechanical
Engineering
Inha University
Incheon
Korea, South

Editorial Office
MDPI
St. Alban-Anlage 66
4052 Basel, Switzerland

This is a reprint of articles from the Special Issue published online in the open access journal *Micromachines* (ISSN 2072-666X) (available at: www.mdpi.com/journal/micromachines/special_issues/micromixer_v2).

For citation purposes, cite each article independently as indicated on the article page online and as indicated below:

LastName, A.A.; LastName, B.B.; LastName, C.C. Article Title. *Journal Name* **Year**, *Volume Number*, Page Range.

ISBN 978-3-0365-6174-5 (Hbk)
ISBN 978-3-0365-6173-8 (PDF)

Contents

About the Editor

Kwang-Yong Kim

Kwang-Yong Kim received his B.S. degree from Seoul National University, and his M.S. and Ph.D. degrees from the Korea Advanced Institute of Science and Technology (KAIST), South Korea. He is currently an Emeritus Professor in the Department of Mechanical Engineering and was a Dean of Engineering College of Inha University, Incheon, Korea. He served as an associate editor of *ASME Journal of Fluids Engineering*, a co-editor-in-chief of the *International Journal of Fluid Machinery and Systems*, the editor-in-chief of the *Transactions of Korean Society of Mechanical Engineers*, the president of Korean Society for Fluid Machinery, and the chairman of the Asian Fluid Machinery Committee. He is currently an associate editor of *ASME Open Journal of Engineering* and an editorial board member of *Micromachines*. He is a fellow of the Korean Academy of Science and Technology, a fellow of National Academy of Engineering of Korea, a fellow of the American Society of Mechanical Engineers (ASME), an associate fellow of the American Institute of Aeronautics and Astronautics (AIAA), and a recipient of order of service merit, "Yellow Stripes"and order of science and technology merit, "Doyak Medal"from Republic of Korea. He is interested in applications of the numerical optimization techniques using various surrogate models and computational fluid dynamics to the designs of micro mixers, micro heat sinks, micro fuel cells, fluid machinery, heat-transfer augmentation devices, etc.

Preface to "Analysis, Design and Fabrication of Micromixers, Volume II"

Micromixers are an important component in micrototal analysis systems and lab-on-a-chip platforms which are widely used for sample preparation and analysis, drug delivery, and biological and chemical synthesis. The successful operation of microfluidic devices requires fast and adequate mixing, but mixing is a challenging task due to the laminar feature of the flow at the microscale. Mixing in laminar flows relies on diffusion and requires a longer channel to achieve complete mixing due to the slow process compared to that in turbulent flows. Hence, it is crucial to overcome this challenge to improve the mixing performance. Based on their mixing mechanism, micromixers are classified into two types: active and passive. Passive micromixers are easy to fabricate and generally use geometry modification to cause chaotic advection or lamination to promote the mixing of fluid samples, unlike active micromixers, which use moving parts or some external agitation/energy for the mixing. This reprint covers new mechanisms, numerical and/or experimental mixing analysis, design, and fabrication of various micromixers, which were reported by the Special Issue 'Analysis, Design and Fabrication of Micromixers II' published in *Micromachines*.

Kwang-Yong Kim
Editor

 *micromachines*

Editorial

Editorial for the Special Issue on Analysis, Design and Fabrication of Micromixers II

Kwang-Yong Kim 

Department of Mechanical Engineering, Inha University, Incheon 22212, Republic of Korea; kykim@inha.ac.kr; Tel.: +82-010-3641-3096

Citation: Kim, K.-Y. Editorial for the Special Issue on Analysis, Design and Fabrication of Micromixers II. *Micromachines* **2022**, *13*, 2176. https://doi.org/10.3390/mi13122176

Received: 4 December 2022
Accepted: 7 December 2022
Published: 8 December 2022

Micromixers are important components of lab-on-a-chip systems, and also have many biological and chemical applications [1]. Micromixers are generally classified into two categories, i.e., active and passive micromixers, depending on the mechanism of mixing enhancement. Our previous Special Issue (Special Issue on Analysis, Design and Fabrication of Micromixers I) [2] reported mostly on passive micromixers, in addition to two electrokinetic micromixers. However, except for two review papers [3,4], the current Special Issue places a greater emphasis on active mixing, reporting on five active [5–9] and six passive [10–15] micromixers. Three of the active micromixers have an electrokinetic driving force [5–7], but the other two are activated by mechanical mechanisms [8] and acoustic streaming [9]. Three studies [5,13,14] employ non-Newtonian working fluids, one of which deals with nano-non-Newtonian fluids [13]. Most of the cases investigate micromixer design, except for ref. [5]. Additionally, most of the passive micromixers have three-dimensional (3D) obstructions [10,12,14] or 3D channel shapes [11,13]. Seven studies [5,6,9,12–15] are numerical, two [7,10] are experimental, and the other two are experimental with numerical validation [8] and numerical with experimental validation [11].

Hejazian et al. [3] reviewed previous studies on microfluidic mix-and-jet devices, which are efficient for sample delivery in biomolecular applications. They introduced designs, characteristics, and fabrication techniques. Experimental techniques for the analyses of mixing and jetting are also summarized. They expect innovative designs of these devices to appear in applications in many other fields, such as basic studies on physics and chemistry, polymer fabrication, and the kinetics of nanoparticles. The review paper by Sahadevan et al. [4], which was selected as a feature paper and an Editor's Choice paper by this journal, reports on recent progress and future perspectives on microfluidic applications of artificial cilia, which have good capabilities in mixing, pumping, and particle handling. Microfluidics using artificial cilia can be an efficient tool to control fluid flow with high precision in lab-on-a-chip and other biomedical applications. They suggest that effective and economic manufacturing techniques must be developed to make artificial cilia more practical in these applications.

Banos et al. [5] examined the effects of combinations of non-Newtonian fluids, slippage, and finite-sized ions on the mixing of fluids in an oscillatory electroosmotic flow. They proposed a numerical model for the analysis of the mixing. The results revealed that the effects enhanced the mixing by about 90%. Shi et al. [6] proposed a novel electrokinetic micromixer with staggered electrodes based on light-actuated AC electroosmosis, and analyzed the mixing of fluids numerically using the finite element method. They performed a parametric study using parameters such as the width, length, and spacing of the electrodes, as well as the channel height to identify effects on mixing performance, and thus optimal micromixer geometry. Yang et al. [7] also proposed a new electrokinetic micromixer with a quasi T-channel and electrically conductive sidewalls. An experiment was carried out to find the effects of Reynolds number, AC voltage and frequency, and electric conductivity ratio on mixing performance. They suggested that the proposed micromixer showed higher mixing performance compared to the conventional micromixers with the electrodes located at the channel outlet.

Koike and Takayama [8] introduced a novel active micromixer design for the concentration control of the fluids. They developed a mechanical mechanism to produce concentration gradients in multiple main chambers, arranged parallel to the body channel. Each main chamber is surrounded by a driving chamber, and agitation is caused in the main chamber due to the expansion or contraction of the driving chamber. The fluid moving back and forth in the channel connecting the main chamber to the body channel controls the concentration. The concentration gradient was determined experimentally, but numerical simulation was also performed to confirm the mechanism of the micromixer. Although they could produce a concentration gradient, the concentration control was not so successful. An active micromixer using acoustic streaming was proposed by Endaylalu and Tien [9]. The acoustic streaming in a Y-junction microchannel was generated by introducing triangular structures at the channel walls. The mixing performance was evaluated numerically. A parametric study of the mixing performance was performed using the inlet velocity, triangular structure's vertex angle, and oscillation amplitude. The results suggested that installing the triangular structure at the junction enhanced mixing by causing acoustic streaming. Additionally, lower inlet velocity, higher oscillation amplitude, and smaller vertex angle also improved the mixing performance.

Oevreeide et al. [10] proposed a novel passive micromixer design with double-curved structures located on a surface of the microchannel. In the experiment, confocal imaging was used to find the flow patterns. Additionally, analyzing the fluorescence pattern monitored the development of homogenization. They suggested that introducing a second curved structure to the simple curved structure had a remarkable effect on mixing performance. Rouhi et al. [11] investigated a passive micromixer with a non-planar spiral microchannel with various cross-sections. The mixing performance was evaluated numerically in a Reynolds number (Re) range of 0.001–50. An experiment was also performed to validate the numerical solution. In the numerical test, a large-angle outward trapezoidal cross-section showed the best mixing performance of up to 95%. Juraeva and Kang [12] presented a passive micromixer design with eight mixing units. Each mixing unit was composed of four alternatively arranged baffles. The mixing units were stacked in a cross-flow direction, and four different arrangements were tested: one-, two-, four-, and eight-step stacking. The mixing was analyzed numerically in a range of Re from 0.5 to 50. The numerical results suggested that the highest mixing was obtained with the eight-step stacking for Re > 2, but with the two-step stacking for Re < 2.

Tayeb et al. [13] numerically analyzed the characteristics of nano-non-Newtonian fluids in a novel passive micromixer with a 3D channel structure. Mixing and heat transfer between two heated fluids were investigated. How the concentration of Al_2O_3 nanoparticles affects the pressure drop and thermal mixing for a Re range from 0.1 to 25 was tested. The results revealed that higher nanofluid concentration and stronger chaotic advection improved the hydrodynamic and thermal performances over the whole Re range remarkably. Mahammedi et al. [14] also numerically examined the mixing of non-Newtonian fluids in Kenics micromixers for Re = 0.1–500. The Y-shaped micromixer consisted of six helical elements arranged alternately. They suggested that for the carboxymethyl cellulose solutions with power–law indices of 0.49–1, the Kenics mixer showed high mixing for both low and high Reynolds numbers. Juraeva and Kang [15] reported on the mixing performance of a modified Tesla micromixer. The design was modified by introducing a tip clearance above the wedge-shaped divider. The mixing performance was analyzed numerically for Re = 0.1–80. In order to measure the mixing performance, the degree of mixing at the micromixer outlet and the pressure drop through the mixer were used. They suggested that the tip clearance introduced in the modified design enhanced the mixing performance in a wide Re range.

I appreciate all the authors who contributed to this Special Issue. Additionally, thanks also go to the reviewers for the submitted papers and the editorial staff who conducted the review process. Further contribution to this topic can be made to the Topical Collection "Micromixers: Analysis, Design and Fabrication" of this journal.

Conflicts of Interest: The author declares no conflict of interest.

References

1. Afzal, A.; Kim, K.-Y. *Analysis and Design Optimization of Micromixers*; SpringerBriefs in Computational Mechanics; Springer Nature Singapore: Singapore, 2021.
2. Kim, K.-Y. Editorial for the Special Issue on Analysis, Design and Fabrication of Micromixers. *Micromachines* **2021**, *12*, 533.
3. Hejazian, M.; Balaur, E.; Abbey, B. Recent Advances and Future Perspectives on Microfluidic Mix-and-Jet Sample Delivery Devices. *Micromachines* **2021**, *12*, 531.
4. Sahadevan, V.; Panigrahi, B.; Chen, C.-Y. Microfluidic Applications of Artificial Cilia: Recent Progress, Demonstration, and Future Perspectives. *Micromachines* **2022**, *13*, 735.
5. Baños, R.; Arcos, J.; Bautista, O.; Méndez, F. Steric and Slippage Effects on Mass Transport by Using an Oscillatory Electroosmotic Flow of Power-Law Fluids. *Micromachines* **2021**, *12*, 539.
6. Shi, L.; Ding, H.; Zhong, X.; Yin, B.; Liu, Z.; Zhou, T. Mixing Mechanism of Microfluidic Mixer with Staggered Virtual Electrode Based on Light-Actuated AC Electroosmosis. *Micromachines* **2021**, *12*, 744.
7. Yang, F.; Zhao, W.; Kuang, C.; Wang, G. Rapid AC Electrokinetic Micromixer with Electrically Conductive Sidewalls. *Micromachines* **2022**, *13*, 34.
8. Koike, F.; Takayama, T. Generation of Concentration Gradients by a Outer-Circumference-Driven On-Chip Mixer. *Micromachines* **2022**, *13*, 68.
9. Endaylalu, S.A.; Tien, W.-H. A Numerical Investigation of the Mixing Performance in a Y-Junction Microchannel Induced by Acoustic Streaming. *Micromachines* **2022**, *13*, 338.
10. Oevreeide, I.H.; Zoellner, A.; Stokke, B.T. Characterization of Mixing Performance Induced by Double Curved Passive Mixing Structures in Microfluidic Channels. *Micromachines* **2021**, *12*, 556.
11. Rouhi, O.; Bazaz, S.R.; Niazmand, H.; Mirakhorli, F.; Mas-hafi, S.; Amiri, H.A.; Miansari, M.; Warkiani, M.E. Numerical and Experimental Study of Cross-Sectional Effects on the Mixing Performance of the Spiral Microfluidics. *Micromachines* **2021**, *12*, 1470.
12. Juraeva, M.; Kang, D.-J. Mixing Performance of a Passive Micro-Mixer with Mixing Units Stacked in Cross Flow Direction. *Micromachines* **2021**, *12*, 1530.
13. Tayeb, N.T.; Hossain, S.; Khan, A.H.; Mostefa, T.; Kim, K.-Y. Evaluation of Hydrodynamic and Thermal Behaviour of Non-Newtonian-Nanofluid Mixing in a Chaotic Micromixer. *Micromachines* **2022**, *13*, 933.
14. Mahammedi, A.; Tayeb, N.T.; Kim, K.-Y.; Hossain, S. Mixing Enhancement of Non-Newtonian Shear-Thinning Fluid for a Kenics Micromixer. *Micromachines* **2021**, *12*, 1494.
15. Juraeva, M.; Kang, D.-J. Mixing Performance of the Modified Tesla Micromixer with Tip Clearance. *Micromachines* **2022**, *13*, 1375.

Review

Recent Advances and Future Perspectives on Microfluidic Mix-and-Jet Sample Delivery Devices

Majid Hejazian, **Eugeniu Balaur** and **Brian Abbey** *

ARC Centre of Excellence in Advanced Molecular Imaging, Department of Chemistry and Physics, La Trobe Institute for Molecular Sciences, La Trobe University, Melbourne, VIC 3086, Australia; m.hejazian@latrobe.edu.au (M.H.); e.balaur@latrobe.edu.au (E.B.)

* Correspondence: b.abbey@latrobe.edu.au

Abstract: The integration of the Gas Dynamic Virtual Nozzle (GDVN) and microfluidic technologies has proven to be a promising sample delivery solution for biomolecular imaging studies and has the potential to be transformative for a range of applications in physics, biology, and chemistry. Here, we review the recent advances in the emerging field of microfluidic mix-and-jet sample delivery devices for the study of biomolecular reaction dynamics. First, we introduce the key parameters and dimensionless numbers involved in their design and characterisation. Then we critically review the techniques used to fabricate these integrated devices and discuss their advantages and disadvantages. We then summarise the most common experimental methods used for the characterisation of both the mixing and jetting components. Finally, we discuss future perspectives on the emerging field of microfluidic mix-and-jet sample delivery devices. In summary, this review aims to introduce this exciting new topic to the wider microfluidics community and to help guide future research in the field.

Keywords: microfluidics; micro-mixer; micro-jet; XFEL; molecular imaging; sample delivery

Citation: Hejazian, M.; Balaur, E.; Abbey, B. Recent Advances and Future Perspectives on Microfluidic Mix-and-Jet Sample Delivery Devices. *Micromachines* **2021**, *12*, 531. https://doi.org/10.3390/mi12050531

Academic Editor: Kwang-Yong Kim

Received: 19 April 2021
Accepted: 4 May 2021
Published: 7 May 2021

Publisher's Note: MDPI stays neutral with regard to jurisdictional claims in published maps and institutional affiliations.

1. Introduction

The three-dimensional structure determination of biological molecules is a critical step for understanding the dynamics of biological reactions and is essential for rational drug design [1]. The emergence of X-ray Free-Electron Lasers (XFELs) has facilitated the measurement of complex protein structures and the associated dynamics of biomolecular systems with atomic resolution [2]. These experiments require rapid and precise delivery of the liquid sample to the X-ray interaction region in order to capture the structural changes that occur in biomolecules on sub-microsecond to millisecond timescales [3]. The use of microfluidic mix-and-jet devices capable of triggering reactions and delivering liquid samples to the X-ray beam via a free-standing jet has become a reliable technique for solving the structure of biomolecules. The free-standing jet provides a continuous supply of liquid sample solution to the high-intensity X-ray pulses whilst minimising background diffraction noise and radiation damage [4]. Over the past decade, innovative fabrication techniques have led to numerous efficient sample delivery solutions using microfluidic technology capable of both rapid mixing and the creation of a free-standing liquid jet [5]. Additionally, cryogenic Electron Microscopy (cryo-EM) is another well-established experimental technique for studying the structure of biomolecules and their dynamic conformational changes [6]. Microfluidic sample delivery devices have also been employed for pre-mixing and deposition of liquid samples onto cryo-EM grids for time-resolved studies [7].

The Gas Dynamic Virtual Nozzle (GDVN) is currently the most commonly employed method for focusing and accelerating liquid sample streams and creating free-standing liquid micro-jets [8]. Injectors that take advantage of the GDVN principle of flow focusing can be categorised based on their fabrication method, i.e., capillary, lithography-based microfluidics, and 3D printed nozzles. A capillary GDVN usually comprises two co-axial capillaries, e.g., fused silica and ceramic hollow capillaries. The inner capillary carries the liquid sample, and the end of the outer capillary is tapered to further focus the gas flow and create a liquid micro-jet [4]. Lithography-based microfluidics nozzles, which often employ high resolution and replicable lithography techniques in their fabrication, have also recently been of interest. Lithography-based microfluidics technology offers greater flexibility in terms of microchannel geometry design beyond conventional capillary-based approaches, allowing for multiple microfluidic components to be integrated onto a single chip. By implementing this method for microfluidic jetting, the GDVN nozzle is fabricated as one of the features of the lithography-based microfluidics alongside other components, e.g., a micro-mixer, on the same chip.

The recently developed 3D printed microfluidic technology offers a host of advantages over standard methods. It enables the creation of low-cost and rapid prototyping of microfluidic devices with intricate 3D designs, which can be readily adjusted at minimal additional effort [9]. Despite the current fabrication challenges [10,11], 3D printing for microfluidic device fabrication has been rapidly moving toward becoming the dominant microfluidic fabrication method for numerous biochemical and biomedical research projects [12–14].

Microfluidic mixers can be classified as either 'active' or 'passive' mixers. Passive micromixers often use complex channel geometries in order to amplify the chaotic advection effect. Passive micromixers are usually integrated into microfluidic sample delivery devices to rapidly mix the solutions and to trigger a reaction before the mixed solution is delivered via the liquid microjet. The chaotic advection effect in passive micromixers maximises the contact surface for mass transfer between the mixed solutions, and consequently increases the overall mixing efficiency [15]. A variety of different designs for passive microfluidic mixers have been reported in the literature [16–18]. For example, Lee et al. [19] have previously systematically reviewed the most common passive micromixer designs and summarised their operational principles and mixing performance. Recently, Raza et al. [18] presented a comparative review based on quantitative analyses of a wide range of different types of passive micromixers, which included looking at their mixing efficiencies, pressure drops, and fabrication costs.

Mix-and-jet microfluidics is an emerging field that has shown a promising capability for implementation as a sample delivery platform for molecular imaging applications. Here, we review the most recent trends in microfluidic jetting, particularly on-chip microfluidics. We first discuss the main parameters for designing the GDVN component. Next, we discuss each of the leading fabrication approaches' limitations and challenges, namely: capillary, on-chip, and 3D printed microfluidics. We then explore different experimental methods for the characterisation of both free-standing liquid jets and integrated micromixers. Finally, we highlight the future potential and opportunities for microfluidic jetting, specifically in the context of molecular imaging applications.

2. Design Considerations

The principle of the GDVN is based on hydrodynamic focusing, which relies on squeezing a continuous fluid stream using a sheath flow, with a different velocity, as depicted in Figure 1. The surrounding sheath fluid, which is injected around the core stream, shapes the core fluid meniscus into a steady micro or nano-jet, which has a smaller size than the outlet microchannel [20]. This section of the review introduces the main design parameters that should be considered when designing GDVN nozzles and passive micro-mixers.

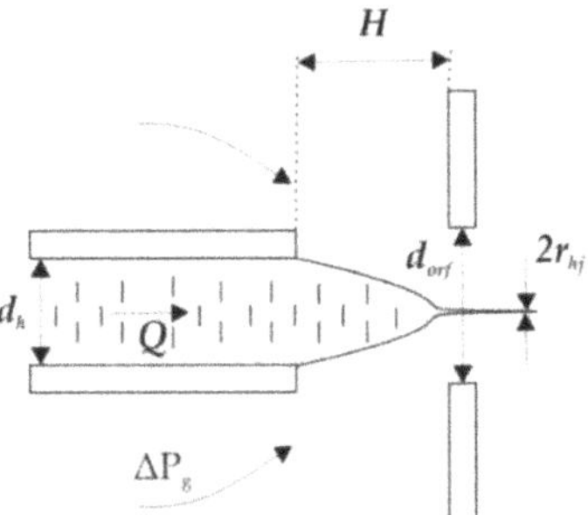

Figure 1. Schematic of the microfluidic nozzle illustrating the main design parameters (reproduced with permission from Reference [21]).

2.1. Main Parameters for Nozzle Design

Ganan-Calvo et al. [21–23] have described the relative effect of the principle geometrical and flow parameters on jetting stability. Aside from the fluid property parameters, the key parameters that determine the jetting regime with a GDVN nozzle are shown in Figure 1. Liquid with a flow rate of Q is injected through the sample microchannel, which has a hydrodynamic diameter of d_h. The liquid meniscus is accelerated via a pressure drop in the gas stream and is hydrodynamically focused in order to form a jet, exiting an orifice with a hydrodynamic diameter of d_{orf}. The distance from the sample microchannel to the orifice is H, and the hydrodynamic radius of the jet is r_{hj}.

The velocity of the jet can be expressed as

$$V_{jet} = \frac{4Q}{\pi r_{hj}{}^2} \tag{1}$$

and the pressure drop of the gas sheath flow is given by

$$\Delta P_g = \frac{\rho_l V_{jet}^2}{2} \tag{2}$$

where ρ_l is the density of the liquid. The Reynolds number (Re) is the ratio of the inertial forces versus the viscous forces within the liquid stream and is defined as

$$Re = \frac{\rho_l Q}{\pi r_{hj} \mu} \tag{3}$$

where μ is the viscosity of the liquid.

The Weber number (We), which is the dimensionless ratio of inertial forces to surface tension forces, is expressed as

$$We = \frac{\rho Q^2}{\pi^2 r_{hj}{}^3 \sigma} \tag{4}$$

where σ is the surface tension of the liquid. The Weber number must be >1 in order to produce a stable jet [24].

Vega et al. [21] experimentally and numerically investigated the effect of these key parameters on jetting stability and mapped the regions of stability and instability using the dimensionless We and Re numbers. From their stability/instability maps, stable jetting tends to occur at relatively higher We and Re numbers, i.e., where the inertial force is dominant. They reported that the transition from unstable to stable regions is mainly determined by the jet dynamics rather than the geometrical parameters. The relative effect of the fluid property parameters, i.e., density, viscosity, and surface tension, are more dominant when working with viscous liquids and very thin liquid jets. Vega et al. also reported that the optimum values for H and Q increase with increasing orifice diameter.

2.2. Main Parameters for Mixer Design

Imaging the dynamics of biomolecules requires sample delivery devices that incorporate micro-mixers capable of efficient mixing with sub-millisecond and millisecond mixing times [25]. The main parameters that are used for the design and evaluation of efficient mixers for integration into mix-and-jet sample delivery devices are the Reynolds number (Re), Peclet number (Pe), and the mixing efficiency (η_{mixing}) [19].

The Peclet number is the ratio of convective mass transport rate to the diffusion mass transport rate and is given by:

$$Pe = \frac{ul}{D} \tag{5}$$

where l is the length of the mixing path, and D is the diffusion coefficient. The chaotic advection effect that is induced by the geometry of the passive mixer microchannels leads to local increases in the velocity and, consequently, an increase in the Pe values.

The evaluation of mixing in microchannels is usually achieved by measuring the degree of mixing at different cross-sections within the mixing channel. This can be quantified using the normalised concentration, c^*, defined as

$$c^* = \frac{c - c_{min}}{c_{max} - c_{min}} \tag{6}$$

where c is the concentration of the species in solution, and the subscripts indicate the minimum (min) and maximum (max) concentration values.

The mixing efficiency (η_{mixing}) is defined as

$$\eta_{mixing} = 1 - \sqrt{\frac{1}{N}\sum_{i=1}^{N}\left(\frac{c_i^* - c_m^*}{c_i^*}\right)^2} \tag{7}$$

where N is the number of sampling points, c_i^* is the normalised concentration at point i, and c_m^* is the mean normalised concentration.

3. Fabrication Methods

There are numerous techniques that are used for the fabrication of integrated mix-and-inject devices. The materials used to fabricate these devices are specially chosen in order to address the key challenges of sample solution compatibility and mechanical stability. Based on the particular fabrication method, the devices can be fabricated with either planar or circular microchannels.

3.1. Co-Axial Capillary Devices

The use of co-axial capillary nozzles for the acceleration of a laminar liquid stream to create microscopic free-standing jet flows was first introduced by Ganan-Calvo [22]. Since then, various innovative developments in GDVNs have been reported for their implementation as sample delivery devices for serial femtosecond crystallography (SFX) using XFELs [8].

The capillary devices typically consist of two co-axial glass capillaries, which are co-aligned to form a GDVN. For a typical capillary-based GDVN [26], the liquid sample capillary has an outer diameter of about 50 µm and is tapered at the end. The surrounding co-axial capillary that allows the gas sheath stream to pass has an inner diameter of around 70 µm; with an average liquid sample flowrate of around 10 µL/min, this results in the creation of a liquid jet with a diameter of 4 µm. The conventional fabrication of the GDVN nozzles involves fabricating the individual nozzles by hand, which can result in inconsistent device characteristics. An example of the lengthy 6-step fabrication procedure, shown in Figure 2, was described in detail by Calvey et al. [27,28], which involves multiple flattening, polishing, tapering, and centring steps that require access to custom-designed chucks and jigs. However, using glass capillaries has the significant advantage of high pressure and solution pH resistance and, as a result, has been the most commonly used method for sample delivery for molecular imaging with XFEL [8,29].

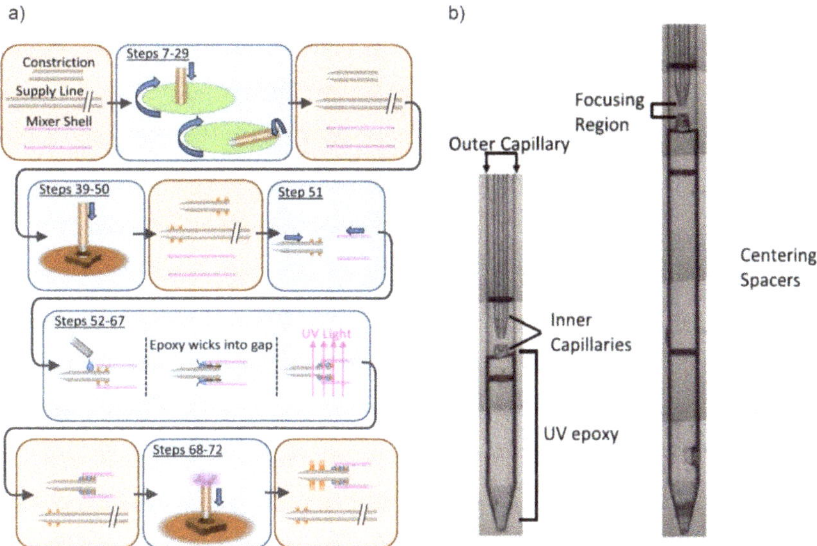

Figure 2. Glass capillary-based GDVN devices. (**a**) overview of the multi-step device fabrication process (reprinted with permission from Reference [28]). (**b**) Optical composite images of the completed devices (reproduced with permission from Reference [27] and Reference [28]).

Beyerlein et al. [30] introduced an easier and faster manufacturing technique for fabricating capillary nozzles based on ceramic micro-injection moulding. Their method offers a higher resolution (~1 μm) and reproducibility whilst having the advantage of also working at lower flow rates and being stable with respect to high pressures, making them compatible with SFX experiments. Zahoor et al. [31,32] performed comprehensive Computational Fluid Dynamics (CFD) studies of ceramic GVDN's based on the Volume of Fluid (VOF) and Finite Volume Method (FVM). Their simulation covers a wide parameter space of liquid Reynolds numbers within the ranges of 17–1222, different geometrical parameters, and Weber numbers in the range of 3–320, which can be used for adjustment of the nozzle geometry design and operating conditions for specific liquid samples. However, the ceramic nozzle moulding method also has significant disadvantages, including the high manufacturing cost of the micro-injection moulding tools and misalignment of the inner capillary.

3.2. Lithography-Based Microfluidics

Soft lithography using Polydimethylsiloxane (PDMS) is another conventional method for the fabrication of microfluidic devices. This fabrication approach is fast, high resolution, reproducible, and cost-effective, and allows for the fabrication of high aspect ratio microchannels. Trebbin et al. [33] first reported mix-and-inject microfluidic devices fabricated using a 3-layer bonding PDMS technique, as shown in Figure 3. The technique enables the fabrication of microchannels with different depths, integrating the GDVN nozzle onto a single microfluidic chip, and producing nozzle arrays. Their microfluidic chip could generate liquid jets with diameters ranging between 0.9 and 20 μm. They reported a range of jet diameter, jet length, and the operating conditions under which their devices were able to produce stable jetting under both atmospheric and vacuum conditions.

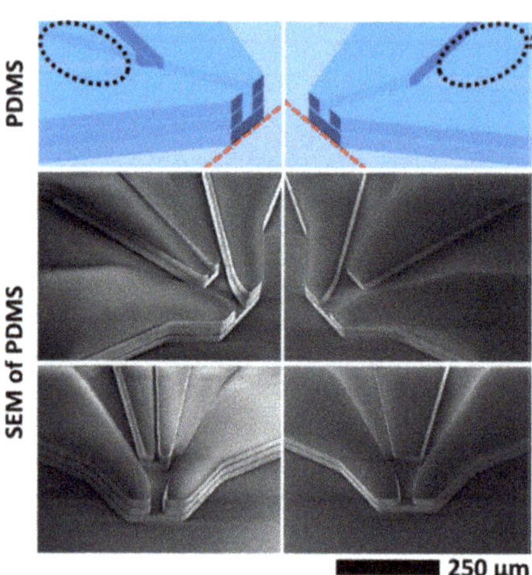

Figure 3. Illustration of the multilayer 3D PDMS microfluidic GDVN devices (reproduced with permission from Reference [33]).

Feng et al. [34] reported the fabrication of a microfluidic sprayer based on a two-layer PDMS technique for use as an alternative to the conventional pipetting/blotting method of cryo-EM for depositing liquid sample droplets on the EM grid. They reported that by changing the sprayer-grid distance and gas pressure, the ice thickness of the droplets could be controlled. Their proposed micro sprayer has the potential to be implemented for solving the structure of apoferritin using single-particle cryo-EM at high resolution.

Microfluidic GDVN nozzles have also been implemented for the production of microfibers. Zhao et al. [35] used soft lithography with PDMS to fabricate a double flow-focusing nozzle microfluidic chip for the production of microfibers. The double-nozzle technique, using DI-water as a sheath flow, prevents drop formation near the exit of the nozzle and generates a continuous stream of microfibers into the atmosphere. Hofmann et al. [36] implemented the same multilayer PDMS bonding technique used by Trebbin et al. [33] to fabricate a microfluidic nozzle device for the generation of ultrafine fibres. Their approach takes advantage of the GDVN principle, which leads to the creation of a steady and continuous stream of uniform microfibers. Precise control over the microfiber diameter and morphology could be achieved by adjusting the air pressure and solution flow rate.

Devices made using the soft lithography method with PDMS suffer from low solvent and pressure resistance, which are significant disadvantages when they are employed for molecular imaging using synchrotron and XFELs, compared to the original glass capillary-based GDVNs. Marmiroli et al. [37] presented a micromachining technique using X-ray lithography to engrave 60 μm thick channels into polymethyl-methacrylate (PMMA) slides, as demonstrated in Figure 4. They used finite element simulations to optimise the geometrical parameters in order to combine a micromixer with a free-standing liquid jet for time-resolved molecular studies at sub-0.1 ms resolution. Their microfluidic injectors were employed for synchrotron small-angle X-ray scattering (SAXS) measurements studying the formation of calcium carbonate from calcium chloride and sodium carbonate. The fastest recorded dynamics that they were able to track occurred on a timescale of just 75 μs.

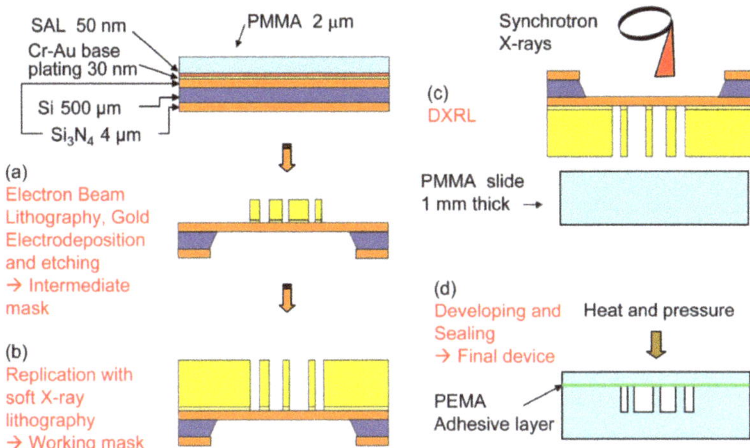

Figure 4. Schematic of the micromixer fabrication process from Marmiroli et al. [37]. (**a**) The production of an intermediate X-ray mask using Electron beam lithography, (**b**) replication of the mask by soft X-ray lithography, (**c**) fabrication of deep micromixer channels on PMMA using Deep X-Ray Lithography (DXRL), (**d**) adhesive bonding of the device. (reproduced with permission from Reference [37]).

Koralek et al. [38] proposed a microfluidic glass chip fabricated using standard hard lithography to create sub-micron liquid sheets, as depicted in Figure 5. They performed optical, infrared, and X-ray spectroscopies to measure the thickness of the liquid sheet, which was found to range from approximately 20 nm to around 1 μm. The liquid sheet was stable for flow rates between 150 and 250 μL/min and a gas flow rate of around 100 SCCM. The nanometer-thick sheet could have transformative potential for applications in infrared, X-ray, electron spectroscopy studies.

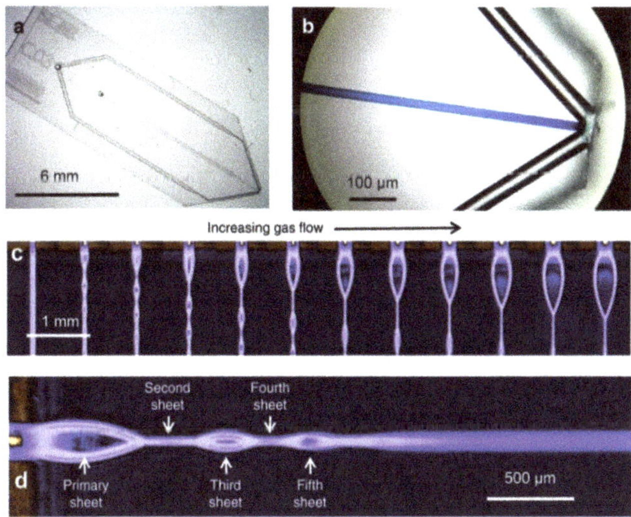

Figure 5. Microfluidic GDVN for ultrathin liquid sheet generation. (**a**) The microfluidic chip (6 × 19 mm) with gas and liquid ports incorporated, (**b**) liquid and gas microchannel can be distinguished via the introduction of blue dye into the liquid channel, (**c**) the jet regime varies as a function of gas pressure, (**d**) a detailed view of the alternating orthogonal liquid sheet structure (reproduced with permission from Reference [38]).

Hejazian et al. [39–41] have proposed a novel SU8 on glass technique to fabricate mix-and-inject devices suitable for experiments at both the synchrotron and XFEL. The use of SU8 for the fabrication of the microchannels provides high chemical inertness and X-ray stability, which is further supported by a glass body to increase the mechanical rigidity making it suitable for enduring high pressures. The microchannels were made using high-resolution photolithography, which offers reproducibility and facilitates the fabrication of serpentine-shaped mixer microchannel structures. The integration of a planar passive micromixer demonstrated a superior mixing performance compared to a straight channel micromixer. Schematics of the 3D design of the jig, the liquid jetting, and the mixing component are demonstrated in Figure 6. Furthermore, they observed three distinct jetting regimes, including the ultrathin liquid sheets reported by Koralek et al. [38], which are achievable by only adjusting the operating conditions with a single device.

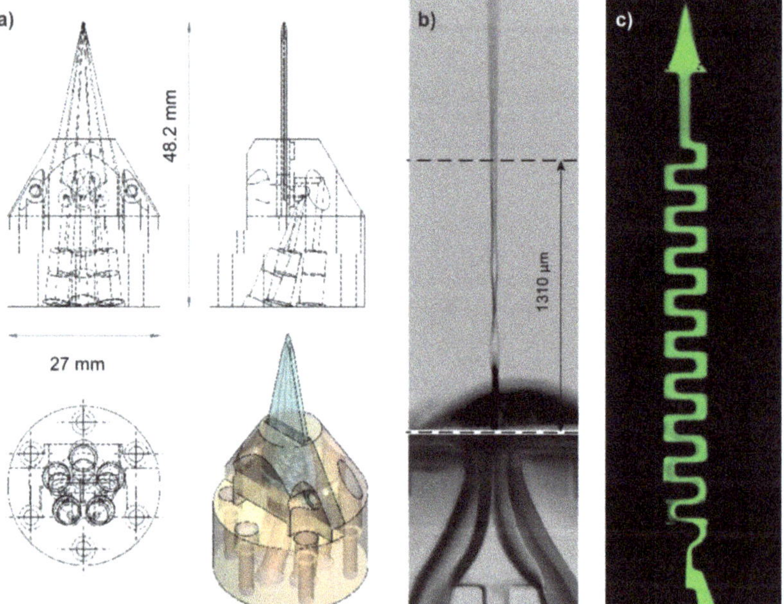

Figure 6. The SU8 on glass microfluidic mix-and-jet devices, (**a**) 3D schematics showing how the microfluidic chip is interfaced to tubing using a custom-made jig, (**b**) the ribbon regime created by the microfluidic mix-and-jet devices under gas flow rates ranging from 162 to 234 mg/min and liquid flow rates of 80 to 100 µL/min, (**c**) mixing of water and a diluted fluoresceine salt solution in the serpentine mixing component (reproduced with permission from Reference [39]).

Vakili et al. [42] presented a prototyping technique based on laser ablation of Kapton® polyimide foils for the fabrication of a microfluidic chip GDVN, shown in Figure 7. Kapton® foils of 125 µm thickness were micromachined using a 193 nm argon fluoride (ArF) excimer laser and bonded to each other using hot embossing. The use of Kapton® sheets has the advantages of having high chemical inertness and x-ray transparency which makes these devices ideal for serial crystallography experiments at synchrotrons and XFELs as well as SAXS measurements at these facilities.

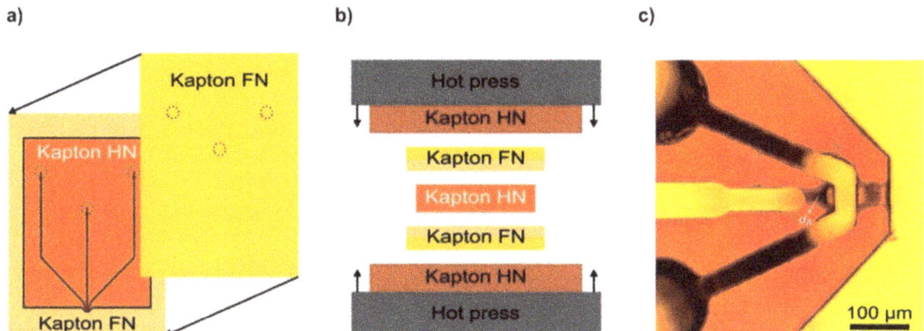

Figure 7. Schematic of the fabrication process of Kapton® GDVN devices. (**a**) Alignment and hot embossing bonding of the Kapton® foils, (**b**) stacking order of the bonding procedure, (**c**) microscopic image of the finished GDVN device showing gas and liquid microchannels (reproduced with permission from Reference [42]).

3.3. Three-Dimensional Printed Microfluidic Devices

The fabrication processes discussed above typically involve time-consuming manual steps and have only limited capability for making complex true 3D micro-features. The 3D-printing fabrication technique has recently gained attention as a fully digital and automated rapid-prototyping method for producing small batches of customised microfluidic devices [43,44]. The technique also reduces assembly work due to the capability of printing chip holders and the chip-to-tubing connections [45].

Despite the current challenges in microfluidic 3D printing [11], for example, the comparatively low throughput, there have been numerous successful reports on the utilisation of submicron resolution 2-photon polymerisation (2PP) 3D printing techniques and using IP-S resist (Nanoscribe GmbH, Karlsruhe, Germany) printing material for the fabrication of mix-and-inject devices. Nelson et al. [46] introduced a 3D printed GDVN sample delivery device for time-resolved studies using an XFEL. They implemented a submicron resolution 2PP 3D printing technique to fabricate nozzle tips, which were glued to gas and liquid capillaries. The off-axis jetting of their first device was corrected by adjusting the design of the nozzle tip, characterised using X-ray tomography, achieving a straight jet. The 3D printing method was able to overcome the geometrical constraints of conventional fabrication methods, and their device was able to achieve stable jetting with a gas pressure lower than for glass GDVNs.

Galinis et al. [47] introduced 3D printed nozzles to create a stable thin liquid sheet jet in a vacuum, using high resolution (0.2 µm) direct two-photon laser writing. They used a custom-made plate holder for batch printing of the nozzles and the average printing time for each nozzle was around 2 h. The devices could withstand pressures of up to 8 bars and achieve a jet thickness within the range of 1.02–4.58 µm at 9.1 mL/min under both vacuum and normal atmospheric conditions. Wiedorn et al. [48] used the 3D printed nozzle design first reported in Nelson et al. [46] for high-resolution structure determination of hen egg-white lysozyme (HEWL) microcrystals (6–8 µm in diameter) using megahertz serial femtosecond crystallography (SFX) at the SPB/SFX beamline at the European XFEL. The 3D printed devices were able to deliver the sample using high-speed liquid jets with a 1.8 µm diameter at speeds of between 50 and 100 m/s to match the megahertz repetition rate, which is equivalent to a total of 150–1200 pulses per second. They found that the high-speed jet speeds produced by the 3D printed sample delivery device combined with the megahertz beamline could significantly reduce sample consumption and the data acquisition time.

Bohne et al. [49] have reported on a hybrid fabrication method consisting of 2PP 3D printing the nozzle head onto a 2D microfluidic silicon-glass chip fabricated via lithography. The method omits the assembly steps for connecting the nozzle tip to the liquid sample and gas channels, which previously required gluing of the capillaries to the nozzle tip. The device was capable of generating stable jets under atmospheric and vacuum conditions, with a 1.5 μm diameter at a liquid flow rate of 1.5 μL/min, and a more than 20 μm diameter at a flow rate of 100 μL/min. The hybrid method allowed for integrating multiple microfluidic components on a single chip to make custom-designed sample delivery devices to suit a particular sample's characteristics.

Nazari et al. [50] employed 2PP 3D printing with an IP-S material to fabricate a GDVN nozzle which had an asymmetric design. Their method was able to achieve submicron resolution printing with a printing time of between 35 min to a few hours. The device could establish stable jets with speeds greater than 170 m/s, which is suitable for MHz XFEL experiments. They systematically characterised the liquid jets produced by the device and reported the jet diameter, length, speed, and Weber number as a function of the gas sheath flow rate by using a dual-pulsed nanosecond image acquisition and analysis method.

Knoska et al. [51] reported an optimised 2-photon stereolithography 3D printing technique achieved by adjusting the print resolution during fabrication to reduce the printing time for mix-and-inject sample delivery devices to minutes. The designed assembly and the liquid jet created by the device are demonstrated in Figure 8. The devices could achieve submicron jets with jet speeds higher than 200 m/s, suitable for megahertz time-resolved structural biology studies at XFELs. They also fabricated and tested a double-orifice nozzle for creating narrower jets with a reduced sample consumption for liquid jet samples. They introduced the X-ray microtomography technique for the characterisation of their 3D millisecond mixer component of the devices. Their 3D integrated micromixer consisted of a series of 180° turn helical elements that facilitate high mixing efficiencies, minimising inertial forces to avoid damaging the microcrystals in the liquid sample whilst maintaining a constant cross-section to prevent blockage of the device. The fabrication methods and their advantages and limitations that were discussed in Section 3 are summarised in Table 1.

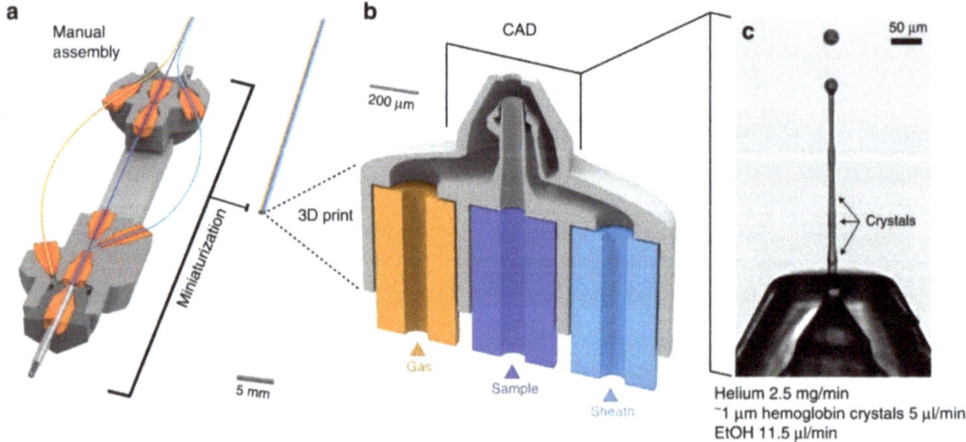

Figure 8. Three-dimensional printed double-flow-focusing GDVN. (**a**) The design assembly consisting of inserting three glass capillaries into a machined 10 cm long aluminium body, (**b**) 3D schematics of the gas orifice with three capillary inlets for gas, liquid sample, and sheath flow, (**c**) the 3D printed nozzle in operation jetting a solution containing 3 μm Hemoglobin crystals (reproduced with permission from [51]).

Table 1. Summary of fabrication methods (see Section 3) along with their pros and cons.

Fabrication Method	Pros	Cons
Co-axial capillary devices fabricated via glass extrusion [27,28]	High pressure and solution pH resistance and uses well-established fabrication methods.	Arduous manual intervention required during fabrication and assembly; poor reproducibility.
Co-axial capillary devices fabricated via ceramic micro-injection moulding [30]	Good reproducibility and reduced fabrication complexity compared to glass co-axial capillary devices.	Manual intervention required during fabrication, processing, and device assembly.
Microfluidic injector devices fabricated in PDMS [33–36]	Straight forward fabrication protocols, reproducible results, high spatial resolution.	Lack of mechanical stability and chemical inertness. Can only handle low pressures.
Deep X-Ray Lithography (DXRL) in PMMA [37]	Reproducible fabrication and high resolution.	Requires access to a synchrotron beamline; low PH resistance due to using PMMA.
Microfluidic glass chip fabrication using hard lithography [38]	High spatial resolution and reproducibility. Chemically and mechanically robust.	Costly manufacturing processes involving a high degree of complexity.
Microfluidic SU8 on glass lithographic fabrication [39–41]	Simple fabrication achieving high resolution combined with chemical and mechanical inertness and design lexibility.	Requires additional micromachining to produce the device inlet and outlet.
Laser ablation of Kapton® polyimide fims [42]	High resolution, and high chemical and mechanical inertness.	Manual alignment required during fabrication employing laser micromachining.
Microfluidic devices fabricated via 3D nanoprinting [46–51]	Automated rapid-prototyping, high spatial resolution, and reproducibility possible.	Requires manual assembly and use of glass capillaries, limited flexibility in terms of geometry.

4. Characterisation Techniques

The fabricated devices require lab testing and calibration before being implemented as sample delivery devices for applications such as molecular imaging at XFEL facilities. In this section, we summarise the standard methods that were used for the characterisation of mixing and jetting within integrated microfluidic mix-and-jet sample delivery devices.

4.1. Jetting Analysis

The analysis of jetting is mostly conducted through microscopic imaging of the microjet to map the stable and unstable regions as a function of the operating parameters, e.g., gas pressure and liquid flow rate. Vega et al. [21] used a Complementary Metal-Oxide-Semiconductor (CMOS) high-speed video camera (Photonfocus MV-D1024-160F, Photonfocus AG, Lachen, Switzerland) to image the fluid meniscus and the jet. The nozzle was illuminated using an optical fibre connected to a light source. An auxiliary charge-coupled devices (CCD) camera, positioned perpendicularly with respect to the CMOS camera, was used to assess the asymmetricity of the flow focusing by acquiring images of the liquid meniscus. The imaging setup was mounted on an optical table with a pneumatic antivibration isolation system for analysing both the stability of the liquid jets and the behaviour of the liquid meniscus.

Galinis et al. [47] measured the thickness of thin liquid sheet jet flows created by their 3D printed nozzles using white light interferometry using a 633 nm He–Ne laser. An optical fibre was used to guide the light into a spectrometer (OceanOptics HR4000, 200–1100 nm, Ocean Optics, Inc., Largo, FL, USA), and the peak values of the focused white light spectral interferograms were used to determine the absolute thickness and flatness of the liquid sheet under both atmospheric pressure and in vacuum conditions.

Beyerlein et al. [30] examined jet stability and break up using a Photron FASTCAM SA4 camera (Photron PTY Ltd., Tokyo, Japan), with a frame rate of 500,000 frames per second and a shutter speed of 1 microsecond, for imaging. The high-speed camera was equipped with a 12× Ultra-Zoom motorised lens and a 10× objective lens to provide a resolution ranging from 0.3 to 3 μm/pixel to image both the jetting (with speeds of up to 20 m/s) and the 10 μm droplets, created after the jet break-up. Schematics of the experimental setup are depicted in Figure 9. Their fast-imaging setup takes advantage of an illumination source consisting of a Karl Storz xenon lamp generating a uniform background, which was coupled to a pulsed laser source to improve the time resolution. Bohne et al. [49] used the same setup as Beyerlein et al. [30] for measurements of liquid jets created by their 3D printed device under atmospheric pressure conditions. They used an environmental scanning electron microscope (SEM, EVO MA 25, Carl Zeiss AG, Oberkochen, Germany) to conduct the measurements under near-vacuum conditions of 100 Pa.

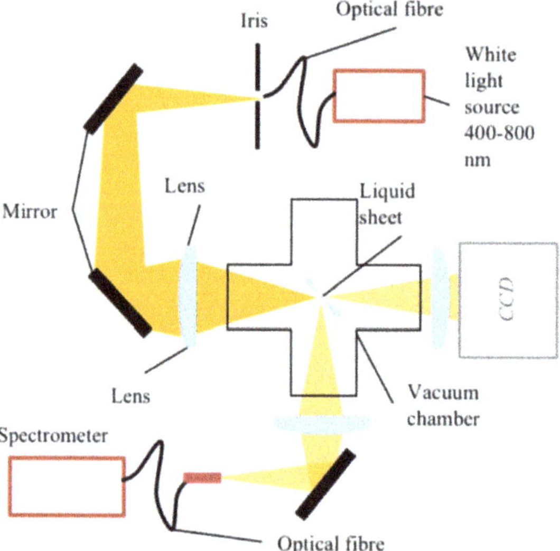

Figure 9. The layout of the testing station used to characterise the nozzle jetting performance under vacuum (reproduced with permission from Reference [30]).

Knoska et al. [51] carried out submicron jet diameter measurements, with jet speeds of over 200 m/s, using a nanosecond double flash imaging. The laser light generated by the dual-pulse laser system (Nano S 50-20 PIV, Litron Lasers, Rugby, Warwickshire, England, UK) illuminated Rhodamine 6G dye (252433, Sigma–Aldrich, St. Louis, MO, USA) to determine the jet velocities directly from the recorded images. They measured jet diameters as small as 536 nm, at a liquid flow rate of 2.4 μL/min, and a gas flow rate of 22.5 mg/min, using their imaging setup.

4.2. Mixing Analysis

The most common and straightforward method for investigating mixing in microfluidic devices is using a Confocal Fluorescence Microscope (CFM) for fluorescent imaging and then analysing the fluorescent intensity profiles to determine the mixing efficiency. Fang et al. [52] implemented a CFM for imaging and quantifying the 3D mixing patterns in microfluidic mixer devices, as shown in Figure 10. They used fluorescent intensity analysis to quantify the mixing efficiencies of the micro-mixers. In addition, they captured clear fluorescent images of the mixing patterns, which demonstrate flow advection and mass exchange. Inguva et al. [53] proposed a high-speed velocimetry technique for measuring fluid speeds of up to 10 m/s in microchannels to study chaotic mixing in microfluidic devices. They implemented a CFM equipped with a water-immersed Olympus UPLSAPO 60XW (Olympus Corp, Shinjuku City, Tokyo, Japan) objective to image a diffraction-limited confocal volume. Velocity profiles were established from analysing the data acquired at different depths within the micro-mixer. Their experimental method could measure fluid speeds with a 20% margin of error.

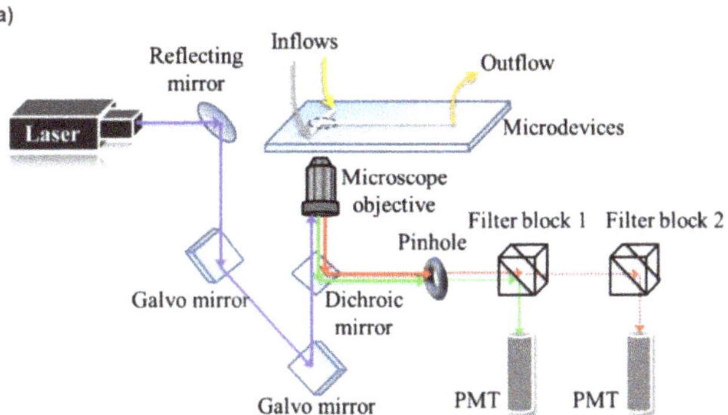

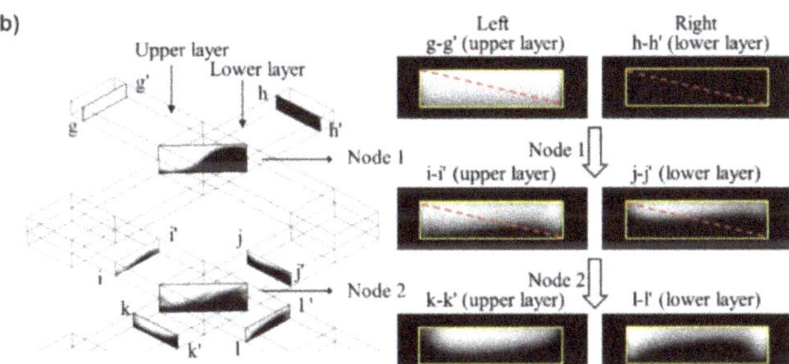

Figure 10. Confocal Fluorescence Microscope (CFM) for fluorescent imaging for mixing analysis. (**a**) Schematic diagram of the CFM system used by Fang et al., (**b**) schematics of the microfluidic mixer channel, and the cross-sectional fluorescent images depicting the progression of chaotic mixing along the mixer (reproduced with permission from Reference [52]).

Xi et al. [54] used Optical Coherence Tomography (OCT) to examine and compare mixing efficiencies of three micro-mixers: a Y channel mixer, a 3D serpentine mixer, and a vortex mixer. They reported that significantly more accurate estimations of mixing efficiencies and flow velocity profiles could be achieved using the OCT method. The visual overlap of fluid flows when using confocal microscopy results in more accurate estimations of mixing efficiencies. Jiang et al. [55] used two-photon fluorescence lifetime imaging microscopy to visualise and study millisecond chaotic mixing dynamics inside microdroplets in an integrated droplet-based microfluidic serpentine mixer device. A flow rate of 1 μL/min was used for the sample streams, and 1.5 μL/min for the sheath flows with a 50×40 μm^2 microchannel cross-section. Their fluorescent intensity analysis results show that the mixing efficiency inside the droplets can reach up to 80% after 18 ms.

Witkowski et al. [56] utilised micro-Particle Image Velocimetry (micro-PIV) to map the velocity profiles within passive micromixers. The images were acquired using an inverted laboratory microscope equipped with a 5.5-megapixel resolution camera. A laser light beam illuminates the fluorescent particles with a diameter of 1 μm, suspended in the carrier liquid flowing through the mixer microchannel. Yang et al. [57] proposed a method to simultaneously determine both the velocity and concentration profiles in microfluidic devices using micro-PIV and particle counting. They used a confocal fluorescence microscope for imaging the flow of microparticles inside the mixer microchannel. They used two distinct algorithms to track the displacement of microparticles for velocity profile determination whilst counting particles of different colours for resolving the concentration distribution.

Huyke et al. [58] investigated both small time scales of mixing and homogenous residence times of a co-axial hydrodynamic focusing mixer using a fluorescein–iodide quenching reaction. A 50 mM fluoresceine sample was first hydrodynamically focused by a buffer sheath and then focused again by a 500 mM KI sheath, as shown in Figure 11. All solutions contained 20 mM Tris and 10 mM HeCl (Sigma–Aldrich, St. Louis, MO, USA) at a measured pH of 8. The mixing times for the fast-quenching fluorescent reaction were determined by fluorescent imaging using an inverted microscope equipped with a suitable illumination source. The mixing efficiencies were then quantified by analysing the fluorescent intensity of the acquired images. The mixing and jetting characterisation techniques that were discussed in Section 4 are summarised in Tables 2 and 3.

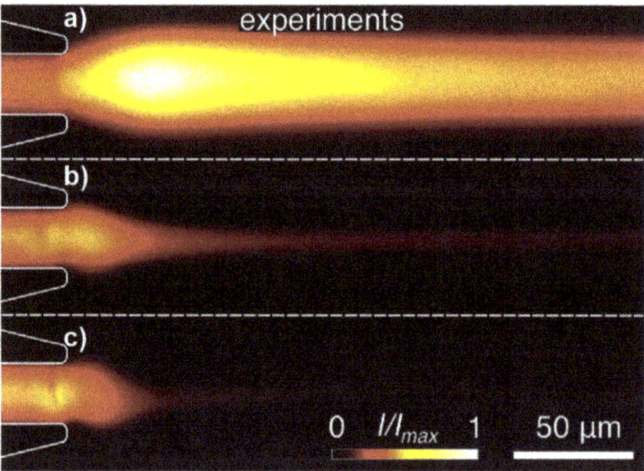

Figure 11. Experimental images of a hydrodynamic focusing mixer evaluated with the fluorescein-iodide quenching reaction technique for three sheath flow rates (Q_{sh}) to sample flow rate (Q_{sa}) ratios (Q_{sh}/Q_{sa}), (**a**) flow rate ratio of 100, (**b**) flow rate ratio of 1000, (**c**) flow rate ratio of 5000. At lower sheath to sample flow rate ratios, the sheath species diffuse into the sample stream (reprinted with permission from Reference [58]).

Table 2. Summary of the techniques used to characterise liquid jetting (see Section 4).

Method	Schematic	Comments
Complementary Metal-Oxide-Semiconductor (CMOS) high-speed video camera [21]	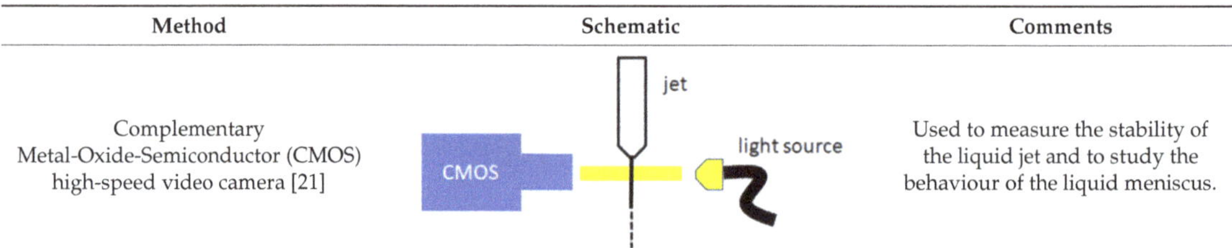	Used to measure the stability of the liquid jet and to study the behaviour of the liquid meniscus.

Table 2. *Cont.*

Method	Schematic	Comments
White-light interferometry [47]	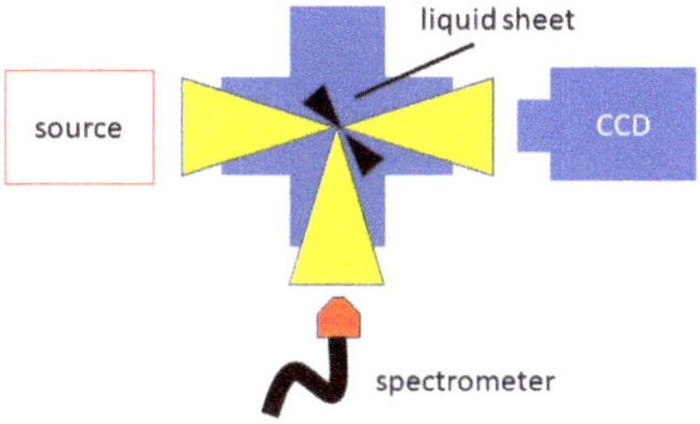 source; liquid sheet; CCD; spectrometer	Measures the absolute thickness and 'flatness' of the liquid sheet under both atmospheric pressure and vacuum conditions.
High-speed microscopic imaging [30,49]	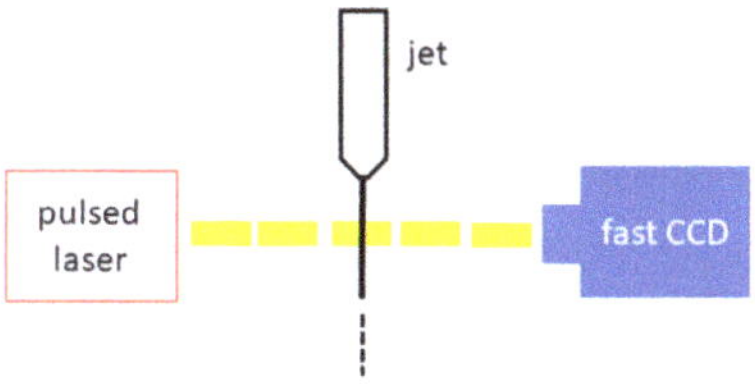 jet; pulsed laser; fast CCD	Used to study the liquid jet stability and the break up of the jet into microdroplets.
Nanosecond double flash imaging [51]	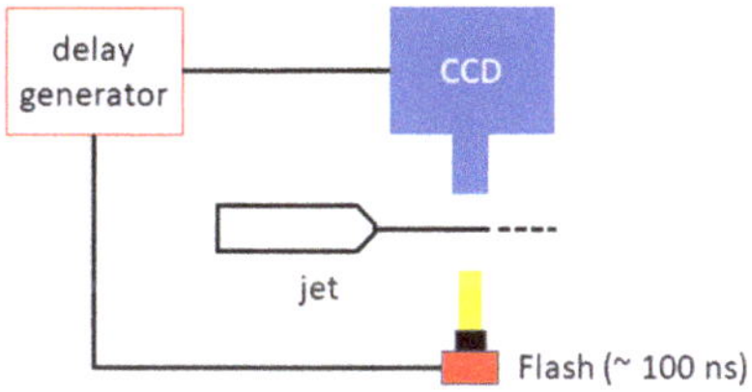 delay generator; CCD; jet; Flash (~ 100 ns)	Used to determine the jet velocity and jet diameter.

Table 3. Summary of the techniques used to characterise microfluidic mixing (see Section 4).

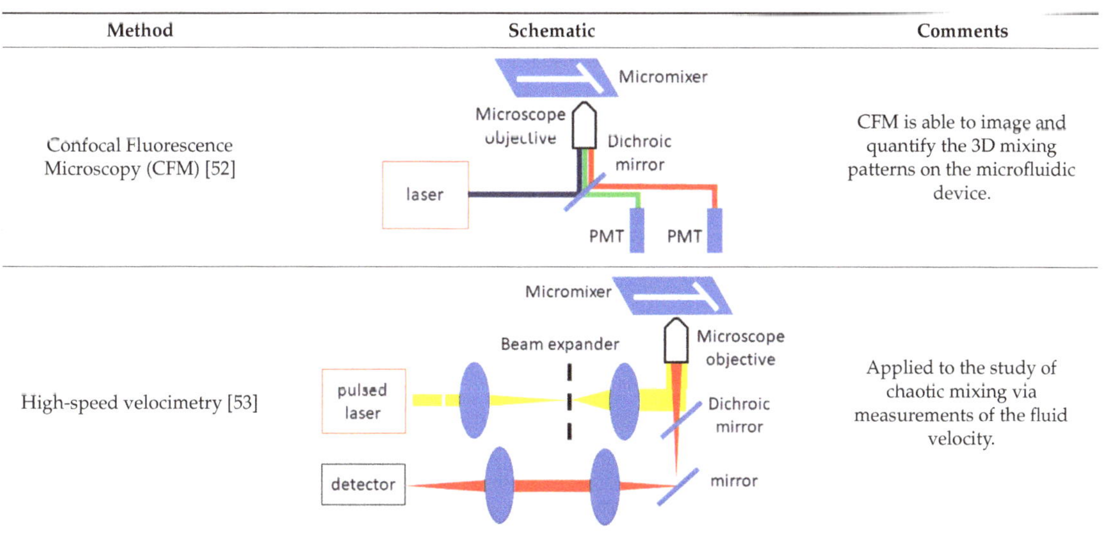

Method	Schematic	Comments
Confocal Fluorescence Microscopy (CFM) [52]	Micromixer; Microscope objective; Dichroic mirror; laser; PMT; PMT	CFM is able to image and quantify the 3D mixing patterns on the microfluidic device.
High-speed velocimetry [53]	Micromixer; Beam expander; Microscope objective; pulsed laser; Dichroic mirror; detector; mirror	Applied to the study of chaotic mixing via measurements of the fluid velocity.

Table 3. *Cont.*

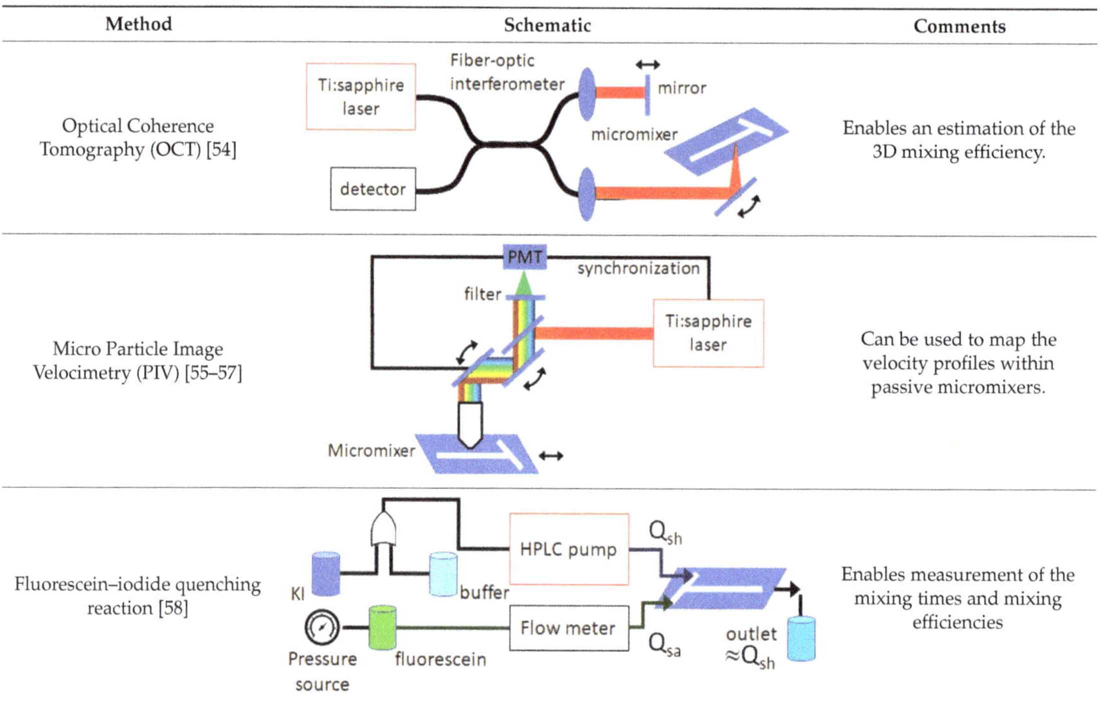

Method	Schematic	Comments
Optical Coherence Tomography (OCT) [54]		Enables an estimation of the 3D mixing efficiency.
Micro Particle Image Velocimetry (PIV) [55–57]		Can be used to map the velocity profiles within passive micromixers.
Fluorescein–iodide quenching reaction [58]		Enables measurement of the mixing times and mixing efficiencies

5. Summary and Perspectives

The recent advances in molecular imaging techniques using cryo-EM, XFEL, and synchrotron facilities necessitates the precise and controlled delivery of mixed solutions. Microfluidic technology has shown promise in addressing the sample delivery needs for molecular imaging technology over recent decades. Here, we have reviewed the recent advances in the emerging field of integrated mix-and-jet microfluidic sample delivery devices.

We introduced the main parameters required for the design of these integrated devices. The nozzle component is mainly designed based on the GDVN principle and integrated into the microfluidic device to generate free-standing liquid jets. The primary dimensionless parameters to be considered for the nozzle design and characterisation of the jet are *We* and *Re*. Passive micromixers are commonly used to trigger biomolecular reactions, taking advantage of chaotic advection and rapid millisecond mixing. The main dimensionless parameters to be considered for the design of a passive mixing component are *Re* and *Pe*, whilst η_{mixing} can characterise the mixing in the mixer microchannel. Additionally, we critically reviewed the techniques used for the fabrication of the mix-and-inject devices. Conventional capillary-based methods for the fabrication of the sample delivery devices are laborious and irreproducible, providing only limited versatility to integrate complex passive micromixers. Numerous techniques for the fabrication of chip-based microfluidic mix-and-inject devices were reported to replace the previous capillary-based techniques. Most of the chip-based planar methods enable the fabrication of rigid and chemically inert devices whilst taking advantage of the design freedom, high resolution, and reproducibility. Recently, 3D printed mix-and-jet microfluidic devices have shown great promise for XFEL single-particle imaging and SFX studies. The new technology facilitates fast and low-cost fabrication of fully 3D mixer and nozzle components that outperform both capillary and on-chip sample delivery devices. Furthermore, we summarised the standard experimental techniques used for the characterisation of both mixing and jetting. For these measurements, both high-speed optical imaging and fluorescent signal analysis were used.

Incorporating GDVN nozzles with microfluidics technology is still a new concept that will open up a host of new applications in many areas, especially in the biological and life sciences. Currently, most of the published references in this field are proof-of-concept of mix-and-inject experiments in which new device architectures and designs are often introduced. In the near future, we can expect to see more reports describing innovative designs and solutions to apply these devices to a range of different fields, including fundamental chemistry and physics, polymer fabrication, the study of the kinetics of nanoparticles, and biomolecular imaging.

Author Contributions: M.H. conceived of the presented idea, developed the structure of the paper, and wrote the manuscript. B.A. and E.B. supervised the work. All authors discussed the review paper and contributed to the final manuscript. All authors have read and agreed to the published version of the manuscript.

Funding: This research was funded by the Australian Research Council Centre of Excellence in Advanced Molecular Imaging, grant number CE140100011.

Data Availability Statement: The data that support the findings of this study are available from the corresponding author upon reasonable request.

Acknowledgments: This work was performed in part at the Melbourne Centre for Nanofabrication (MCN) in the Victorian Node of the Australian National Fabrication Facility (ANFF). The authors would like to acknowledge the support of the Australian Research Council (ARC) Centre of Excellence in Advanced Molecular Imaging.

Conflicts of Interest: The authors declare no conflict of interest.

References

1. Zhao, F.-Z.; Sun, B.; Yu, L.; Xiao, Q.-J.; Wang, Z.-J.; Chen, L.-L.; Liang, H.; Wang, Q.-S.; He, J.-H.; Yin, D.-C. A novel sample delivery system based on circular motion for in situ serial synchrotron crystallography. *Lab Chip* **2020**, *20*, 3888–3898. [CrossRef] [PubMed]
2. Chapman, H.N. X-Ray Free-Electron Lasers for the Structure and Dynamics of Macromolecules. *Annu. Rev. Biochem.* **2019**, *88*, 35–58. [CrossRef] [PubMed]
3. Fenwick, R.B.; Esteban-Martín, S.; Salvatella, X. Understanding biomolecular motion, recognition, and allostery by use of conformational ensembles. *Eur. Biophys. J.* **2011**, *40*, 1339–1355. [CrossRef] [PubMed]
4. Ghazal, A.; Lafleur, J.P.; Mortensen, K.; Kutter, J.P.; Arleth, L.; Jensen, G.V. Recent advances in X-ray compatible microfluidics for applications in soft materials and life sciences. *Lab Chip* **2016**, *16*, 4263–4295. [CrossRef]
5. Grunbein, M.L.; Nass Kovacs, G. Sample delivery for serial crystallography at free-electron lasers and synchrotrons. *Acta Crystallogr. Sect. D* **2019**, *75*, 178–191. [CrossRef]
6. Ognjenović, J.; Grisshammer, R.; Subramaniam, S. Frontiers in Cryo Electron Microscopy of Complex Macromolecular Assemblies. *Annu. Rev. Biomed. Eng.* **2019**, *21*, 395–415. [CrossRef]
7. Banerjee, A.; Bhakta, S.; Sengupta, J. Integrative approaches in cryogenic electron microscopy: Recent advances in structural biology and future perspectives. *iScience* **2021**, *24*, 102044. [CrossRef]
8. Echelmeier, A.; Sonker, M.; Ros, A. Microfluidic sample delivery for serial crystallography using XFELs. *Anal. Bioanal. Chem.* **2019**, *411*, 6535–6547. [CrossRef]
9. He, Y.; Wu, Y.; Fu, J.-Z.; Gao, Q.; Qiu, J.-J. Developments of 3D Printing Microfluidics and Applications in Chemistry and Biology: A Review. *Electroanalysis* **2016**, *28*, 1658–1678. [CrossRef]
10. Nielsen, A.V.; Beauchamp, M.J.; Nordin, G.P.; Woolley, A.T. 3D Printed Microfluidics. *Annu. Rev. Anal. Chem.* **2020**, *13*, 45–65. [CrossRef]
11. Waheed, S.; Cabot, J.M.; Macdonald, N.P.; Lewis, T.; Guijt, R.M.; Paull, B.; Breadmore, M.C. 3D printed microfluidic devices: Enablers and barriers. *Lab Chip* **2016**, *16*, 1993–2013. [CrossRef]
12. Tasoglu, S.; Folch, A. Editorial for the Special Issue on 3D Printed Microfluidic Devices. *Micromachines* **2018**, *9*, 609. [CrossRef]
13. Weisgrab, G.; Ovsianikov, A.; Costa, P.F. Functional 3D Printing for Microfluidic Chips. *Adv. Mater. Technol.* **2019**, *4*, 1900275. [CrossRef]
14. Mehta, V.; Rath, S.N. 3D printed microfluidic devices: A review focused on four fundamental manufacturing approaches and implications on the field of healthcare. *Bio Des. Manuf.* **2021**. [CrossRef]
15. Cai, G.; Xue, L.; Zhang, H.; Lin, J. A Review on Micromixers. *Micromachines* **2017**, *8*, 274. [CrossRef]
16. Suh, Y.K.; Kang, S. A Review on Mixing in Microfluidics. *Micromachines* **2010**, *1*, 82. [CrossRef]

17. Mansur, E.A.; Ye, M.; Wang, Y.; Dai, Y. A State-of-the-Art Review of Mixing in Microfluidic Mixers. *Chin. J. Chem. Eng.* **2008**, *16*, 503–516. [CrossRef]
18. Lee, C.-Y.; Chang, C.-L.; Wang, Y.-N.; Fu, L.-M. Microfluidic mixing: A review. *Int. J. Mol. Sci.* **2011**, *12*, 3263–3287. [CrossRef]
19. Lee, C.-Y.; Wang, W.-T.; Liu, C.-C.; Fu, L.-M. Passive mixers in microfluidic systems: A review. *Chem. Eng. J.* **2016**, *288*, 146–160. [CrossRef]
20. Dziubinski, M. Hydrodynamic Focusing in Microfluidic Devices. In *Advances in Microfluidics*; Kelly, R.T., Ed.; IntechOpen: London, UK, 2012; pp. 29–54. [CrossRef]
21. Vega, E.J.; Montanero, J.M.; Herrada, M.A.; Gañán-Calvo, A.M. Global and local instability of flow focusing: The influence of the geometry. *Phys. Fluids* **2010**, *22*, 64105. [CrossRef]
22. Gañán-Calvo, A.M. Generation of Steady Liquid Microthreads and Micron-Sized Monodisperse Sprays in Gas Streams. *Phys. Rev. Lett.* **1998**, *80*, 285–288. [CrossRef]
23. Gañán-Calvo, A.M. Jetting–dripping transition of a liquid jet in a lower viscosity co-flowing immiscible liquid: The minimum flow rate in flow focusing. *J. Fluid Mech.* **2006**, *553*, 75–84. [CrossRef]
24. Wiedorn, M.O.; Awel, S.; Morgan, A.J.; Ayyer, K.; Gevorkov, Y.; Fleckenstein, H.; Roth, N.; Adriano, L.; Bean, R.; Beyerlein, K.R.; et al. Rapid sample delivery for megahertz serial crystallography at X-ray FELs. *IUCrJ* **2018**, *5*, 574–584. [CrossRef] [PubMed]
25. Lu, Z.; McMahon, J.; Mohamed, H.; Barnard, D.; Shaikh, T.R.; Mannella, C.A.; Wagenknecht, T.; Lu, T.-M. Passive Microfluidic device for Sub Millisecond Mixing. *Sens. Actuators B Chem.* **2010**, *144*, 301–309. [CrossRef]
26. Weierstall, U. Liquid sample delivery techniques for serial femtosecond crystallography. *Philos. Trans. R. Soc. Lond. B Biol. Sci.* **2014**, *369*, 20130337. [CrossRef]
27. Calvey, G.D.; Katz, A.M.; Schaffer, C.B.; Pollack, L. Mixing injector enables time-resolved crystallography with high hit rate at X-ray free electron lasers. *Struct. Dyn.* **2016**, *3*, 054301. [CrossRef]
28. Calvey, G.D.; Katz, A.M.; Pollack, L. Microfluidic Mixing Injector Holder Enables Routine Structural Enzymology Measurements with Mix-and-Inject Serial Crystallography Using X-ray Free Electron Lasers. *Anal. Chem.* **2019**, *91*, 7139–7144. [CrossRef]
29. Cheng, R.K. Towards an Optimal Sample Delivery Method for Serial Crystallography at XFEL. *Crystals* **2020**, *10*, 215. [CrossRef]
30. Beyerlein, K.R.; Adriano, L.; Heymann, M.; Kirian, R.; Knoška, J.; Wilde, F.; Chapman, H.N.; Bajt, S. Ceramic micro-injection molded nozzles for serial femtosecond crystallography sample delivery. *Rev. Sci. Instrum.* **2015**, *86*, 125104. [CrossRef]
31. Zahoor, R.; Bajt, S.; Šarler, B. Influence of Gas Dynamic Virtual Nozzle Geometry on Micro-Jet Characteristics. *Int. J. Multiph. Flow* **2018**, *104*, 152–165. [CrossRef]
32. Zahoor, R.; Belšak, G.; Bajt, S.; Šarler, B. Simulation of liquid micro-jet in free expanding high-speed co-flowing gas streams. *Microfluid. Nanofluid.* **2018**, *22*, 87. [CrossRef]
33. Trebbin, M.; Krüger, K.; DePonte, D.; Roth, S.V.; Chapman, H.N.; Förster, S. Microfluidic Liquid Jet System with compatibility for atmospheric and high-vacuum conditions. *Lab Chip* **2014**, *14*, 1733–1745. [CrossRef]
34. Feng, X.; Fu, Z.; Kaledhonkar, S.; Jia, Y.; Shah, B.; Jin, A.; Liu, Z.; Sun, M.; Chen, B.; Grassucci, R.A.; et al. A Fast and Effective Microfluidic Spraying-Plunging Method for High-Resolution Single-Particle Cryo-EM. *Structure* **2017**, *25*, 663–670. [CrossRef]
35. Zhao, J.; Xiong, W.; Yu, N.; Yang, X. Continuous Jetting of Alginate Microfiber in Atmosphere Based on a Microfluidic Chip. *Micromachines* **2017**, *8*, 8. [CrossRef]
36. Hofmann, E.; Krüger, K.; Haynl, C.; Scheibel, T.; Trebbin, M.; Förster, S. Microfluidic nozzle device for ultrafine fiber solution blow spinning with precise diameter control. *Lab Chip* **2018**, *18*, 2225–2234. [CrossRef]
37. Marmiroli, B.; Grenci, G.; Cacho-Nerin, F.; Sartori, B.; Ferrari, E.; Laggner, P.; Businaro, L.; Amenitsch, H. Free jet micromixer to study fast chemical reactions by small angle X-ray scattering. *Lab Chip* **2009**, *9*, 2063–2069. [CrossRef]
38. Koralek, J.D.; Kim, J.B.; Bruza, P.; Curry, C.B.; Chen, Z.; Bechtel, H.A. Generation and characterization of ultrathin free flowing liquid sheets. *Nat. Commun.* **2018**, *9*, 1353. [CrossRef]
39. Hejazian, M.; Darmanin, C.; Balaur, E.; Abbey, B. Mixing and jetting analysis using continuous flow microfluidic sample delivery devices. *RSC Adv.* **2020**, *10*, 15694–15701. [CrossRef]
40. Hejazian, M.; Balaur, E.; Flueckiger, L.; Hor, L.; Darmanin, C.; Abbey, B. Microfluidic mixing and jetting devices based on SU8 and glass for time-resolved molecular imaging experiments. In Proceedings of the Microfluidics, BioMEMS, and Medical Microsystems XVII, San Francisco, CA, USA, 2–4 February 2019; p. 108750D.
41. Hejazian, M.; Balaur, E.; Abbey, B. A Numerical Study of Sub-Millisecond Integrated Mix-and-Inject Microfluidic Devices for Sample Delivery at Synchrotron and XFELs. *Appl. Sci.* **2021**, *11*, 3404. [CrossRef]
42. Vakili, M.; Vasireddi, R.; Gwozdz, P.V.; Monteiro, D.C.F.; Heymann, M.; Blick, R.H.; Trebbin, M. Microfluidic polyimide gas dynamic virtual nozzles for serial crystallography. *Rev. Sci. Instrum.* **2020**, *91*, 85108. [CrossRef]
43. Ho, C.M.B.; Ng, S.H.; Li, K.H.H.; Yoon, Y.-J. 3D printed microfluidics for biological applications. *Lab Chip* **2015**, *15*, 3627–3637. [CrossRef] [PubMed]
44. Amin, R.; Knowlton, S.; Hart, A.; Yenilmez, B.; Ghaderinezhad, F.; Katebifar, S.; Messina, M.; Khademhosseini, A.; Tasoglu, S. 3D-printed microfluidic devices. *Biofabrication* **2016**, *8*, 022001. [CrossRef] [PubMed]
45. Van den Driesche, S.; Lucklum, F.; Bunge, F.; Vellekoop, M.J. 3D Printing Solutions for Microfluidic Chip-To-World Connections. *Micromachines* **2018**, *9*, 71. [CrossRef]

46. Nelson, G.; Kirian, R.A.; Weierstall, U.; Zatsepin, N.A.; Faragó, T.; Baumbach, T.; Wilde, F.; Niesler, F.B.; Zimmer, B.; Ishigami, I.; et al. Three-dimensional-printed gas dynamic virtual nozzles for x-ray laser sample delivery. *Opt. Express* **2016**, *24*, 11515–11530. [CrossRef]
47. Galinis, G.; Strucka, J.; Barnard, J.C.T.; Braun, A.; Smith, R.A.; Marangos, J.P. Micrometer-thickness liquid sheet jets flowing in vacuum. *Rev. Sci. Instrum.* **2017**, *88*, 083117. [CrossRef]
48. Wiedorn, M.O.; Oberthür, D.; Bean, R.; Schubert, R.; Werner, N.; Abbey, B. Megahertz serial crystallography. *Nat. Commun.* **2018**, *9*, 4025. [CrossRef] [PubMed]
49. Bohne, S.; Heymann, M.; Chapman, H.N.; Trieu, H.K.; Bajt, S. 3D printed nozzles on a silicon fluidic chip. *Rev. Sci. Instrum.* **2019**, *90*, 035108. [CrossRef]
50. Nazari, R.; Zaare, S.; Alvarez, R.C.; Karpos, K.; Engelman, T.; Madsen, C.; Nelson, G.; Spence, J.C.H.; Weierstall, U.; Adrian, R.J.; et al. 3D printing of gas-dynamic virtual nozzles and optical characterization of high-speed microjets. *Opt. Express* **2020**, *28*, 21749–21765. [CrossRef] [PubMed]
51. Knoska, J.; Adriano, L.; Awel, S.; Beyerlein, K.R.; Yefanov, O.; Oberthuer, D.; Murillo, G.E.P.; Roth, N.; Sarrou, I.; Villanueva-Perez, P.; et al. Ultracompact 3D microfluidics for time-resolved structural biology. *Nat. Commun.* **2020**, *11*, 657. [CrossRef] [PubMed]
52. Fang, W.F.; Hsu, M.H.; Chen, Y.T.; Yang, J.T. Characterization of microfluidic mixing and reaction in microchannels via analysis of cross-sectional patterns. *Biomicrofluidics* **2011**, *5*, 014111. [CrossRef]
53. Inguva, V.; Rothstein, J.P.; Bilsel, O.; Perot, B.J. High-speed velocimetry in microfluidic protein mixers using confocal fluorescence decay microscopy. *Exp. Fluids* **2018**, *59*, 177. [CrossRef]
54. Xi, C.; Marks, D.L.; Parikh, D.S.; Raskin, L.; Boppart, S.A. Structural and functional imaging of 3D microfluidic mixers using optical coherence tomography. *Proc. Natl. Acad. Sci. USA* **2004**, *101*, 7516. [CrossRef]
55. Jiang, L.; Zeng, Y.; Zhou, H.; Qu, J.Y.; Yao, S. Visualizing millisecond chaotic mixing dynamics in microdroplets: A direct comparison of experiment and simulation. *Biomicrofluidics* **2012**, *6*, 012810. [CrossRef]
56. Witkowski, D.; Kubicki, W.; Dziuban, J.A.; Jašíková, D.; Karczemska, A. Micro-Particle Image Velocimetry for imaging flows in passive microfluidic mixers. *Metrol. Meas. Syst.* **2018**, *25*, 441–450.
57. Yang, J.-T.; Lai, Y.-H.; Fang, W.-F.; Hsu, M.-H. Simultaneous measurement of concentrations and velocities of submicron species using multicolor imaging and microparticle image velocimetry. *Biomicrofluidics* **2010**, *4*, 014109. [CrossRef]
58. Huyke, D.A.; Ramachandran, A.; Oyarzun, D.I.; Kroll, T.; DePonte, D.P.; Santiago, J.G. On the competition between mixing rate and uniformity in a coaxial hydrodynamic focusing mixer. *Anal. Chim. Acta* **2020**, *1103*, 1–10. [CrossRef]

 micromachines

Review

Microfluidic Applications of Artificial Cilia: Recent Progress, Demonstration, and Future Perspectives

Vignesh Sahadevan [1,†], Bivas Panigrahi [2,†] and Chia-Yuan Chen [1,*]

[1] Department of Mechanical Engineering, National Cheng Kung University, Tainan 701, Taiwan
[2] Department of Refrigeration, Air Conditioning and Energy Engineering, National Chin-Yi University of Technology, Taichung 411, Taiwan; bivas@ncut.edu.tw
* Correspondence: chiayuac@mail.ncku.edu.tw; Tel.: +886-2757575-62169; Fax: +886-2352973
† These authors contributed equally to this work.

Abstract: Artificial cilia-based microfluidics is a promising alternative in lab-on-a-chip applications which provides an efficient way to manipulate fluid flow in a microfluidic environment with high precision. Additionally, it can induce favorable local flows toward practical biomedical applications. The endowment of artificial cilia with their anatomy and capabilities such as mixing, pumping, transporting, and sensing lead to advance next-generation applications including precision medicine, digital nanofluidics, and lab-on-chip systems. This review summarizes the importance and significance of the artificial cilia, delineates the recent progress in artificial cilia-based microfluidics toward microfluidic application, and provides future perspectives. The presented knowledge and insights are envisaged to pave the way for innovative advances for the research communities in miniaturization.

Keywords: artificial cilia; microfluidics; flow manipulation; biological/medical applications

Citation: Sahadevan, V.; Panigrahi, B.; Chen, C.-Y. Microfluidic Applications of Artificial Cilia: Recent Progress, Demonstration, and Future Perspectives. *Micromachines* **2022**, *13*, 735. https://doi.org/10.3390/mi13050735

Academic Editor: Kwang-Yong Kim

Received: 23 March 2022
Accepted: 19 April 2022
Published: 3 May 2022

1. Introduction

Biological cilia are hair-like microscopic structures found on the outer surfaces of nearly every mammalian cell. These microscopic structures allow the cells to interact and sense their surrounding environment [1]. The length of cilia usually varies between 2–15 μm, and they possess complex internal structures comprised of outer doublet microtubules, central microtubules (Axoneme), and dynein arms, which further determines cilia's function as well as their motion. According to their functions, biological cilia can be broadly classified into two major groups: motile and non-motile, or primary cilia [2]. The cilia could be identified, whether motile or primary, using the arrangements of nine pairs of two microtubules with two central pair-structure known as axonemes. In general, if the cilium is with nine pairs of two microtubules but without two-central pair apparatus, the cilium is widely considered as primary cilium. On the other hand, if the cilium has nine pairs of two peripheral microtubules with two central pair microtubules, the arrangement is considered a motile cilium [3,4]. Motile cilia exhibit spatial, temporal, and even oriental asymmetry in their motion to generate flow around them. In contrast, the non-motile cilia are stationary by nature and act as a sensor. Applications of motile cilia are immense. Motile cilia on the outer membrane paramecia help them propel in the fluid 10 times faster than their body length. Additionally, the cilia on the outer surface of the juvenile starfish allow them to select food by creating vortices around them [5,6]. Non-motile cilia can be found on the kidney tubule, where they sense the direction of urine flow and direct the cells accordingly. The motile cilia exhibit a planar motion and beat in a straight path during the forward stroke and roll back to their original position by moving close to the surface in a tangential manner during the recovery stroke. This spatial asymmetry can generate a substantial flow around the cells [1]. Similarly, cilia on the embryo nodal cell exhibit a tilted conical beating path that produces the fluid flow to determine further left-right symmetry in the body [7,8].

Microfluidics is the science of manipulating the amount of fluid with channels, where at least one dimension is generally in a range of 10~1000 μm [9]. Microfluidics exploits its most apparent characteristic, such as its miniaturized size to efficiently handle a small amount of fluid in a minimal time scale. The flow physics of microfluidics is purely laminar, and the viscous force dominates over the inertial force in the flow regime. Hence, it is quite challenging to manipulate the fluid flow in the microfluidic environment due to the effect of the viscous force. As discussed earlier, nature provides an ingenious way for the microorganisms and eukaryotic cells to manipulate the flow around cells by means of cilia. Taking inspiration from the natural cilia, artificial cilia have now been fabricated in the laboratory environment to manipulate the fluid flow within the microfluidic environment. The applications and usages of these artificial cilia-based microfluidic devices are unparalleled.

In the past two decades, the development of microfluidics and artificial cilia research has increased drastically. The existing reviews on microfluidic based devices focused on any specific topics. For instance, the studies were focused on particular actuation techniques like magnetics [10–14] and light [15], specific applications like mixing [12], robotics [16] and particle manipulation [11,17,18], or fabrication techniques like 3D printing [19]. Few excellent reviews are presented in the field of artificial cilia-based microfluidic devices [20,21]. These reviews have extensively discussed the principles of artificial cilia, fabrication processes, actuation mechanisms, modeling of their motion, etc. Along with these fundamental aspects, this review article has put particular emphasis on describing various biological as well as the industrial applications of these artificial cilia-based microfluidic devices. This will not only bridge the existing knowledge gap in the field of artificial-based microfluidics but will also provide a perspective towards future applications and possible research directions. The arrangement of the article is delineated as follows. First, the importance of the natural cilia dynamics was discussed, followed by the description of the fabrication techniques for artificial cilia and the respective actuation methodologies. In the following sections, dynamic beating behaviors of artificial cilia and their applications in microfluidics were also illustrated. Their roles in contemporary applications were discussed. Finally, in view of the general summary, an outlook on conclusions and future perspectives was offered.

2. From Natural Cilia to Artificial Cilia

Natural cilia are hair-like structures that perform the operations such as feeding, swimming, transporting, moving, and sensing functions in almost all cell types [22–25]. In motile natural cilia category, their applications in coral reef [22] and the respiratory tract [23] towards particle transportation is exemplary. The natural cilia in the coral surfaces generate metachronal wave motion to create flows using the asymmetric dynamic beating behavior. Due to this flow behavior, the materials such as oxygen and nutrients are transported without the typical wave streams [22]. The natural cilia shielded on the lumen within the human respiratory tract transport the dust and bacteria to remove towards the oropharynx [23]. Natural cilia in the Cactus spines and trachea are well regarded for their directional transporting ability due to their asymmetric motion and anisotropic surface. Beating natural cilia with amorphous sheets and dense tufts can create the propulsion forces which facilitate the locomotion in many species [26–28]. Depending upon the size of the organisms, the functions of the natural cilia vary. For example, including the surface of starfish larvae, many organismic ciliary structures were evolved towards facilitating their feeding. The non-motile natural cilia perform sensing operations and adapt to the ambient atmosphere [23]. They were used for sensing work to defend the predators and sense the food and atmosphere in the Animalia kingdom. The natural cilia in the spider tarsi legs sense air flow and other vibrations. The external vibration makes the natural cilia bend where the electrical impulses create and reach the spiders [23].

Motivated by these biological cilia, congeners such as artificial cilia have been advanced and exploited in microfluidics and micro/nanorobotics to realize performances of propelling, transporting, moving, and mixing [29–35]. Even though the in-depth information and uses of artificial cilia in transportation, moving, and sensing will be seen in the upcoming chapters, the insights into the evolvement of artificial cilia from natural cilia are explained here by detailing their participation with examples of each significant application. Wang et al. [15] fabricated magnetically actuated cilia using Polydimethylsiloxane (PDMS) and cobalt powder by biomimicking cactus spines and trachea cilia in favor of transportation. The artificial cilia were capable of transporting hydrogel slices directionally using their anisotropic surface and asymmetric motion. The asymmetric stroke of the artificial cilia in the sequential magnetic field drove the hydrogel forward. The asymmetric beating behavior and metachronal wave motion are two intriguing benefits exploited using artificial cilia [16,28,29]. The nonreciprocal or asymmetric beating behavior was achieved by creating the difference in the swept area of forward and recovery strokes of artificial cilia [30]. The nonreciprocal motion was measured by the difference between the swept area of the artificial cilia tip and the swept area of the semicircle. The semicircle was created by the artificial cilia length as the radius. The direction of the particle transportation can be adjusted by changing the levels of forward and recovery strokes [31]. The out-of-phase behavior or the phase difference in the neighboring artificial cilia raised the oscillating waves above the surface of the artificial cilia, known as metachronal wave motion. Another study was reported [34] in favor of transportation, in which the photo-actuated artificial cilia were obtained by mimicking the Paramecium aurelia's complex mechanical functions and surface responsiveness. The light-controllable cilia were fabricated using Diarylethene and are capable of transporting objects, demonstrated to transport the polystyrene bead (PB) of 1 mm in diameter. Recent bioinspired artificial cilia established from natural cilia for transporting ability are reported elsewhere [23,30,31,34,36–40]. Inspired by the starfish surface, researchers designed ciliary bands [29], which can be actuated by ultrasound. The ciliary band was successfully demonstrated for locomotion, trapping polystyrene particles, and transportation of water droplets.

The biomimetic inventions have not ended here—the establishments are also reflected in sensing artificial cilia. For sensing, conceptual-wise, the working principle of the artificial was designed similar to the working principle of natural cilia. Artificial cilia for sensing have two sections. The first section is the cilia part where typical magnetic particles (Carbonyl iron powder (CIP) and, neodymium-iron-boron (NdFeB)), PDMS, glass fibers, and polymer materials were used. The second section is the signal processing section. Graphene nanoplatelets [41], carbon nanotubes (CNTs), SiO_2, carbon nanofiber, iron nanowires, and AgNW were employed as the sensing/signal processing part [41–48]. Signals such as the piezoelectric and piezoresistive types were created due to cilia bending and processed through the signal processing units [41]. For instance, in the study [42], the high aspect ratio (HAR) artificial cilia were fabricated by replicating goldfish's neuromasts and arthropod filiform hair. The HAR cilia sensor was comprised of PDMS cilia structure, and graphene nanoplatelet infused microchannel. The change in resistance of graphene nanoplatelet was realized in the presence of external flow or touch over the PDMS cilia sensor.

3. Fabrication Techniques for Artificial Cilia

3.1. Micro-Molding Fabrication Techniques

The magnetically actuated artificial cilia fabricated by means of micro-molding techniques required a less complex fabrication process and considered precise by nature. This fabrication technique is one of the template-based fabrication techniques. The micro-mold fabrication process involves four critical steps. The first step is to pattern the mold corresponding to the artificial cilia. The second process introduces the polymer material to the pattern. The third step is arranged for the solidification of the pattern. The final step is separating or peeling off the artificial cilia. Chen and his team [49–55] fabricated a wide range of artificial cilia with universality using micro-molding fabrication processes. For instance, Wu et al. [56] demonstrated a micro-molding fabrication process that involves a series of computerized numerical control (CNC) micromachining processes towards preparing the mould for artificial cilia., PDMS-magnetic composite casting was carried out followed by PDMS casting to create the structure of artificial cilia and its microfluidic environment.

Following the PDMS casting, the sample was kept in the hot plate for curing at 90 °C before secluding the artificial cilia [56]. The critical challenge coming up with magnetic artificial cilia is their size, as they are considered relatively oversized than the natural cilia. Recently, the magnetic artificial cilia [57] were showcased of the same size (Radius = 200 nm, Length = 6μm) as their counterpart (i.e., biological cilia). The proposed magnetic artificial cilia were fabricated using a tailored molding process. Figure 1A illustrates the use of micro-molding fabrication process towards the fabrication of multi-segmented magnetic artificial cilia. The fabrication was carried out following processes such as: (i) artificial cilia were patterned in an acrylic sheet, (ii) magnetic and PDMS mixture were poured into the pattern, (iii) the pattern was cured on the hot plate at 85 °C for 48 h, (iv) procedures of peeling-off artificial cilia from the acrylic substrate and the following magnetization of the artificial cilia.

3.2. *Photolithography Fabrication Techniques*

The photolithography fabrication technology was used to fabricate soft patterned thin film on the substrate using light. In general, UV light is a popular alternative to this technique. Still, various lights with different wavelengths, such as X-rays, visible light, and extreme UV rays, were also employed depending on the requirements. In a study [58] where a two-step lithography process was employed to fabricate nickel-iron (Ni-Fe) permalloy-based artificial cilia by the researchers. In the first step, Cu's sacrificial layer was sputtered on the negative photoresist material (NR9 1500Py Futurex), followed by the Ni-Fe layer. The desired thickness of the artificial cilia was defined by the deposition of Ni-Fe. The ciliary structures were formed after removing the photoresist material using acetone. In the second step of lithography, the ciliary structures were pinned on the glass substrate using the Ti anchor to hold the cilia on the substrate. Then, the sacrificial Cu layer was removed by 5% ammonium hydroxide solution.

The photolithography fabrication technique is preferred over some other fabrication techniques for the following reasons. The first reason is that this technique offers high precision. In addition, this method is highly controlled together with high throughput, compared to the bead self-assembly [59,60], which required additional control [58]. The recent discussions and approaches over photolithography processes to fabricate artificial cilia can be found elsewhere [29,44,61,62]. Figure 1B illustrates the photolithography fabrication process depicted fabricating magnetic artificial cilia following processes such as: (i) substrate preparing, (ii) anchor preparing, (iii) cilia body layer preparing, (iv) ciliary shape developing, (v) cilia coating, and (vi) peeling-off.

3.3. *3D/4D/5D Printing Fabrication Techniques*

Previously explained, micro-molding and photolithography fabrication technologies are limited because different molds need to be used for different designs in the micro-molding fabrication process. The photolithography fabrication technology requires photoresist, external light sources, the repeatability of the lithography process, and the cleanroom fabrication setup. In the 3D printing technique eliminates these extensive experiment setups and various sizes of artificial cilia could be made without fabrication complexity. A 3D CAD diagram was used to design the three-dimensional artificial cilium to fabricate it under computer programming [63]. The 3D printing technology for artificial cilia was well illustrated by Liu et al. [64]. Recent discussions and approaches over 3D printing technology (Figure 1C) to fabricate artificial cilia can be found in some recent articles [41,42,46,65].

In 2004, 4D printing technology for multi-material [66] was proposed by by S. Tibbits. In the 4D printing technique, the programmed multi-materials can morph their shape over a period of time, even after they came out of the printer. Recently, 4D printing of artificial cilia was proposed by Tsumori et al. [67]. The printing system printed out the 3D artificial cilia whose anisotropy could be changed simultaneously. Next to 4D printing, 5D-printing was proposed by Tsumori et al. [68] for the magnetic artificial cilia. In this process, three design parameters (x, y, z) were used to fabricate the artificial cilia's shape. Two more parameters (θ, ψ) were utilized in aligning magnetic chain clusters. Five design parameters (x, y, z, θ, ψ) were optimized simultaneously.

3.4. Facile Bottom-Up Approaches

The facile bottom-up approach is one of the template-free fabrication methods which is unconstrained and the artificial cilia made by this process is highly flexible by nature. This approach involves simple fabrication and operating steps. Timonen et al. [69] demonstrated the facile bottom-up approach (Figure 1D) to fabricate the magnetic artificial cilia. The magnetic particle such as cobalt was mixed with the solvent toluene and elastomeric poly (styrene-block-isoprene-block-styrene) polymer. Poly (tetrafluoroethylene) (PTFE) was used as the substrate in the study. The suspension was ultrasonicated for 10 s to have the high aspect magnetic artificial cilia. In addition, the bottom-up approach was used to fabricate the minimal model system [70] comprised of microtubule (MT) bundles and molecular motors. The exemplified minimal model system was demonstrated to resemble the beating behaviors of the eukaryotic cilia and flagella by self-assembling the MTs and molecular motors.

3.5. Roll-Pulling Approaches

The roll-pulling approach was showcased by the researchers in the study [71] where the custom-made setup was comprised of aluminum roll, substrate, and the precursor medium. The aluminum roll was surrounded by magnetic pillars fabricated by soft lithography. The aluminum roll and glass substrate were separated in a particular gap and rotated with a constant line speed with the help of a rigid string to eliminate friction. The glass substrate was carried the precursor medium comprised of magnetic particles and PDMS. The precursor medium was then pulled by the pre-molded magnetic pillars during the roll rotation and became filaments of a certain length before breaking. The dimensions of magnetic filaments can be adjusted by the size of micro-pillars and the line speed of the roll and substrate. The proposed fabrication technique has advantages over other fabrication techniques in terms of scaling down the dimension of artificial cilia. For example, the 3D printing technology [42] for PDMS artificial cilia is challenging to fabricate high aspect ratio cilia. It required additional attention to print out PDMS polymer materials for microchannels and biomimetic structures (with less than 150 μm diameter) due to its size limitation.

3.6. Self-Assembly Fabrication Techniques

Wang et al. [60] demonstrated a new affordable in-situ fabrication process named the self-assembly technique. In this fabrication process, micro sized beads were self-assembled and constructed to form the artificial cilia and encapsulated with soft polymer coatings. The self-assembly approach for the artificial cilia has been illustrated in Figure 1E [59]. The self-assembly approach was comprised of three electromagnets. The three electromagnets were arranged orthogonally to direct the applied magnetic field. The superparamagnetic particles were assembled to create the artificial cilium due to the controlled magnetic field.

3.7. Field-Effect Spinning Approaches

High sensing performance can be exploited by developing artificial cilia with identical sensory functions. The field-effect spinning (FES) method was reported in the study [72] for the fabrication of artificial cilia. In the previously reported conventional approach, the fiber filaments were pulled without direction. In contrast to the above-mentioned approach, the FES method was demonstrated to produce the vertical and uniformly sized artificial cilia.

3.8. Dip-Coating Fabrication Techniques

The dip-coating fabrication technique can be used for the fabrication of large size ciliary structures. High aspect ratio shapes can be fabricated using this approach with the following of four steps. In the first step, to give the shape to the ciliary structure, the metals like stainless steel pins were used along with the dielectric material. In the second step, the fixture, the dielectric material, and the spaces between the cilia were covered by rubber shims. In the third step, the structure was removed from the downside, and the dielectric material followed the structure of the previously placed fixture immediately. In the fourth step, the downside channels were sealed. The step-by-step fabrication processes along flow charts can be found in the study [73].

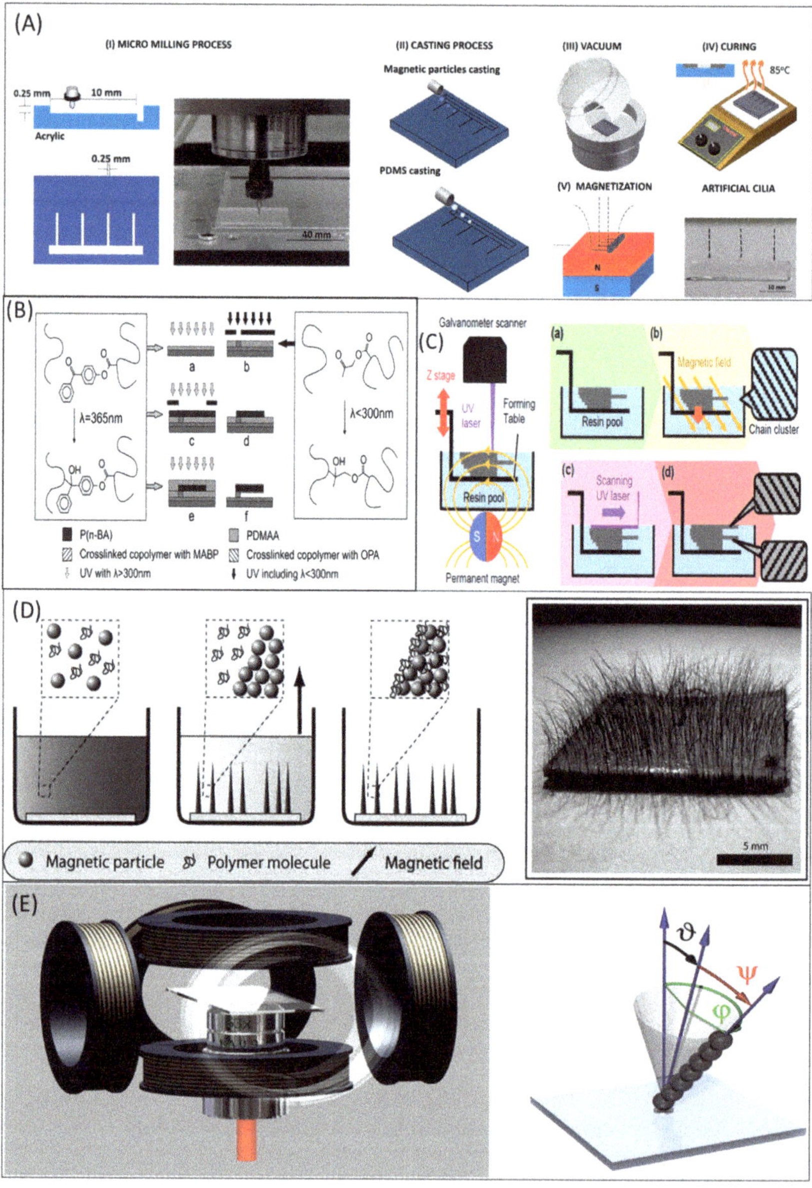

Figure 1. Fabrication techniques for artificial cilia. (**A**) The micro-molding fabrication process demonstrated fabricating multi-segmented magnetic artificial cilia using the following fabrication processes: (**i**) cilia shape patterning in acrylic sheet, (**ii**) magnetic and PDMS mixture pouring in the

pattern, (**iii**) cured in the hot surface plate at 85 °C for 48 h, and (**iv**) procedures of peeling-off artificial cilia from the acrylic substrate and magnetization. The figure was reproduced with permission from [55], under a Creative Commons BY Non-Commercial No Derivative Works (CC BY-NC-ND 4.0) license, published by Elsevier, 2021. (**B**) Photolithography fabrication process depicted fabricating magnetic artificial cilia using the fabrication following processes: (**i**) substrate preparing, (**ii**) anchor preparing, (**iii**) cilia body layer preparing, (**iv**) ciliary shape developing, (**v**) cilia coating and (**vi**) peeling-off. The figure was reproduced with permission from [74], published by John Wiley and Sons, 2011. (**C**) 3D printing fabrication process illustrated to manufacture magnetic artificial cilia.The 3D printing technique was based on stereolithography. Initially, the resin material was poured, and the UV laser of 355 nm was employed to cure. The cilia body were magnetized using a permanent magnet. The surface of the substrate was set at the level of the layer. The substrate table went down to form the next layer after the curing process of the previous layer, and the process was repeated until the desired shape was achieved. The figure was reproduced with permission from [68], published by the Society of Photopolymer Science and Technology (SPST), 2018. (**D**) The facile bottom-up approach in which elastomeric poly (styrene-block-isoprene-block-styrene), toluene, and magnetic particles were mixed, and the mixture was ultrasonicated. A PTFE dish was placed in the aqueous medium (**left**). The setup was kept in the magnetic field. As a result of the magnetic field, the conical structure was formed by magnetic particles, but the polymer was still in the suspensions (**middle**). Toluene evaporated the suspension when the polymer material covered the empty holes between the magnetic particles and artificial cilia were fabricated (**right**). The figure was reproduced with permission from [69], published by the American Chemical Society, 2010. (**E**) Self-assembly approach in which three magnetic coils were arranged orthogonally to render the magnetic field. The artificial cilium was created by assembling the superparamagnetic particles due to the magnetic field (**right**). The figure was reproduced with permission from [59], published by the National Academy of Sciences, 2010.

4. Artificial Cilia Actuation Methodologies

4.1. Optical Actuation Techniques

Optical actuation is the wireless actuation technology preferred in autonomous wireless microsystems [75]. Optically actuated cilia were established from materials consisting of acrylates or methacrylates and liquid-crystal polymer entailing azobenzene dyes using inkjet printing technology [76]. The visible light had a wavelength of 455 550 nm, and ultraviolet light was used to get the desired bending and complex movements of the artificial cilia in the water. The optical actuation method for the artificial cilia has been illustrated in Figure 2A. The figure illustrates the light actuation process demonstrated by Broer et al. [77]. The artificial cilia were made by the cross-linked polymers functionalized with azobenzene. The polymer was responsive to light and prototyped to transport the particles in the aqueous medium.

4.2. Electrostatic Actuation Techniques

Electrostatic actuation is done by the force between the conducting electrically charged objects. The electrostatic force used to actuate the artificial cilia can be attractive or repulsive. Den Toonder et al. [78] demonstrated such artificial cilia, which had a length and width of 100 mm, and 20 mm, respectively, and was made of polyimide (PI) and chromium (Cr). The artificial cilia had been electrostatically actuated for microfluidic applications. The rapid ciliary motion of the produced cilia was suitable for delivering the fluid flow over 0.6 mm/s [78]. The recent straightforward method to find the fluid flow velocity can be found elsewhere [79]. The Electrostatic actuation method for the artificial cilia has been illustrated in Figure 2B [80]. The setup comprised one moving electrode in the center, sandwiched by multiple ciliary electrodes deposited on two fixed electrodes on the outside. The electrostatic force was released in the spaces between moving and ciliary electrodes when the applied voltage was exerted between the fixed and moving electrode. Due to the electrostatic force, the moving electrode displaced towards the fixed electrode in parallel as well as perpendicular manner.

4.3. *pH Actuation Techniques*

The pH actuation is predominantly used for hydrogel actuators. But the actuation of hydrogel cilia was not limited to pH alone [81]. Hydrogels can be swelled and shrank over 10% of their original volume due to specific stimuli such as temperature (T), light, pH, etc., [82]. It gets attention due to the recent development of microfluidics and MEMS, increasing the need for small devices to work in microscale platforms. Hydrogels are the 3D polymer networks with 99 wt % of water. Changing pH intensity was used to stimulus artificial cilia for hydrogel actuation [21,81]. In the study [83], hydrogel-actuated artificial cilia were fabricated using micropost arrays and microfins. The actuators can be bent and upright straight by treating acids and bases. The soft lithography technique is ideal for manufacturing hydrogel artificial cilia [84]. The schematics of the pH actuation process for hydrogel cilia is seen in Figure 2C.

4.4. *Resonance Actuation Techniques*

In the resonance actuation technique, the artificial cilia were actuated under the resonance of the lead– zirconate–titanate (PZT) microstage. The piezoelectric transducer was excited by a signal generator. Photolithography microfabrication technique and deep reactive ion etching (DRIE) were used to fabricate the master mold. PDMS polymer was filled in the mold to manufacture the artificial cilia [85,86]. The resonance-actuated cilia-assisted micromixers can provide an efficient uniform mixing performance than diffusion-and-vibration mixtures [86].

4.5. *Magnetic Actuation*

Many actuation techniques exist to actuate the artificial cilia to harness the desired motions and deliver specific applications. But most actuation techniques have their drawbacks hindering their applications in biological domain. This is the primary reason why the researchers have turned their attention to the magnetic actuation system. For example, the electrostatic actuation system cannot be used for biological fluid because it leads to electrolysis [21,87]. Similarly, water is necessary for the hydrogel actuation of artificial cilia. In the air, water molecules gets absorbed, which leads to a long response time of a few hours even for miniaturized cilia [82,88]. Artificial cilia can be actuated using a permanent magnet [68,89–91] or electromagnet [49,50,92–95]. Both symmetric and asymmetric motions can be obtained from artificial cilia under the influence of the magnetic stimuli [49,58,96–99].

4.5.1. Electromagnetic Actuation

Chen et al. [98] demonstrated the electromagnetic actuation technique for the artificial cilia using the custom-built electromagnetic system with four magnetic coils to achieve real time actuation for various microfluidic applications. The Electromagnetic actuation method for the artificial cilia has been illustrated in Figure 2D [56]. The figure shows the electromagnetic actuation comprised of four solenoidal coils. The complete algorithm to actuate a series of artificial cilia using a single electromagnet can be found elsewhere [95].

4.5.2. Permanent Magnetic Actuation

Rotating a permanent magnet is the alternative method requiring less effort and complexity towards artificial cilia actuation [100]. Hanasoge et al. [58] achieved the asymmetric beating of artificial cilia by rotating the permanent magnet. The study reported that magnetic actuation induced the forward stroke; recovery stroke by releasing the accumulated elastic force. The asymmetric beating was achieved by the interplay of elastic, magnetic, and viscous forces. Figure 2E illustrates permanent magnetic actuation in which the permanent magnet of 0.5 T magnetic field was used to actuate the nano-artificial cilia. The magnetic field was sufficient to generate the deflection up to 7 μm, equivalent to 20° of bending angle of artificial cilia.

4.6. Acoustic Actuation Techniques

The need for simple fabrication and straightforward actuation for the artificial cilia in lab-on-chip applications is in high demand. Orbay et al. [101] fabricated the artificial cilia using initial photolithography and UV polymerization technique which eliminated the complex magnetization process. The artificial cilia were actuated using the piezoelectric transducer (PZT) (81-7BB-27-4L0, Murata Electronics, Japan). The thin epoxy layer was used to connect the PZT to the optical path of the PDMS microchannel. The piezoelectric transducer was driven by the sine waves induced through the function generator. The RF amplifier (25A250A, Amplifier Research, USA) was used to amplify the sine waves. The artificial cilia were tested for the mixing operation where the fluorescein and DI water were used. It was found that the increment of mixing performance and complete mixing can be achieved by the voltage up.

4.7. Electric Stimulation Actuation Techniques

The electric stimulation actuation technique is practically advantageous for the remote and precise control of the artificial cilia. In this approach, the ciliary structures were made of dielectric material. The dielectric materials were polarized due to the dielectrophoresis in the alternating current (AC) electric field when the dielectric nanoparticles were aligned along the electric field direction, leading to the deformation of the artificial cilia. In the study [102], $BaTiO_3$ was used as the dielectric nanoparticles along the PDMS cilia body. Dielectric fiber materials were used in a recent study [73]. By converting external physical cues into electric impulses using the piezoelectric or triboelectric effect principles, this actuation technique can also opt for sensors.

4.8. Induced Charge Electro-Osmosis Using AC Electric Field Techniques

The Induced charge electro-osmosis (ICEO) cilium was fabricated using a basic self-organizing process [103–105] demonstrated by Sugioka and the team. A graphite rod was immersed in the deionized water and intercalated between two Cu electrodes. As shown in Figure 2F, initially, SW_1 was turned off. SW_2 was turned on by applying DC electric voltage (V_o) between two Cu electrodes for duration t_{DC} (30–120 s), which led to the formation of Carbon artificial cilia at position X_o. The fabricated artificial cilia had a fibrous network which was identified using energy-dispersive x-ray spectroscopy (EDS). SW_1 was turned on to actuate the artificial cilia by applying AC electric peak voltage between the graphite rod and the right Cu electrode at time t_{AC}, when the non-fixed end of artificial cilia (X_o) moved with the deflection range h. The prototyped artificial cilia were capable of producing asymmetric motions. Using the same ICEO and implementing AC electric field principle, a metachronal wave motion was showcased by actuating three artificial cilia with different lengths [106]. Recently, ICEO actuated artificial cilia were demonstrated to transport the square-shaped polyethylene object [107].

4.9. Pneumatical Actuation Techniques

The artificial elastomers were actuated using the pressure sources created by electro-pneumatic actuators. Pneumatically actuated artificial cilia were showcased by Milana and the team [108], where six monolithic PDMS cylinders were offset two times to provide large deflection and swept area. Twelve pressure inputs and pneumatic valves were used to control the six elastomers. Pressure inputs and valves were controlled by the Lab view graphical user interface (GUI). The air inside the elastomer could be diffused with glycerol due to the pressure. The artificial cilia were filled with water to avoid this problem. As a result, the pneumatic signal was converted to hydraulic signals. Figure 2G illustrates the experimental setup of pneumatic actuation. In the same group's previous study [109], the step-by-step fabrication process of two pneumatic artificial cilia was shown. The two independent pneumatic actuators were used to deliver two degrees of freedom and spatial asymmetry. Figure 3F illustrates the pneumatic actuation technique in which two pneumatic actuators are actuated using dedicated pressure sources. Each actuator has an inflatable cavity body. Similar pneumatic microactuators which were fabricated using the micro-molding processes are reported here [110,111]. Onck et el. [36] established a study where the mixing, pumping, and transport capability of pneumatically actuated artificial cilia were analyzed numerically and experimentally. The study further reported that the antiplectic metachronal wave motion increases the mixing and transportation above and below the ciliary surfaces.

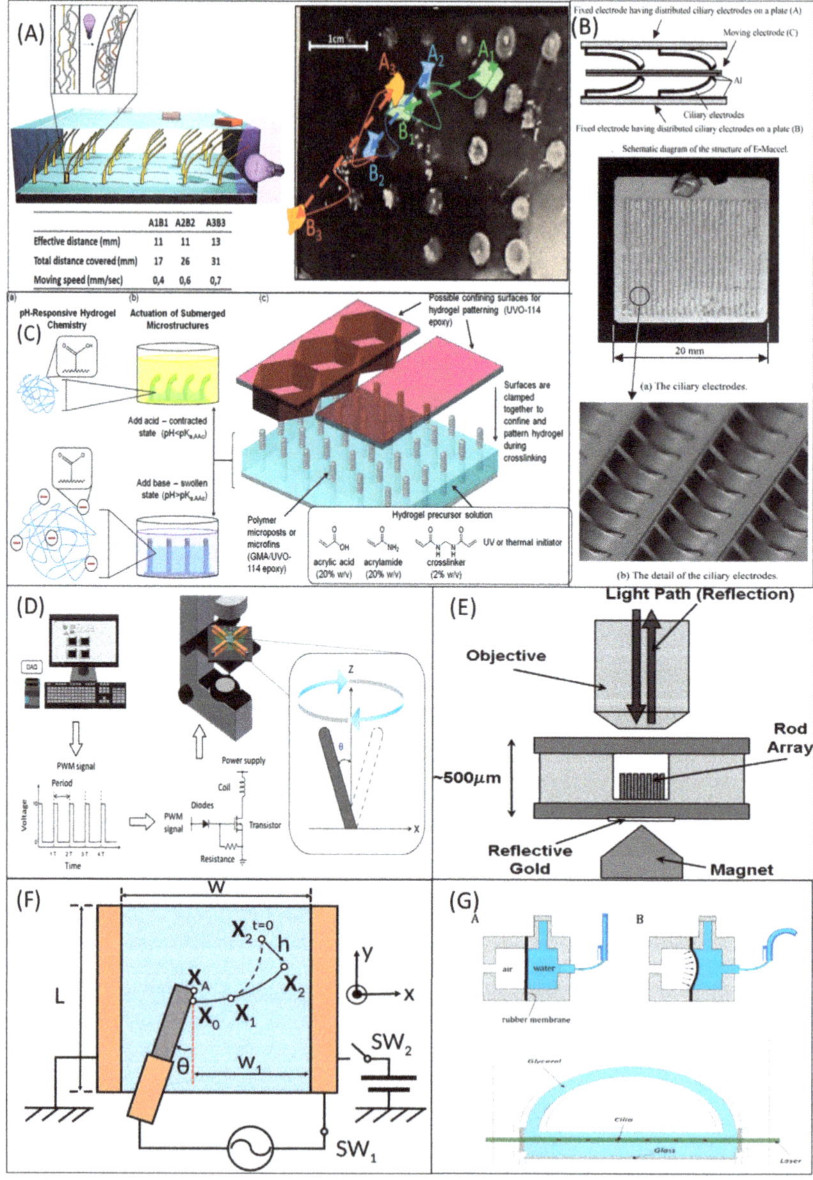

Figure 2. Artificial cilia actuation methodologies. (**A**) Optic actuation in which the photoresponsive polymers transported objects by having bending motion as responsive towards the light (**left**). The curved lines were the trajectory paths on which the objects were transported, and the dashed lines

were the effective linear distance the objects covered (**right**). The figure was reproduced with permission from [77], published by John Wiley and Sons, 2016. (**B**) The electrostatic actuation setup comprised two fixed electrodes, one moving electrode, and multiple ciliary electrodes (**top**). The electrostatic force was released in the spaces between moving and ciliary electrodes when the applied voltage was exerted in between the fixed and moving electrode. Due to the electrostatic force, the moving electrode moved towards the fixed electrode parallel and perpendicularly. The figure was reproduced with permission from [80], published by the Institute of Electrical Engineers, 2004. (**C**) The schematics of pH actuation process for hydrogel artificial cilia. The hydrogel response towards pH was shown (**left**). The figure was reproduced with permission from [83], published by John Wiley and Sons, 2011. (**D**) The schematics of electromagnetic actuation for artificial cilia. The algorithm was fed into the system to create pulse width modulation through which the artificial cilia were actuated using electromagnet coils. Inset: The beating trajectory of the artificial cilia. The figure was reproduced with permission from [56] under a Creative Commons BY (CC BY) license, published by MDPI, 2017. (**E**) Permanent magnetic actuation in which the permanent magnet of 0.5 T magnetic field was used to actuate the nanorod artificial cilia. The magnetic field was sufficient to generate the deflection up to 7 μm, equivalent to 20° of bending angle of artificial cilia. The figure was reproduced with permission from [100], published by the American Chemical Society, 2007. (**F**) ICEO artificial cilia actuation using the electric field (details of the experimental setup were discussed previously in the Induced charge electro-osmosis using AC electric field section). The figure was reproduced with permission from [103] under a Creative Commons BY (CC BY) license, published by AIP, 2020. (**G**) The strategy of pneumatic actuation (details of the experimental setup were discussed previously in the pneumatic actuation section). The figure was reproduced with permission from [108], under a Creative Commons Attribution NonCommercial License 4.0 (CC BY-NC) license, published by the American Association for the Advancement of Science (AAAS), 2020.

4.10. Thermal Actuation Techniques

Recent studies demonstrated that the self-swinging motion of the pendulum could be made by the asymmetrical heat transfer from the pendulum to the immersed medium [112]. Based on this principle, nichrome wire was used as the heat engine to actuate the cilium by Sugioka et al. [113]. The nichrome wire was used as a resistance wire. The artificial cilium was self-swinged in the nucleate boiling regime due to the asymmetrical heat transfer. DC electric charge was sent to the U-shaped wire immersed in the deionized water to set up the nucleate boiling regime. The same group achieved the metachronal wave motion in the low Reynolds number regime [114] using pendulums of different lengths.

4.11. Actuation Techniques for Multi-Responsive Artificial Cilia

The ability to respond to multiple stimuli makes artificial cilia known for a wide range of locomotive potential, advanced applications, and functions. The multi-responsive ability leads artificial cilia to next-generation intelligent robots. The cilia body comprises more than one responsive material to response multiple stimuli [115,116]. For instance, Mendes and team [117] fabricated the artificial cilia using electro-responsive gel and magneto responsive nanoparticles. The body of the artificial cilia can deform in the different pH concentrations. As a result of this ability, artificial cilia can be used to detect the pH. Similarly, another artificial cilia array was reported [118], fabricated using the PDMS and CrO_2 nanoparticles. The study has discussed the multi-responsive technique [119], where apart from actuating artificial cilia through the magnetic field, the shape can be further reconfigured through photo-thermal heating.

5. Dynamic Beating Behaviors of Artificial Cilia

The flowing nature created by the beating of artificial cilia can be manipulated by the beating trajectory [51,120]. The dynamic beating behaviors also took an important place in the nonreciprocal nature of cilia, resulting in spatial asymmetry and fluid transportation [7,8]. The beating trajectories were configured and optimized to improve the mixing can be found elsewhere [49,78,92]. The dynamic beating behaviors can be either symmetric or asymmetric by nature. The structures and motion of the motile and primary cilia were illustrated in Figures 3A and 3B, respectively, followed by the symmetric and asymmetric dynamic beating behaviors of artificial cilia.

5.1. Symmetric Dynamic Beating Behaviors

Wu et al. [56] demonstrated the symmetric trajectory beating of artificial cilia to achieve both mixing and micropropulsion operations. The observed mixing efficiency and fluid flow rate of micropropulsion operation were 0.84 and 0.089 μL/min, respectively. Chen et al. [50] analyzed the mixing performance for three different trajectories. They are 1. circular, 2. back-and-forth oscillation, and 3. figure-of-eight pattern of artificial cilia. Out of the above three trajectories, figure-eight efficiently provided the mixing performance in the highly viscous fluids. The mixing performance was increased to 0.86 from 0.79 [121]. All the reported beating trajectories were great examples of symmetric dynamic beating behaviors. The symmetric dynamic beating trajectory of the artificial cilium is illustrated in Figure 3C. Figure 3D illustrates two different symmetric trajectories which have been tested to analyze the mixing operation.

5.2. Asymmetric Dynamic Beating Behaviors

In contrast to symmetric motion, the studies [59,122] showed that asymmetric motion and improper coordination lead to complex fluid flow. The study of the asymmetric dynamic beating behaviors proved that conical activity is the best and simplest asymmetric nonreciprocal motion [123]. Vilfan et al. [59] demonstrated a nonreciprocal beating by inducing the conical rotation using magnetically actuated artificial cilia. The artificial cilia rotate at the angular frequency of $\omega = \varphi/t$. By changing the semi-cone angle, the pumping efficiency of the cilia could be adjusted. Figure 3E illustrates the tilted conical path, which is the best example of the asymmetric motion of the self-assembled cilia.

In low-Re number flow regimes, the fluid propulsion was achieved by the motion asymmetric beating of cilia. Out of a wide range of motion asymmetry, the spatial asymmetry was induced by the difference in the effective and recovery strokes trajectory is challenging to achieve and difficult to control. Milana et al. [109] attained this type of asymmetry using two pneumatic actuators with individual control. The performance of the artificial cilia was tested with asymmetric beating trajectories, as shown in Figure 3F Flow measurements have shown that the two degrees of freedom could generate fluid propulsion in the low Re number regime. It was found that the mixing using the asymmetric motion of artificial cilia was 1.34 times higher than the mixing obtained using the symmetric movement of the artificial cilia [98]. The recent discussions on asymmetric actuation trajectory induced by the artificial cilia can be found elsewhere [58,98].

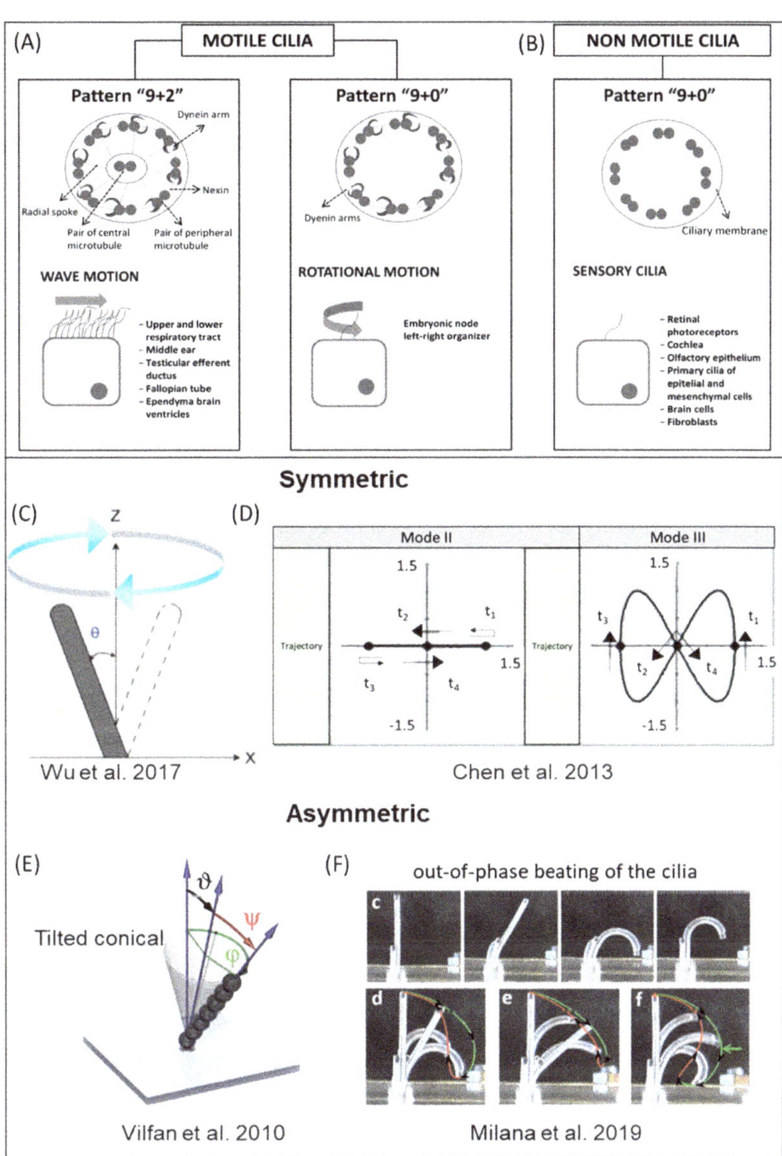

Figure 3. The dynamic beating behaviors of artificial cilia. (**A**) Structure and motion of motile biological cilia. The figure was adapted from [124] under a Creative Commons BY (CC-BY 4.0) license, published by IBIMA Publishing, 2015. (**B**) Structure and motion of primary biological cilia. The figure was adapted from [124] under a Creative Commons BY (CC-BY 4.0) license, published by IBIMA Publishing, 2015. The dynamic beating trajectory behaviors for artificial cilia were depicted as symmetric beating behavior [50,56] and asymmetric beating behavior [59,109] in schematics. (**C**) The symmetric dynamic beating trajectory of the artificial cilium has been illustrated by the 2D schematic. The figure was adapted from [56] under a Creative Commons BY (CC BY) license, published by MDPI, 2017. (**D**) Two different symmetric trajectories have been tested to analyze the mixing operation. The figure was reproduced with permission from [50], published by the Royal Society of Chemistry, 2013. (**E**) The superparamagnetic colloidal particles were assembled to create artificial cilia in the magnetic

field. The artificial cilium was actuated under the influence of the homogeneous magnetic field. The tilted conical path observed by the artificial cilium was the most uncomplicated asymmetric motion used to pump the fluid. The figure was reproduced with permission from [59], published by the National Academy of Sciences, 2010. (**F**) The performance of the artificial cilia was tested with asymmetric beating trajectories. The figure was reproduced with permission from [109], published by John Wiley and Sons, 2019.

6. Artificial Cilia for Microfluidic Applications

The significant property of microfluidics is that they are capable of delivering practical applications such as mixing, separating, trapping, transporting, and pumping, which involves a small amount of fluid. Passive microfluidic devices such as capillary flow devices might be exploited to get the aforementioned processes done. However, the size and efficiency of such devices are debatable. Integrating an artificial cilia-like setup within microfluidic environment leads to further miniaturization of microfluidic devices. The flexibility in their structural rigidity, actuation capability, and flow generation capability makes artificial cilia as necessary components for the microfluidic devices. Applications of artificial cilia range from lab-on–chip, sensing, particle manipulation, soft robotics, and biomedical devices etc. Several prominent applications in the aforementioned areas are discussed underneath.

6.1. Flow Propulsion

Microfluidic channels/devices have difficulties with biochemical reactions due to contamination and limited flow rates. The artificial cilia were positioned within the microchannels to generate regulated fluid flow by actuating them with respect to the external stimuli. For instance, in the study [99], varying the magnetic particle distribution during the artificial cilia fabrication process led to four kinds of versatile flows: circulatory fluid flows, direction-reversible flows, oscillating flows, and pulsatile flows. Besides solving the problems of limited flow rate and contamination, the proposed artificial cilia pump increased the options of fluids to manipulate. In intelligent robots and microfluidic devices, integrating both functional device capable of both sensing and pumping was challenging. Kong et al. [125] demonstrated the self-adaptive magnetic photonic nano-chain cilia arrays to address this issue. The importance of the metachronal wave motion and the asymmetric dynamic beating behavior were briefly explained earlier. The artificial cilia were indulged in the different experimental setups to achieve those characteristics for better pumping performance. Surprisingly, the strategy is different from that in natural cilia. Toonder et al. [62,126] demonstrated two different ways to achieve metachronal wave motion, and their unique benefits were also discussed. First [126], a custom-made magnetic actuation setup provided a non-uniform magnetic field. The metachronal wave was achieved by causing adjacent cilia to move out of phase in the non-uniform magnetic field. The combined effect of inertia force, metachronal and asymmetric wave motion of artificial cilia created the velocity of 3000 μm/s in water medium and 60 μm/s in pure glycerol in the low Reynolds number of 0.05. In the second approach [62], the magnetic artificial cilia were actuated in the uniform magnetic field. However, during the curing process, the magnetic artificial cilia (MAC) array was kept on the group of rod-shaped magnets with an alternating dipole orientation with each neighboring magnets for the remanent magnetization and paramagnetic particle distribution. The metachronal motion was achieved using this strategy during magnetic actuation in the uniform magnetic fields. Apart from pumping capability, the reported MAC, could climb in the slopes from 0° to 180° and carry weighs of 10 times higher than the MAC array. The proposed artificial cilia shed light on the applications of swimming microrobots, biofouling, and on-chip micropumps. In addition, recently, an explicit study [127] established a scientific investigation on the metachronal coordination of the artificial cilia towards the generated fluid flow. In the study, M.Sitti and his team concluded that the antiplectic metachronal coordination waves of the artificial cilia improved the fluid flow. The magnetically actuated artificial cilia of same size as the natural cilia (~10 μm) were fabricated using the microfabrication approach for this study. Concerning the viscosity of the fluid, magnetic torque, and elastic forces of the infinitesimal segment of the artificial cilium, the scaling analysis on the cilia length of the single cilium were conducted. The established study not only proved that the induced flow was enhanced by the integration of the non-reciprocal motion and metachronal coordination but also unveiled the implications of boundary surfaces of artificial cilia, locations of the defects, properties of fluid environment towards the resulted fluid flow in the Re number as same as the biological counterparts of the artificial cilia.

The presence of the artificial cilia in the electroosmotic pumps needs to be noticed very well for future bioinspired thermal micro/nanofluidic technologies. Recently, Saleem et al. [128] numerically analyzed the transportation of thermally radiated nanofluid in the microchannel in which the inside layer was settled with artificial cilia. In similar physical conditions, parameters such as electric potential, directional flow, velocity, pressure, temperature, and entropy generation were numerically analyzed by the same group [129–131]. Outside osmotic pumps, a numerical study [132] was reported to examine the heat transfer in the rectangular channel under the influence of the mechanical stirrer or artificial cilia. The study was proposed to find the heat increment or reduction over the increasing Re and Peclet number.

6.2. Mixing

The nature of the microfluidic regime is highly different from the microfluidic regime, due to the viscous force dominates the inertia force. Hence, mixing any reagent with the fluid medium is difficult in microfluidic devices. The challenge is to create a perfect/complete mixing in this microfluidic regime. Artificial cilia are well-known actuators for mixing and considered as an active microfluidic device. They have outperformed diffusion and vibration-induced mixers. Studies further reveal that the artificial cilia-based mixers enable the mixing operation more than the passive micromixers [133]. Researchers investigated artificial cilia to improve the mixing performance; as a result, fields from hydrogen production to microalgae growth can benefit from it. In terms of hydrogen production, a study stated the importance of artificial cilia mixing in photocatalytic activity [134]. Photocatalytic activity of g-C_3N_4 involves a recycling problem and the efficiency of hydrogen production is low. Researchers incorporated artificial cilia in the photocatalytic process in which graphene oxide (GO) was used as a bridge between the ciliary array and g-C_3N_4. Figure 4A illustrates the relationship between the increasing mixing performance and the actuation frequency of artificial cilia over time. Surprisingly, hydrogen production was improved under the influence of actuated artificial cilia by 75% compared with the control group with a static state. Moreover, the problem of recycling was resolved by employing artificial cilia. More on the photocatalytic activity improvement using optimized artificial cilia is discussed later in the photocatalysis section.

Artificial cilia with multifunctional abilities were made to develop bioinspired systems through which a broad spectrum of applications were made [135]. Figure 4B illustrates the result of the mixing experiment conducted by the artificial cilia in metachronal and nodal-like synchronous motions. Significant mixing performance was achieved by nodal-like synchronous motion and the overall performance barely affected by the arrangement.

In biochemical analyses, droplet-based digital microfluidics and medical diagnosis, fluid droplet manipulations such as directional transportation and mixing are necessary. However high-speed mixing, and high-volume droplet transportation are challenging. Zhou et al. [136] fabricated artificial cilia with super-hydrophobicity which were actuated by permanent magnets. Such artificial cilia were demonstrated to achieve complete mixing in ~1.5 s, and it is 60 times improvement compared with typical diffusional mixing. Figure 4C illustrates the mixing performance of the magnetic responsive film shielded by the micro-level artificial cilia and the relationship between mixing performance and frequency. Thanks to the permanent magnet that generated superhydrophobicity without any additional surface salinization, the superhydrophobicity lasted even after all the tests such as pressing, mechanical abrasion, chemical reactions, and sand abrasion. In the concept of acquainting both superhydrophobicity and wettability in the artificial cilia, a novel fabrication approach [30] was introduced by researchers. Figure 4D illustrates chemical reactions such as transportation and mixing of starch and iodine droplets by artificial cilia.

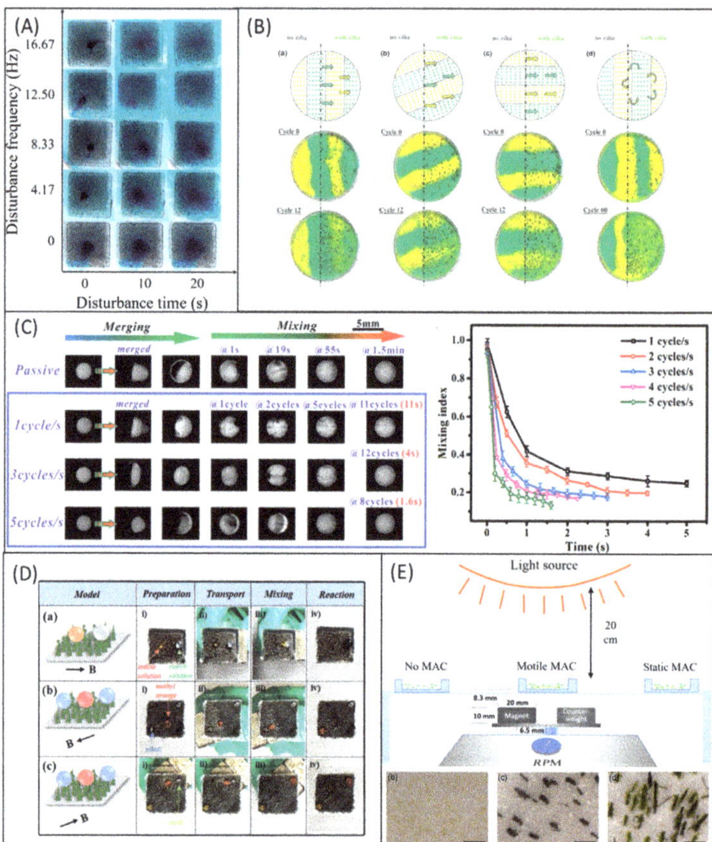

Figure 4. Artificial cilia for mixing. (**A**) Artificial cilia were actuated by magnetic fields in different frequencies to diffuse the blue ink in the water medium. The figure was reproduced with permission from [134], published by the American Chemical Society, 2020. (**B**) The mixing experiment was conducted by the artificial cilia in metachronal motion and nodal-like synchronous motion. Significant mixing performance was achieved by nodal-like synchronous motion barely affected by the arrangement. The figure was reproduced with permission from [135], published by John Wiley and Sons, 2021. (**C**) The mixing performance of the magnetic responsive film shielded by the microlevel artificial cilia was reported here. Complete mixing of fluorescent droplets was achieved in 1.6 s. The relationship between mixing performance and the frequency was plotted (right). The figure was reproduced with permission from [136], published by the American Chemical Society, 2021. (**D**) Under the influence of the magnetic field, microchemical reactions such as transportation and mixing of starch and iodine droplets were demonstrated by artificial cilia. The figure was reproduced with permission from [30], published by John Wiley and Sons, 2021. (**E**) The experimental setup used for microalgae culture embedded with artificial cilia mixing subsets: microalgae growth under (**b**) no cilia (**c**) static magnetic artificial cilia (**d**) motile magnetic artificial cilia. The figure was reproduced with permission from [137], under a Creative Commons BY Non-Commercial No Derivative Works (CC BY-NC-ND 4.0) license, published by John Wiley and Sons, 2021.

Microalgae such as Scenedesmus subspicatus is an essential product in food, biofuel, and many bio-products. In recent days, microalgae have been cultured in microfluidic channels. A study [137] reported use of magnetic artificial cilia to improve the growth rate of microalgae by creating flow and mixing. Using plasma treatment, hydrophilic artificial cilia were made. The mixing performance induced by the hydrophilic and hydrophobic artificial cilia improved the microalgae growth by ten times and two times, respectively, compared with the control group. The experimental setup used for microalgae culture embedded with artificial cilia mixing was illustrated in Figure 4E. In the figure, the subsets showcased the microalgae growth under (b) no cilia, (c) static artificial cilia (d) motile artificial cilia.

To improve mixing, magnetic artificial cilia were facilitated in the microchannel by researchers, and numerous physical attributes were analyzed both theoretically and experimentally [50,98,138,139]. However, mixing various reagents in the microsized platforms requires real-time adaption in the beating trajectory of artificial cilia. As a result of the issue, researchers intended to control the artificial cilia using fingertip drawing through remote control [138]. The artificial cilia setup was tested with four different trajectories. The built-in method was a perfect figure of eight pattern. The observed results showed the mixing performance of approximately 0.8 in 8 s, which is practically advantageous for a highly viscous medium. In general, Newtonian fluids [140], water and ink [86], silicone oils [78], high viscous dyed solutions [50], DI water and a fluorescent dye [141], colored dye solutions [133], and aqueous glycerol solution [49] were used to calculate the mixing performance by researchers.

6.3. Sensing

Several types of research have been conducted towards the design development of artificial cilia sensors [48,142–152]. The artificial cilia sensors were designed to be used in the marine system's sensors [153,154], flow sensors [48,155,156], force sensors [150], and tactile and texture sensors. The hydrophone is the ideal example of artificial cilia-based sensors designed and fabricated to use in marine system sensors. Annulus-shaped ciliary hydrophone [153] was recently established by mimicking the neuromasts in lateral fish lines by the researchers. The reported ciliary hydrophone can be used to determine the ship's motion. The proposed annulus ciliary-shaped sensor outperformed the previously reported bionic cilium-shaped sensor by the same group [154]. In response to the need for flow sensors, Alfadhel et al. 2014 [48] designed and developed a magnetic nanocomposite artificial cilium-like structure on a magnetoimpedance (GMI) thin-film sensor. The ciliary pillars have high elasticity and deflect corresponding to external flow resulting in a net change in mean magnetic intensity value perturbing the impedance of the GMI sensor. By relating the change in impedance to the flow property, the relation was established to quantify the flow. This sensor can detect air and water flow with the sensitivity of 24 mΩ (mm)$^{-1}$ s and 0.9 mΩ (mm)$^{-1}$ s, respectively [48]. The ciliary nanowires can be used as the tactile sensor to sense the touch and flow manipulations which were connected to the giant magneto-impedance (GMI) sensor. When the artificial cilia were touched by flow or humans, the magnetic intensity on the GMI sensor was disturbed, and thus the force could be detected [142,157]. Ribeiro et al. utilized a giant magnetoresistance sensor (GMR) and devised a simulation model to fabricate ciliary force sensors for various robotic applications [146]. Figure 5A illustrates the tactile Sensor for harsh environmental conditions. The stray magnetic field of the giant magnetoresistive (GMR) sensor was changed when the cilia were deflected due to touching. The changes in the magnetic field led to the variation in the resistance by which external force was sensed [152]. In robotics, the skins of robots were expected to have sensitivity with high accuracy; as a result, the robots can have comparable quality to the human skins. A study was proposed where the electronic cilia (EC) [158] could sense both magnetic field and pressure. The reported electronic cilia (EC) were designed and demonstrated to use in e-skin. Researchers claimed that the benefit of the integrated pressure-magnetic field sensor EC is more remarkable than piezoresistive sensors [45,159] and magnetic sensors. Figure 5B illustrates that the cilia-inspired flow sensor comprised the artificial cilium and the mini shaker. Dielectric material was connected to the mini shaker, and it was kept at the ciliary tip in the rest position with the tip displacement response pointing towards the flow. Triplicate measurements of the tip displacement were conducted when the dielectric was actuated at the constant frequency of 35 Hz [45]. Nanotubular cilia with polyimide substrate on the nonpatterned Anode Aluminum Oxide (AAO) templates and Cr and Au deposition by sputtering created a non-harmful flexible temperature sensor which was successfully tested on the eggshell without electrical failure (Figure 5C) [160]. Another study [44] was recently established where the magnetized nanocomposite artificial cilia were embedded with working magnetoresistive sensors. The fabricated sensors sensed the mature level of the strawberries and blueberries nondestructively. Small-sized nano ciliary arrays are sensitive and accurate to transport the vibration with the triboelectric effect. Due to this feature, nanometer ciliary arrays were demonstrated to be used as vibration sensors which were illustrated in Figure 5D as ubiquitous surfaces, human-machine interfaces, and buttonless keyboards such as touching and tapping keyboards for computers and other electronic devices [161]. The sensing principles were summarized along with fabrication and actuation methods in Table S1: Classifications of technologies and actuation/sensing mechanisms in Supplementary Materials.

6.4. Contemporary Emerging Applications

6.4.1. Zebrafish Research

Zebrafish have a similar genetic structure to humans, and researchers are interested in taking up zebrafish research to the next level by employing artificial cilia for zebrafish research. For instance, the artificial cilia were embedded in the microchannel which was facilitated with a moving wall feature using shape memory alloy to rotate and control the zebrafish stepwise for the benefit of hemodynamic screening [162]. The shape memory alloy-based miniaturized actuator's detailed fabrication and the controlling process can be found elsewhere [163]. Other recent findings on artificial cilia-assisted zebrafish research can be found elsewhere [52,94,95]. For instance, the artificial cilia in the microfluidic platforms used to activate sperm are shown in Figure 6A [94].

6.4.2. Minimal Robots/Microrobots and Soft Robots

Artificial ciliary arrays have been utilized as soft robots and encoded with both symplectic and antiplectic metachronal wave like motion. Recent research [65] by Nelson et al. 2020 showed (Figure 6B) that the ciliary arrays can be used for particle manipulation in the fluid medium and soft robots in the air medium. The soft robots from the same study can crawl and roll, which can be controlled accurately by the magnetic field and inspired by the giant African millipede. Other recent discussions over artificial cilia embedded robotic engineering can be found elsewhere [164–166]. A computational model on Belousov–Zhabotinsky (BZ) cilia to create soft robots oscillating using light for the fluid environments was reported by Balazs et al. [167]. A multi-legged soft millirobot was recently reported by Lu et al. [165]. The magnetic field-guided assembly approach, a template-free fabrication technique, was used to fabricate the robot. In this approach, the magnetic particle was mixed with PDMS and hexane. Under the magnetic field, the tapered feet structures of artificial cilia were produced on the polystyrene substrate. Further, the substrate was kept at 80 °C for 1 h before removing the ciliary robot from the polystyrene substrate to fabricate the robot.

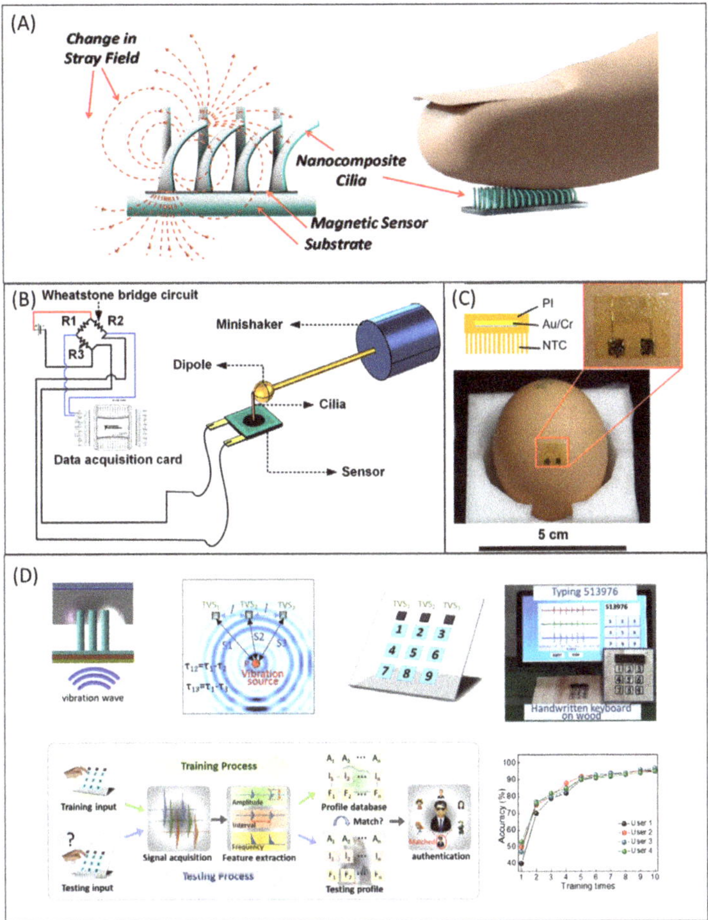

Figure 5. Artificial cilia as sensors. (**A**) Tactile Sensor for harsh environmental conditions. The stray magnetic field of the giant magnetoresistive (GMR) sensor was changed when the artificial cilia were deflected due to touching. The changes in the magnetic field led to the variation in the resistance by which external force was sensed. The figure was reproduced with permission from [152] under a Creative Commons BY (CC-BY) license, published by MDPI, 2016. (**B**) The cilia-inspired flow sensor was comprised of the artificial cilium and the mini shaker. The dielectric was connected to the mini shaker, and it was kept at the ciliary tip in the rest position to find the tip displacement response towards the flow. Triplicate measurements of the tip displacement were conducted when the dielectric was actuated at the constant frequency of 35 Hz. The figure was reproduced with permission from [45], under a Creative Commons BY (CC BY) license, published by MDPI, 2020. (**C**) Photographic image of temperature sensor created and assembled using nanotubular cilia (NTC). PI/Au/Cr/PI with NTC assembled temperature sensor used to find the temperature on the egg (PI: Polyamide, Au: gold, and Cr: Chromium) (**top left**). For comparison, a conventional sensor tested the egg's temperature when heated using the oven (**bottom**). Images reproduced with permission from [160], published by the American Chemical Society, 2020. (**D**) The schematic illustration of the triboelectric vibration sensor (**top left**). The artificial cilia sensors were demonstrated to be a keyboard (**top left**). The black diagram of the training and testing process for authentication was shown (**bottom left**). The graph plotted training times and the accuracy percentage (**bottom left**). Images reproduced with permission from [161], published by Elsevier, 2019.

6.4.3. Wearable Devices/Electro-Devices

The modern wearable electro-devices are not heavy but deformable. It requires non-toxic solvents and glue to make "sticky and play" kits. As a result of this specific requirement, the study demonstrated membrane-type flexible and nanotubular cilia to increase the spatial interfacial adhesion on complex shapes such as stone, bark, and textiles [160]. Recently, wearable artificial cilia [168] were fabricated using polymer materials in the form of a microneedle array. The fabricated microneedle array was demonstrated in the mice for psoriasis to be a drug delivery system for transdermal treatments.

6.4.4. Artificial Cilia with Wettability and Hydrophobicity

Hydrophiles/wettability and hydrophobicity are the two physical properties that are endeavored in artificial cilia surfaces. By acquiring these two properties, artificial cilia can manipulate droplets and solid particles in all kinds of mediums and facilitate droplets' pinning and no pinning options. For example, the study [169] has shown that oil droplets could be transported due to superoleophobicity. The properties can be achieved by various methods such as lubricating, characterizations, and topography modifications by physical and chemical methods. Both permanent and switchable options were encoded [39,170–173]. The unidirectional wetting properties of the artificial cilia are shown in Figure 6C.

6.4.5. Energy Harvesting

Electromagnetic induction coils were used to harvest the electrical energy transformed from ambient vibrations. The electromagnetic induction coil can have either a stationary coil and moving magnetic flux or a stationary magnetic flux and moving coil. A recent study proposed an artificial cilia embedded unique structure to harvest the energy (Figure 6D), and known as the magnetic composite energy harvester [174]. The energy harvester was comprised of two main components: (i) the microfabricated coils on the plane and (ii) high aspect ratio (HAR) artificial cilia. The modulus of elasticity of proof mass and HAR of the artificial cilia were endeavored to have the setup to respond low frequencies. In a recent study [175], magnetic actuated artificial cilia were prototyped to showcase the waste energy conversion process. The artificial cilia were fabricated using ZnO, which is a piezoelectric material. The artificial cilia harvested the mechanical energy from the flow. The piezoelectric catalysis reaction was triggered by the rotating artificial cilia. As a result of the piezoelectric catalytic performance, Ag nanoparticles were dispersed on the ZnO cilia. The catalytic performance was improved by this process. The mechanical energy was converted into chemical energy, and the study paved the path towards other energy conversions such as hydrogen energy conversion from waste energy.

6.4.6. Antifouling or Self-Cleaning

Fouling is the accumulation of unwanted microparticles in the liquid surfaces, water quality analyzers, marine sensors, and lab-on-a-chip applications. Actuation of the artificial in the microfluidic chip removed 99% of algae, and the study [176] (Figure 6E) shows artificial cilia can be used for antifouling. The level of cleanliness can be found using the simple formula as follows,

$$Cleanliness = 1 - \left(\frac{A_{Heavy} + \frac{1}{2} A_{Normal}}{A_{Total}} \right) \tag{1}$$

In the total observation area (A_{total}), A_{heavy} refers to a heavily affected zone, A_{normal} refers to a commonly affected zone [176]. In the existing literature on antifouling or self-cleaning, the magnetic artificial cilia employed were several hundred micrometers in length mostly. The dimension limits the application of magnetic artificial cilia to a large scale in microfluidic chips. Recently, an investigation [177] was reported in which the magnetic artificial cilia were sized the same as their biological counterpart which could provide a high degree of cleanliness from 69% in the worst scenarios to almost 100% in the frequency of 40Hz. The other recent discussions over self-cleaning by artificial cilia can be found elsewhere [169,178,179].

6.4.7. Photocatalysis

Photocatalysis attempts a promising approach to create a sustainable society. Some examples are photocatalytic electrolysis of water, solar power generation, carbon capture technology, and pollutant degradation. So, researchers decided to enhance the photocatalytic activity using motile ciliary films. They found that the observed performance is better than the conventional stationery photocatalyst films. Many studies have exposed that artificial cilia with various sizes and dynamic beating behaviors improved photocatalytic activity [134,180]. They address issues such as environmental contamination and energy supply problems. For example, a study [181] shows that the synergic effects of the magnetic artificial cilium, ZnO nanowires, and CdS quantum dots enhanced the H_2 generation considerably. Several materials, such as TiO_2, ZnO, CdS, MoS_2, and $BiVO_4$, have been reported as efficient photocatalyists through testing them in artificial cilia embedded device. It is well known that powder photocatalyst inherently benefits from interior mass transfer in quick succession. As discussed earlier, the artificial cilia-based micromixer can mix rapidly. Hence, the artificial cilia enhanced photocatalytic performance by improving the internal mass transfer [180,182–185]. The study (Zhang et al.) discussed that researchers combined magnetically actuated artificial cilia and motile photocatalyst film which led to a three-fold improvement in photocatalytic activity compared to traditional planar film [180]. ZnO is a well-known photocatalyst, promised material due to its nontoxicity and thermal stability. Peng et al. established a study where ZnO nanorods were converted to ZnO nanosheets by only increasing the seeding time. The ZnO nanosheets with exposed (001) facets were inserted into the motile inner film by seed-mediated hydrothermal growth strategy for the first time. In the study, (001) facets exposed ZnO nanosheet arrays on the magnetic artificial cilia, and ZnO nanorod arrays film were tested for photocatalytic efficiency. It was realized that ZnO nanosheet arrays with active (001) facets film-coated cilia have 2.4 fold better photolytic performance compared to ZnO nanorods [185]. The schematic (Figure 6F) has illustrated the strategy of artificial cilia for the photocatalysis process. Another study reported that 2D TiO_2 nanosheet film was associated with 3D magnetic artificial cilium. Under the influence of the magnetic field rotation of 800 rpm yielded RhB degradation considerably [183]. Peng et al. further designed a photocatalytic film that integrated the $BiVO_4$, ZnO, and magnetic artificial cilia to increase mass transfer, light absorption, and photocatalytic activity [182]. Another recent study [93] showed the possibility of expanding the photodegradation capability by changing the arrangement of artificial cilia.

6.4.8. Particle Manipulation

Particle manipulation is the process where single or multiple solid particles or liquid droplets are transported in the sub-microscale fluid medium by artificial cilia. Droplet-based microfluidics involves the manipulations of (a) water droplets and (b) oil droplets. Strong observations of droplet generation, merging, separation, and sorting are required for such droplet manipulation. On the other hand, solid particle-based microfluidics involves the manipulations of (c) polymer or viscoelastic particles.

(a) Water droplets manipulation

In water droplet manipulation, single or multiple water droplets were transported in the liquid or air medium by the ubiquitous established strategies by the artificial cilia. Magnetically responsive artificial cilia were primarily implemented to direct the water droplets in the microfluidic device. Tilting angle and contact angle from the surface are the predominant parameters of the artificial cilia to control the droplets. Those angles by means of bending deflections were adjusted using the magnetic field. In addition, the ciliary body was enhanced with advanced necessity surface properties such as hydrophobic and wettability, optical properties, and intrinsic body materials to control the droplets. The investigation [172] discussed a novel droplet transporting using the ferromagnetic artificial cilia. In the study, the artificial cilia were upgraded with the switchable hydrophobicity wettability, and the surface properties. The wettability and adhesion can be switched reversely by applying the magnetic field. The on/off control of the magnetic field was utilized to bring back the adhesion and surface sliding consecutively to transport the droplets. If the magnetic field was switched off, the surface would become water-repellent and if the magnetic field was switched on the surface would become water-adhesive.

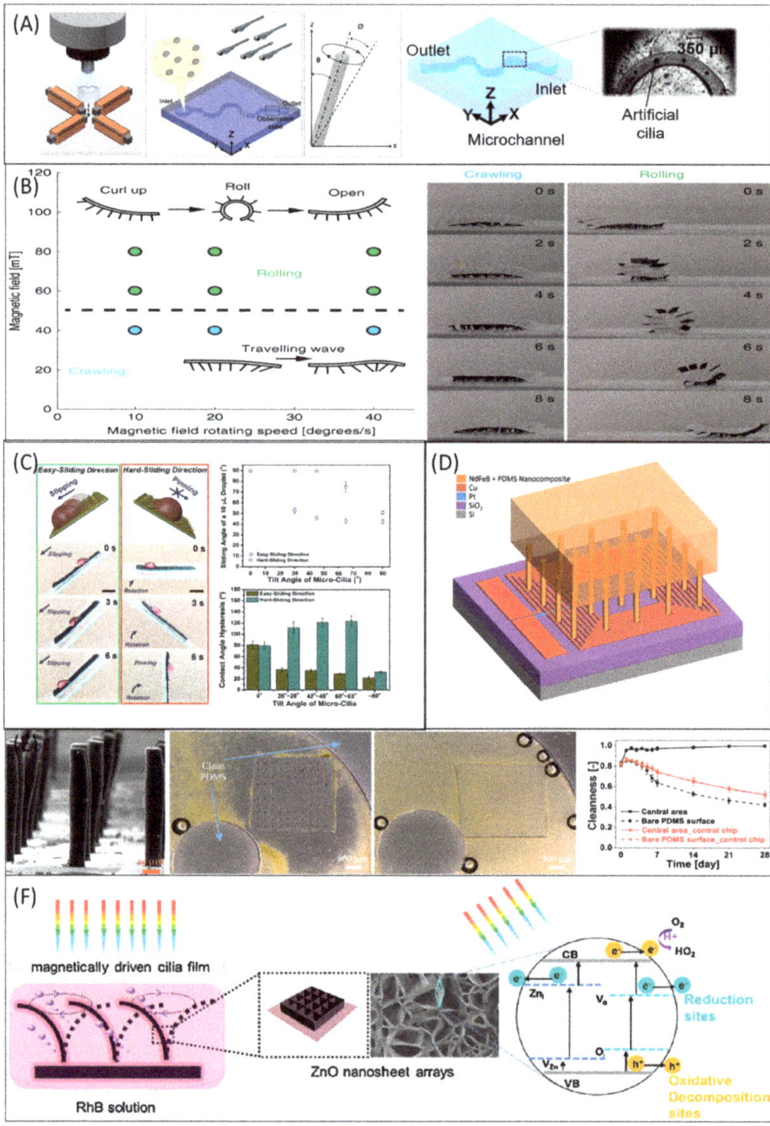

Figure 6. Contemporary emerging applications. (**A**) Illustration of artificial cilia used for Zebrafish research. The magnetic actuation setup was used to control the cilia (**left**). The microfluidic channel embedded the artificial cilia to activate the zebrafish sperms (**second left**). SEM picture of the observation area (**right**). The figure was reproduced with permission from [94], under a Creative Commons BY (CC BY 4.0) license, published by Springer Nature, 2018. (**B**) Artificial cilia embedded

minimal robots/Microrobots and soft robots. The photographic images show the crawling and rolling movement of the soft robot (**right**). The graph plotted the magnetic field rotational speed and the magnetic field (**left**). The figure was reproduced with permission from [65], under a Creative Commons BY (CC BY 4.0) license, published by Springer Nature, 2020. (**C**) Artificial cilia surface with wettability. For the sake of wettable property, the slipping and pinning of the droplet on the ciliary array were showcased (**right**). The figure was reproduced with permission from [170], published by John Wiley and Sons, 2017. (**D**) The energy harvester comprised two main components: (**i**) the microfabricated coils on the plane and (**ii**) high aspect ratio (HAR) artificial cilia. The modulus of elasticity of proof mass and HAR of the artificial cilia were endeavored to have the setup to respond to low frequencies. The figure was reproduced with permission from [174], published by John Wiley and Sons, 2018. (**E**) Artificial cilia for antifouling and self-cleaning. The ciliated area was clean compared with the control group (**second from left**). Non-actuated cilia led to fouling, captured after 28 days using fluorescent microscopy (third one from left). The figure was reproduced with permission from [176], under a Creative Commons BY Non-Commercial No Derivative Works (CC BY-NC-ND) license, published by the American Chemical Society, 2020. (**F**) The schematic has illustrated the strategy of artificial cilia for the photocatalysis process. Inner-motile ZnO nanofilm (001) facets were actuated to increase the mass transfer and mixing performance and improve the photocatalytic process. The figure was reproduced with permission from [185], published by Elsevier, 2017.

Instead of depending upon the magnetic field ultimately, hydrophobicity and slippery properties were achieved on the surfaces of artificial cilia by infusing them in the lubricants. For instance, the artificial cilia [186] were fabricated using PDMS and carbonyl iron microspheres. The polystyrene (PS) nanoparticles were deposited on the ciliary body. Thus, it was comprised of a hierarchical structure. The water droplet was pinned by the hydrophobic property of the ciliary tip. The infused lubricant (perfluorinated oils) indulged in the slipperiness of the droplet when the ciliary body was tilted due to the exerted magnetic field. In the same year, a similar slippery approach was used in another study [170]. Figure 7A illustrates the transportation of the water droplet by artificial ciliary lubricated surfaces. The lubricated surface of the magnetic responsive arrays was achieved by filling in the silicone oil. In the uniform magnetic field, the entire ciliary array responded to the magnetic field.

On the other hand, a single row or column of artificial cilia responded to the dual magnets comprised junction induced magnetic field, which can create precise unidirectional wave motion. Double the magnet junctions have been introduced as a spectacular milestone in droplet manipulation [37,39]. Song et al. [39] demonstrated that the artificial cilia that moved the water droplet using dual magnets comprised junctions induced unidirectional wave motion using a single column of the ciliary array. The single droplet was identified to travel in the straight line and arc orbit whereas two droplets were moved in the parallel trajectories and mixed in the single orbit.

(b) Oil droplets manipulation

Oil droplet manipulation has been getting attention recently in material processes, antifouling, and self-cleaning. For example, Zhang and his team [171] fabricated artificial cilia with mushroom head microstructures. The artificial cilia were showcased the amphiphilic ability by manipulating oil droplets in the water medium and water droplets in the air medium by switching reverse wettability. Importantly, no lubricant treatment was required for such artificial cilia. Another study [169] demonstrated the oleophobic ability of the artificial microcilia by transporting oil droplets in air and underwater by transforming the surface into hydrophobic.

(c) Polymer or viscoelastic particles manipulation

Over many years, the transporting mechanics of the artificial cilia were improved drastically, yet there is a lack of innovation towards the transportation of the viscous/polymer particles. The conical trajectory movement of the beating cilia, changing shapes, and the frictional forces of the beating artificial cilia are the strategies behind the manipulations of the viscous/polymer particles [38,178]. The relationship between particle transportation and the ratio of particle size to cilia pitch was reported in the study [187]. A recent study [38] ascertained the manipulation of polylactic acid (PLA) particle transportation in air and water using the magnetically actuated artificial cilia. It was concluded that the adhesive and the frictional forces between the artificial cilia and the moving particles are necessary for particle transportation. The transportation was achieved in the direction of the tilted conical movement and the effective stroke direction [38]. Figure 7B showcased the transport and trapping capability of artificial ciliary bands. By changing the acoustic-electric field, both transporting and trapping beads in the aqueous medium were achieved by the same microarchitecture of the ciliary band. Pacheco et al. [188] proposed a study where the modular materials such as L-Mu^{3}Gel and CF-Mu^{3}Gel, equivalent to the physiological and CF mucus, respectively, were transported successfully with the magnetic artificial cilia. The artificial mucus models L-Mu^{3}Gel and CF-Mu^{3}Gel can be created using readily available materials. The study helps to understand the drug processing time in the mucus clearance ciliary region. Figure 7C illustrates the modeling of human mucus transportation.

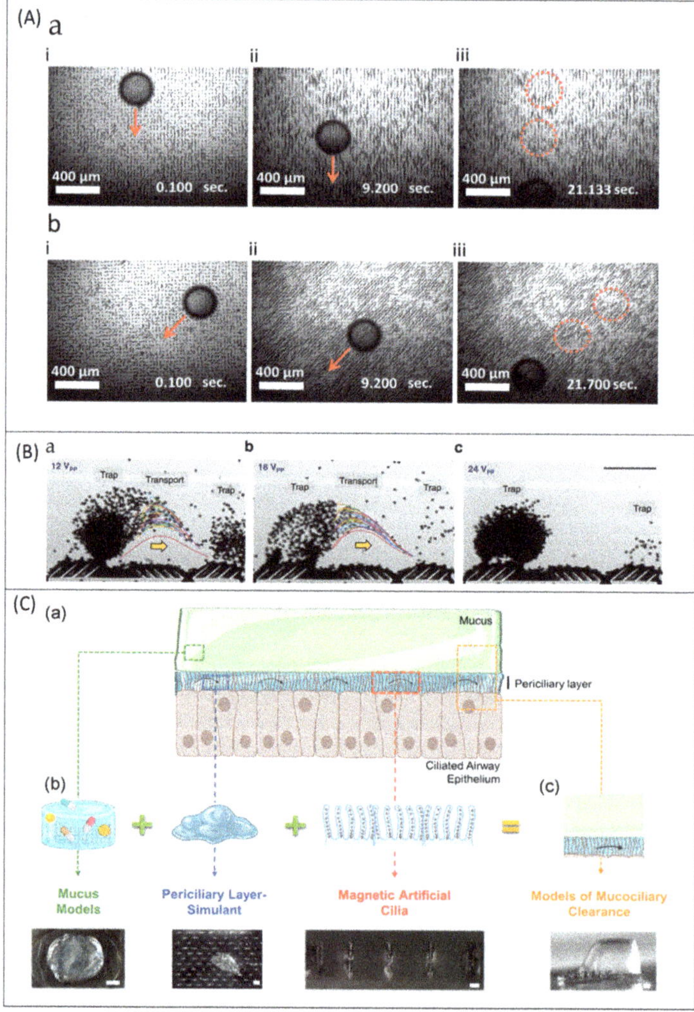

Figure 7. Artificial cilia for particle manipulation. **(A)** The transportation of the water droplet by artificial

ciliary lubricated surfaces was shown in the schematics. Filling in the silicone oil led to the lubricated surfaces of the magnetic responsive arrays. The figure was reproduced with permission from [189], under a Creative Commons BY (CC BY) license, published by AIP Publishing, 2020. (**B**) The transport and trapping capability of artificial ciliary bands were showcased. By changing the acoustic-electric field, both transporting and trapping beads in the aqueous medium were achieved by the same microarchitecture of the ciliary band. The figure was reproduced with permission from [29], under a Creative Commons BY (CC BY 4.0) license, published by Springer Nature, 2021. (**C**) Illustration of modeling human mucus transportation. As a milestone in the viscoelastic particle transportation by artificial cilia, physiological and pathological mucus were modeled using L-Mu^{3}Gel and CF-Mu^{3}Gel and achieved transportation in the respiratory airway model by artificial cilia. The figure was reproduced with permission from [188], published by John Wiley and Sons, 2021.

7. Conclusions and Future Directions

Artificial cilia have been developed to be implicit dominant tools for microfluidic applications. Certain research groups have sought to increase artificial cilia's efficiency, especially in providing micromixing, pumping, and particle handling applications for microfluidic purposes. In addition, artificial cilia have been enriched with the ability to carry, sense, communicate and locomote. This review briefly discussed the concepts from the current developments, fabrication processes, actuation strategies, dynamic beating behaviors, functionalities, uniqueness, the potential applications, deliverables, remaining challenges, and the future trends by biomimetic cilia. The advantages and disadvantages of each strategy have been discussed explicitly.

Bringing artificial cilia to commercial applications has been challenging since the work was started. Upgrading the artificial cilia abilities has been exceptionally vital for understanding the mechanisms that can be utilized in the future. Cost-effective and efficient manufacturing strategies must be set up to deliver artificial cilia to more actual utilization in lab-on-a-chip gadgets. Even though the natural cilia are highly different from today's biomimetic cilia, advances in technology will undoubtedly produce artificial cilia that are close to the natural cilia and generate a large variety of new and exciting achievements in the future.

Supplementary Materials: The following are available online at https://www.mdpi.com/article/10.3390/mi13050735/s1, Table S1: Classifications of technologies and actuation/sensing mechanisms.

Author Contributions: C.-Y.C. conceived of the presented idea. V.S. and B.P. wrote the manuscript. C.-Y.C. revised the manuscript and supervised the work. All authors have read and agreed to the published version of the manuscript.

Funding: This study was supported through the Ministry of Science and Technology of Taiwan under Contract No. MOST 108-2221-E-006-221-MY4 (to Chia-Yuan Chen). This work would not be possible without the facility provided by the Center for Micro/Nano Science and Technology, National Cheng Kung University. This research was supported in part by Higher Education Sprout Project, Ministry of Education to the Headquarters of University Advancement at National Cheng Kung University (NCKU).

Conflicts of Interest: The authors declare no conflict of interest.

References

1. Sleigh, M.A. *Cilia and Flagella*; Elsevier Science & Technology Books: Amsterdam, The Netherlands, 1974.
2. Mitchison, H.M.; Valente, E.M. Motile and non-motile cilia in human pathology: From function to phenotypes. *J. Pathol.* **2017**, *241*, 294–309. [CrossRef] [PubMed]
3. Ibañez-Tallon, I.; Heintz, N.; Omran, H. To beat or not to beat: Roles of cilia in development and disease. *Hum. Mol. Genet.* **2003**, *12*, R27–R35. [CrossRef] [PubMed]
4. Rosenbaum, J.L.; Witman, G.B. Intraflagellar transport. *Nat. Rev. Mol. Cell Biol.* **2002**, *3*, 813–825. [CrossRef] [PubMed]
5. Gilpin, W.; Prakash, V.N.; Prakash, M. Vortex arrays and ciliary tangles underlie the feeding–swimming trade-off in starfish larvae. *Nat. Phys.* **2017**, *13*, 380. [CrossRef]
6. Mayne, R.; Whiting, J.G.; Wheway, G.; Melhuish, C.; Adamatzky, A. Particle sorting by *Paramecium cilia* arrays. *Biosystems* **2017**, *156*, 46–52. [CrossRef]

7. Hirokawa, N.; Tanaka, Y.; Okada, Y.; Takeda, S. Nodal flow and the generation of left-right asymmetry. *Cell* **2006**, *125*, 33–45. [CrossRef]
8. Smith, D.; Blake, J.; Gaffney, E. Fluid mechanics of nodal flow due to embryonic primary cilia. *J. R. Soc. Interface* **2008**, *5*, 567–573. [CrossRef]
9. Whitesides, G.M. The origins and the future of microfluidics. *Nature* **2006**, *442*, 368–373. [CrossRef]
10. Su, P.; Ren, C.; Fu, Y.; Guo, J.; Member, I.; Guo, J.; Yuan, Q. Magnetophoresis in microfluidic lab: Recent advance. *Sens. Actuators. A Phys.* **2021**, *332*, 113180. [CrossRef]
11. Cao, Q.L.; Fan, Q.; Chen, Q.; Liu, C.T.; Han, X.T.; Li, L. Recent advances in manipulation of micro- and nano-objects with magnetic fields at small scales. *Mater. Horiz.* **2020**, *7*, 638–666. [CrossRef]
12. Shanko, E.S.; van de Burgt, Y.; Anderson, P.D.; den Toonder, J.M.J. Microfluidic magnetic mixing at low reynolds numbers and in stagnant fluids. *Micromachines* **2019**, *10*, 731. [CrossRef]
13. Galera, A.C.; San Miguel, V.; Baselga, J. Magneto-mechanical surfaces design. *Chem. Rec.* **2018**, *18*, 1010–1019. [CrossRef] [PubMed]
14. Yunas, J.; Mulyanti, B.; Hamidah, I.; Mohd Said, M.; Pawinanto, R.E.; Ali, W.; Subandi, A.; Hamzah, A.A.; Latif, R.; Yeop Majlis, B. Polymer-based MEMS electromagnetic actuator for biomedical application: A review. *Polymers* **2020**, *12*, 1184. [CrossRef]
15. Mehta, K.; Peeketi, A.R.; Liu, L.; Broer, D.; Onck, P.; Annabattula, R.K. Design and applications of light responsive liquid crystal polymer thin films. *Appl. Phys. Rev.* **2020**, *7*, 041306. [CrossRef]
16. El-Atab, N.; Mishra, R.B.; Al-Modaf, F.; Joharji, L.; Alsharif, A.A.; Alamoudi, H.; Diaz, M.; Qaiser, N.; Hussain, M.M. Soft actuators for soft robotic applications: A review. *Adv. Intell. Syst.* **2020**, *2*, 2000128. [CrossRef]
17. Zhang, S.Z.; Wang, Y.; Onck, P.; den Toonder, J. A concise review of microfluidic particle manipulation methods. *Microfluid. Nanofluid.* **2020**, *24*, 1. [CrossRef]
18. Xuan, X.C. Recent advances in direct current electrokinetic manipulation of particles for microfluidic applications. *Electrophoresis* **2019**, *40*, 2484–2513. [CrossRef]
19. You, S.; Li, J.; Zhu, W.; Yu, C.; Mei, D.; Chen, S. Nanoscale 3D printing of hydrogels for cellular tissue engineering. *J. Mater. Chem. B* **2018**, *6*, 2187–2197. [CrossRef]
20. Zhang, X.; Guo, J.; Fu, X.; Zhang, D.; Zhao, Y. Tailoring flexible arrays for artificial cilia actuators. *Adv. Intell. Syst.* **2020**, *3*, 2000225. [CrossRef]
21. den Toonder, J.M.J.; Onck, P.R. Microfluidic manipulation with artificial/bioinspired cilia. *Trends Biotechnol.* **2013**, *31*, 85–91. [CrossRef]
22. Shapiro, O.H.; Fernandez, V.I.; Garren, M.; Guasto, J.S.; Debaillon-Vesque, F.P.; Kramarsky-Winter, E.; Vardi, A.; Stocker, R. Vortical ciliary flows actively enhance mass transport in reef corals. *Proc. Natl. Acad. Sci. USA* **2014**, *111*, 13391–13396. [CrossRef]
23. Ben, S.; Yao, J.J.; Ning, Y.Z.; Zhao, Z.H.; Zha, J.L.; Tian, D.L.; Liu, K.S.; Jiang, L. A bioinspired magnetic responsive cilia array surface for microspheres underwater directional transport. *Sci. China-Chem.* **2020**, *63*, 347–353. [CrossRef]
24. Lim, D.; Lahooti, M.; Kim, D. Inertia-driven flow symmetry breaking by oscillating plates. *AIP Adv.* **2019**, *9*, 105119. [CrossRef]
25. Chateau, S.; Favier, J.; Poncet, S.; D'Ortona, U. Why antiplectic metachronal cilia waves are optimal to transport bronchial mucus. *Phys. Rev. E* **2019**, *100*, 042405. [CrossRef]
26. Gilpin, W.; Bull, M.S.; Prakash, M. The multiscale physics of cilia and flagella. *Nat. Rev. Phys.* **2020**, *2*, 74–88. [CrossRef]
27. Gemmell, B.J.; Colin, S.P.; Costello, J.H.; Sutherland, K.R. A ctenophore (comb jelly) employs vortex rebound dynamics and outperforms other gelatinous swimmers. *R. Soc. Open Sci.* **2019**, *6*, 181615. [CrossRef] [PubMed]
28. Arnellos, A.; Keijzer, F. Bodily complexity: Integrated multicellular organizations for contraction-based motility. *Front. Physiol.* **2019**, *10*, 1268. [CrossRef] [PubMed]
29. Dillinger, C.; Nama, N.; Ahmed, D. Ultrasound-activated ciliary bands for microrobotic systems inspired by starfish. *Nat. Commun.* **2021**, *12*, 6455. [CrossRef] [PubMed]
30. Fan, Y.; Li, S.; Wei, D.; Fang, Z.; Han, Z.; Liu, Y. Bioinspired Superhydrophobic Cilia for droplets transportation and microchemical reaction. *Adv. Mater. Interfaces* **2021**, *8*, 2101408. [CrossRef]
31. Zhang, T.Z.; Wang, Y.F.; Dai, B.; Xu, T.L. Bioinspired transport surface driven by air flow. *Adv. Mater. Interfaces* **2020**, *7*, 2001331. [CrossRef]
32. Furusawa, M.; Maeda, K.; Azukizawa, S.; Shinoda, H.; Tsumori, F. Bio-mimic motion of elastic material dispersed with hard-magnetic particles. *J. Photopolym. Sci. Technol.* **2019**, *32*, 309–313. [CrossRef]
33. Luo, Z.R.; Evans, B.A.; Chang, C.H. Magnetically actuated dynamic iridescence inspired by the neon tetra. *Acs Nano* **2019**, *13*, 4657–4666. [CrossRef] [PubMed]
34. Nishimura, R.; Fujimoto, A.; Yasuda, N.; Morimoto, M.; Nagasaka, T.; Sotome, H.; Ito, S.; Miyasaka, H.; Yokojima, S.; Nakamura, S.; et al. Object transportation system mimicking the cilia of *Paramecium aurelia* making use of the light-controllable crystal bending behavior of a photochromic diarylethene. *Angewandte Chemie-Int. Ed.* **2019**, *58*, 13308–13312. [CrossRef] [PubMed]
35. Whiting, J.G.H.; Mayne, R.; Melhuish, C.; Adamatzky, A. A Cilia-inspired closed-loop sensor-actuator array. *J. Bionic Eng.* **2018**, *15*, 526–532. [CrossRef]
36. Zhang, R.; den Toonder, J.; Onck, P.R. Transport and mixing by metachronal waves in nonreciprocal soft robotic pneumatic artificial cilia at low Reynolds numbers. *Phys. Fluids* **2021**, *33*, 092009. [CrossRef]

37. Demirors, A.F.; Aykut, S.; Ganzeboom, S.; Meier, Y.A.; Hardeman, R.; Graaf, J.; Mathijssen, A.; Poloni, E.; Carpenter, J.A.; Unlu, C.; et al. Amphibious transport of fluids and solids by soft magnetic carpets. *Adv. Sci.* **2021**, *8*, 2102510. [CrossRef]
38. Zhang, S.Z.; Zhang, R.J.; Wang, Y.; Onck, P.R.; den Toonder, J.M.J. Controlled multidirectional particle transportation by magnetic artificial cilia. *ACS Nano* **2020**, *14*, 10313–10323. [CrossRef]
39. Song, Y.G.; Jiang, S.J.; Li, G.Q.; Zhang, Y.C.; Wu, H.; Xue, C.; You, H.S.; Zhang, D.H.; Cai, Y.; Zhu, J.G.; et al. Cross-species bioinspired anisotropic surfaces for active droplet transportation driven by unidirectional microcolumn waves. *ACS Appl. Mater. Interfaces* **2020**, *12*, 42264–42273. [CrossRef]
40. Wang, Y.F.; Chen, X.D.; Sun, K.; Li, K.; Zhang, F.L.; Dai, B.; Shen, J.; Hu, G.Q.; Wang, S.T. Directional transport of centimeter-scale object on anisotropic microcilia surface under water. *Sci. China-Mater.* **2019**, *62*, 236–244. [CrossRef]
41. Kamat, A.M.; Pei, Y.; Kottapalli, A.G. Bioinspired cilia sensors with graphene sensing elements fabricated using 3D printing and casting. *Nanomaterials* **2019**, *9*, 954. [CrossRef]
42. Kamat, A.M.; Pei, Y.; Jayawardhana, B.; Kottapalli, A.G.P. Biomimetic soft polymer microstructures and piezoresistive graphene mems sensors using sacrificial metal 3D printing. *ACS Appl. Mater. Interfaces* **2021**, *13*, 1094–1104. [CrossRef] [PubMed]
43. Dehdashti, E.; Reich, G.W.; Masoud, H. Collective sensitivity of artificial hair sensors to flow direction. *AIAA J.* **2021**, *59*, 1135–1141. [CrossRef]
44. Carvalho, M.; Ribeiro, P.; Romão, V.; Cardoso, S. Smart fingertip sensor for food quality control: Fruit maturity assessment with a magnetic device. *J. Magn. Magn. Mater.* **2021**, 168116. [CrossRef]
45. Sengupta, D.; Trap, D.; Kottapalli, A.G.P. Piezoresistive carbon nanofiber-based cilia-inspired flow sensor. *Nanomaterials* **2020**, *10*, 211. [CrossRef] [PubMed]
46. Kamat, A.M.; Zheng, X.; Jayawardhana, B.; Kottapalli, A.G.P. Bioinspired PDMS-graphene cantilever flow sensors using 3D printing and replica moulding. *Nanotechnology* **2020**, *32*, 095501. [CrossRef]
47. Zhou, Q.; Ji, B.; Wei, Y.Z.; Hu, B.; Gao, Y.B.; Xu, Q.S.; Zhou, J.; Zhou, B.P. A bio-inspired cilia array as the dielectric layer for flexible capacitive pressure sensors with high sensitivity and a broad detection range. *J. Mater. Chem. A* **2019**, *7*, 27334–27346. [CrossRef]
48. Alfadhel, A.; Li, B.; Zaher, A.; Yassine, O.; Kosel, J. A magnetic nanocomposite for biomimetic flow sensing. *Lab Chip* **2014**, *14*, 4362–4369. [CrossRef]
49. Chen, C.-Y.; Hsu, C.-C.; Mani, K.; Panigrahi, B. Hydrodynamic influences of artificial cilia beating behaviors on micromixing. *Chem. Eng. Process Process Intensif.* **2016**, *99*, 33–40. [CrossRef]
50. Chen, C.-Y.; Chen, C.-Y.; Lin, C.-Y.; Hu, Y.-T. Magnetically actuated artificial cilia for optimum mixing performance in microfluidics. *Lab Chip* **2013**, *13*, 2834–2839. [CrossRef]
51. Chen, C.-Y.; Cheng, L.Y.; Hsu, C.C.; Mani, K. Microscale flow propulsion through bioinspired and magnetically actuated artificial cilia. *Biomicrofluidics* **2015**, *9*, 034105. [CrossRef]
52. Huang, P.-Y.; Panigrahi, B.; Lu, C.-H.; Huang, P.-F.; Chen, C.-Y. An artificial cilia-based micromixer towards the activation of zebrafish sperms. *Sens. Actuators. B Chem.* **2017**, *244*, 541–548. [CrossRef]
53. Lu, C.-H.; Tang, C.-H.; Ghayal, N.; Panigrahi, B.; Chen, C.-Y.; Chen, C.-Y. On the improvement of visible-responsive photodegradation through artificial cilia. *Sens. Actuators. A Phys.* **2019**, *285*, 234–240. [CrossRef]
54. Wu, Y.-A.; Panigrahi, B.; Chen, C.-Y. Hydrodynamically efficient micropropulsion through a new artificial cilia beating concept. *Microsyst. Technol.* **2017**, *23*, 5893–5902. [CrossRef]
55. Panigrahi, B.; Vignesh, S.; Chen, C.-Y. Shape-programmable artificial cilia for microfluidics. *iScience* **2021**, *24*, 103367. [CrossRef]
56. Wu, Y.A.; Panigrahi, B.; Lu, Y.H.; Chen, C.Y. An integrated artificial cilia based microfluidic device for micropumping and micromixing applications. *Micromachines* **2017**, *8*, 260. [CrossRef]
57. Ul Islam, T.; Bellouard, Y.; den Toonder, J.M.J. Highly motile nanoscale magnetic artificial cilia. *Proc. Natl. Acad. Sci. USA* **2021**, *118*. [CrossRef]
58. Hanasoge, S.; Ballard, M.; Hesketh, P.J.; Alexeev, A. Asymmetric motion of magnetically actuated artificial cilia. *Lab Chip* **2017**, *17*, 3138–3145. [CrossRef]
59. Vilfan, M.; Potočnik, A.; Kavčič, B.; Osterman, N.; Poberaj, I.; Vilfan, A.; Babič, D. Self-assembled artificial cilia. *Proc. Natl. Acad. Sci. USA* **2010**, *107*, 1844–1847. [CrossRef]
60. Wang, Y.; Gao, Y.; Wyss, H.; Anderson, P.; den Toonder, J. Out of the cleanroom, self-assembled magnetic artificial cilia. *Lab Chip* **2013**, *13*, 3360–3366. [CrossRef]
61. Bryan, M.T.; Martin, E.L.; Pac, A.; Gilbert, A.D.; Ogrin, F.Y. Metachronal waves in magnetic micro-robotic paddles for artificial cilia. *Commun. Mater.* **2021**, *2*, 14. [CrossRef]
62. Zhang, S.; Cui, Z.; Wang, Y.; den Toonder, J. Metachronal μ-cilia for on-chip integrated pumps and climbing robots. *ACS Appl. Mater. Interfaces* **2021**, *13*, 20845–20857. [CrossRef] [PubMed]
63. Tyagi, M.; Spinks, G.M.; Jager, E.W. Fully 3D printed soft microactuators for soft microrobotics. *Smart Mater. Struct.* **2020**, *29*, 085032. [CrossRef]
64. Liu, F.; Alici, G.; Zhang, B.; Beirne, S.; Li, W. Fabrication and characterization of a magnetic micro-actuator based on deformable Fe-doped PDMS artificial cilium using 3D printing. *Smart Mater. Struct.* **2015**, *24*, 035015. [CrossRef]
65. Gu, H.; Boehler, Q.; Cui, H.; Secchi, E.; Savorana, G.; De Marco, C.; Gervasoni, S.; Peyron, Q.; Huang, T.-Y.; Pane, S. Magnetic cilia carpets with programmable metachronal waves. *Nat. Commun.* **2020**, *11*, 2637. [CrossRef] [PubMed]

66. Tibbits, S. 4D printing: Multi-material shape change. *Archit. Des.* **2014**, *84*, 116–121. [CrossRef]
67. Azukizawa, S.; Shinoda, H.; Tsumori, F. 4D-printing system for elastic magnetic actuators. In Proceedings of the 2019 IEEE 32nd International Conference on Micro Electro Mechanical Systems (MEMS), Seoul, Korea, 27–31 January 2019; pp. 248–251.
68. Azukizawa, S.; Shinoda, H.; Tokumaru, K.; Tsumori, F. 3D printing system of magnetic anisotropy for artificial cilia. *J. Photopolym. Sci. Technol.* **2018**, *31*, 139–144. [CrossRef]
69. Timonen, J.V.; Johans, C.; Kontturi, K.s.; Walther, A.; Ikkala, O.; Ras, R.H. A facile template-free approach to magnetodriven, multifunctional artificial cilia. *ACS Appl. Mater. Interfaces* **2010**, *2*, 2226–2230. [CrossRef]
70. Sanchez, T.; Welch, D.; Nicastro, D.; Dogic, Z. Cilia-like beating of active microtubule bundles. *Science* **2011**, *333*, 456–459. [CrossRef]
71. Wang, Y.; den Toonder, J.; Cardinaels, R.; Anderson, P. A continuous roll-pulling approach for the fabrication of magnetic artificial cilia with microfluidic pumping capability. *Lab Chip* **2016**, *16*, 2277–2286. [CrossRef]
72. Jeong, W.; Jeong, S.M.; Lim, T.; Hang, C.Y.; Yang, H.; Lee, B.W.; Park, S.Y.; Ju, S. Self-emitting artificial cilia produced by field effect spinning. *ACS Appl. Mater. Interfaces* **2019**, *11*, 35286–35293. [CrossRef]
73. Becker, K.P.; Chen, Y.F.; Wood, R.J. Mechanically programmable dip molding of high aspect ratio soft actuator arrays. *Adv. Funct. Mater.* **2020**, *30*, 1908919. [CrossRef]
74. Belardi, J.; Schorr, N.; Prucker, O.; Rühe, J. Artificial cilia: Generation of magnetic actuators in microfluidic systems. *Adv. Funct. Mater.* **2011**, *21*, 3314–3320. [CrossRef]
75. Gaudet, M.; Arscott, S. Optical actuation of microelectromechanical systems using photoelectrowetting. *Appl. Phys. Lett.* **2012**, *100*, 224103. [CrossRef]
76. Van Oosten, C.L.; Bastiaansen, C.W.; Broer, D.J. Printed artificial cilia from liquid-crystal network actuators modularly driven by light. *Nat. Mater.* **2009**, *8*, 677–682. [CrossRef]
77. Gelebart, A.H.; Mc Bride, M.; Schenning, A.; Bowman, C.N.; Broer, D.J. Photoresponsive fiber array: Toward mimicking the collective motion of cilia for transport applications. *Adv. Funct. Mater.* **2016**, *26*, 5322–5327. [CrossRef]
78. den Toonder, J.; Bos, F.; Broer, D.; Filippini, L.; Gillies, M.; de Goede, J.; Mol, T.; Reijme, M.; Talen, W.; Wilderbeek, H. Artificial cilia for active micro-fluidic mixing. *Lab Chip* **2008**, *8*, 533–541. [CrossRef] [PubMed]
79. Danis, U.; Rasooli, R.; Chen, C.-Y.; Dur, O.; Sitti, M.; Pekkan, K. Thrust and hydrodynamic efficiency of the bundled flagella. *Micromachines* **2019**, *10*, 449. [CrossRef]
80. Minami, K.; Yano, S. Electrostatic micro actuator with distributed ciliary electrodes (E-Maccel). *IEEJ Trans. Sens. Micromach.* **2004**, *124*, 381–386. [CrossRef]
81. Glazer, P.; Leuven, J.; An, H.; Lemay, S.; Mendes, E. Hydrogel-based multi-stimuli responsive cilia. In Proceedings of the 2013 NSTI Nanotechnology Conference and Expo, Nanotech 2013, Washington, DC, USA, 12–16 May 2013; pp. 138–141.
82. Ionov, L. Hydrogel-based actuators: Possibilities and limitations. *Mater. Today* **2014**, *17*, 494–503. [CrossRef]
83. Zarzar, L.D.; Kim, P.; Aizenberg, J. Bio-inspired design of submerged hydrogel-actuated polymer microstructures operating in response to pH. *Adv. Mater.* **2011**, *23*, 1442–1446. [CrossRef]
84. Pokroy, B.; Epstein, A.K.; Persson-Gulda, M.C.; Aizenberg, J. Fabrication of bioinspired actuated nanostructures with arbitrary geometry and stiffness. *Adv. Mater.* **2009**, *21*, 463–469. [CrossRef]
85. Oh, K.; Chung, J.-H.; Devasia, S.; Riley, J.J. Bio-mimetic silicone cilia for microfluidic manipulation. *Lab Chip* **2009**, *9*, 1561–1566. [CrossRef] [PubMed]
86. Oh, K.; Smith, B.; Devasia, S.; Riley, J.J.; Chung, J.H. Characterization of mixing performance for bio-mimetic silicone cilia. *Microfluid. Nanofluid.* **2010**, *9*, 645–655. [CrossRef]
87. Baltussen, M.; Anderson, P.; Bos, F.; den Toonder, J. Inertial flow effects in a micro-mixer based on artificial cilia. *Lab Chip* **2009**, *9*, 2326–2331. [CrossRef] [PubMed]
88. O'Grady, M.L.; Kuo, P.-l.; Parker, K.K. Optimization of electroactive hydrogel actuators. *ACS Appl. Mater. Interfaces* **2010**, *2*, 343–346. [CrossRef]
89. Tsumori, F.; Saijou, A.; Osada, T.; Miura, H. Development of actuation system for artificial cilia with magnetic elastomer. *Jpn. J. Appl. Phys.* **2015**, *54*, 06FP12. [CrossRef]
90. Tsumori, F.; Marume, R.; Saijou, A.; Kudo, K.; Osada, T.; Miura, H. Metachronal wave of artificial cilia array actuated by applied magnetic field. *Jpn. J. Appl. Phys.* **2016**, *55*, 06GP19. [CrossRef]
91. Shinoda, H.; Tsumori, F. Development of micro pump using magnetic artificial cilia with metachronal wave. In Proceedings of the 2020 IEEE 33rd International Conference on Micro Electro Mechanical Systems (MEMS), Vancouver, BC, Canada, 18–22 January 2020; pp. 497–500.
92. Chen, C.-Y.; Lin, C.-Y.; Hu, Y.-T.; Cheng, L.-Y.; Hsu, C.-C. Efficient micromixing through artificial cilia actuation with fish-schooling configuration. *Chem. Eng. J.* **2015**, *259*, 391–396. [CrossRef]
93. Li, T.-C.; Panigraphi, B.; Chen, W.-T.; Chen, C.-Y.; Chen, C.-Y. Hydrodynamic benefits of artificial cilia distribution towards photodegradation processes. *Sens. Actuators. A Phys.* **2020**, *313*, 112184. [CrossRef]
94. Panigrahi, B.; Lu, C.H.; Ghayal, N.; Chen, C.Y. Sperm activation through orbital and self-axis revolutions using an artificial cilia embedded serpentine microfluidic platform. *Sci. Rep.* **2018**, *8*, 4605. [CrossRef]
95. Chen, C.-Y.; Chang Chien, T.-C.; Mani, K.; Tsai, H.-Y. Axial orientation control of zebrafish larvae using artificial cilia. *Microfluid. Nanofluid.* **2016**, *20*, 12. [CrossRef]

96. Hanasoge, S.; Hesketh, P.J.; Alexeev, A. Metachronal motion of artificial magnetic cilia. *Soft Matter* **2018**, *14*, 3689–3693. [CrossRef] [PubMed]
97. Hanasoge, S.; Hesketh, P.J.; Alexeev, A. Microfluidic pumping using artificial magnetic cilia. *Microsyst. Nanoeng.* **2018**, *4*, 11. [CrossRef]
98. Chen, C.-Y.; Lin, C.-Y.; Hu, Y.-T. Inducing 3D vortical flow patterns with 2D asymmetric actuation of artificial cilia for high-performance active micromixing. *Exp. Fluids* **2014**, *55*, 1765. [CrossRef]
99. Zhang, S.Z.; Wang, Y.; Lavrijsen, R.; Onck, P.R.; den Toonder, J.M.J. Versatile microfluidic flow generated by moulded magnetic artificial cilia. *Sens. Actuators. B-Chem.* **2018**, *263*, 614–624. [CrossRef]
100. Evans, B.; Shields, A.; Carroll, R.L.; Washburn, S.; Falvo, M.; Superfine, R. Magnetically actuated nanorod arrays as biomimetic cilia. *Nano Lett.* **2007**, *7*, 1428–1434. [CrossRef]
101. Orbay, S.; Ozcelik, A.; Bachman, H.; Huang, T.J. Acoustic actuation of in situ fabricated artificial cilia. *J. Micromech. Microeng.* **2018**, *28*, 025012. [CrossRef]
102. Dai, B.; Li, S.H.; Xu, T.L.; Wang, Y.F.; Zhang, F.L.; Gu, Z.; Wang, S.T. Artificial asymmetric cilia array of dielectric elastomer for cargo transportation. *ACS Appl. Mater. Interfaces* **2018**, *10*, 42979–42984. [CrossRef]
103. Sugioka, H.; Ishikawa, M. Artificial carbon cilium using induced charge electro-osmosis. *AIP Adv.* **2020**, *10*, 055302. [CrossRef]
104. Sugioka, H.; Ishikawa, M.; Kado, T. Selective carbon self-wiring from a graphite rod in water under a DC electric field. *J. Phys. Soc. Jpn.* **2020**, *89*. [CrossRef]
105. Sugioka, H.; Nakano, N.; Mizuno, Y. High-speed periodic beating motion of a spiral gold thread using induced charge electro-osmosis with a two-electrode structure. *J. Phys. Soc. Jpn.* **2019**, *88*, 084801. [CrossRef]
106. Sugioka, H.; Yoshijima, H. Metachronal motion of artificial cilia using induced charge electro-osmosis. *Coll. Surf. Physicochem. Eng. Aspects* **2021**, *626*, 127023. [CrossRef]
107. Sugioka, H.; Mizuno, Y.; Nambo, Y. Experimental demonstration of catching and releasing functions of artificial cilia using induced charge electro-osmosis. *J. Phys. Soc. Jpn.* **2020**, *89*, 054401. [CrossRef]
108. Milana, E.; Zhang, R.; Vetrano, M.R.; Peerlinck, S.; De Volder, M.; Onck, P.R.; Reynaerts, D.; Gorissen, B. Metachronal patterns in artificial cilia for low Reynolds number fluid propulsion. *Sci. Adv.* **2020**, *6*, eabd2508. [CrossRef] [PubMed]
109. Milana, E.; Gorissen, B.; Peerlinck, S.; De Volder, M.; Reynaerts, D. Artificial soft cilia with asymmetric beating patterns for biomimetic low-Reynolds-number fluid propulsion. *Adv. Funct. Mater.* **2019**, *29*, 1900462. [CrossRef]
110. Gorissen, B.; De Volder, M.; Reynaerts, D. Pneumatically-actuated artificial cilia array for biomimetic fluid propulsion. *Lab Chip* **2015**, *15*, 4348–4355. [CrossRef]
111. Gorissen, B.; Vincentie, W.; Al-Bender, F.; Reynaerts, D.; De Volder, M. Modeling and bonding-free fabrication of flexible fluidic microactuators with a bending motion. *J. Micromech. Microeng.* **2013**, *23*, 045012. [CrossRef]
112. Sugioka, H.; Kubota, M. Strong self-propelled swing motion due to the growing instability on heat transfer. *J. Phys. Soc. Jpn.* **2020**, *89*, 064402. [CrossRef]
113. Sugioka, H.; Kubota, M.; Tanaka, M. High-speed asymmetric motion of thermally actuated cilium. *J. Phys. Soc. Jpn.* **2020**, *89*, 114402. [CrossRef]
114. Sugioka, H.; Tomita, W.; Tanaka, M. Metachronal motion of a thermally actuated double pendulum driven by self-propulsion caused by spontaneous asymmetrical heat transfer. *J. Appl. Phys.* **2021**, *129*, 244701. [CrossRef]
115. Yao, Y.X.; Waters, J.T.; Shneidman, A.V.; Cui, J.X.; Wang, X.G.; Mandsberg, N.K.; Li, S.C.; Balazs, A.C.; Aizenberg, J. Multiresponsive polymeric microstructures with encoded predetermined and self-regulated deformability. *Proc. Natl. Acad. Sci. USA* **2018**, *115*, 12950–12955. [CrossRef] [PubMed]
116. Raak, R.J.H.; Houben, S.J.A.; Schenning, A.; Broer, D.J. Patterned and collective motion of densely packed tapered multiresponsive liquid crystal cilia. *Adv. Mater. Technol.* **2022**, *7*, 2101619. [CrossRef]
117. Glazer, P.J.; Leuven, J.; An, H.; Lemay, S.G.; Mendes, E. Multi-stimuli responsive hydrogel cilia. *Adv. Funct. Mater.* **2013**, *23*, 2964–2970. [CrossRef]
118. Li, M.; Kim, T.; Guidetti, G.; Wang, Y.; Omenetto, F.G. Optomechanically actuated microcilia for locally reconfigurable surfaces. *Adv. Mater.* **2020**, *32*, 2004147. [CrossRef] [PubMed]
119. Liu, J.A.C.; Evans, B.A.; Tracy, J.B. Photothermally reconfigurable shape memory magnetic cilia. *Adv. Mater. Technol.* **2020**, *5*, 2000147. [CrossRef]
120. Wang, Y.; Gao, Y.; Wyss, H.M.; Anderson, P.D.; den Toonder, J.M. Artificial cilia fabricated using magnetic fiber drawing generate substantial fluid flow. *Microfluid. Nanofluid.* **2015**, *18*, 167–174. [CrossRef]
121. Fang, Y.; Ye, Y.; Shen, R.; Zhu, P.; Guo, R.; Hu, Y.; Wu, L. Mixing enhancement by simple periodic geometric features in microchannels. *Chem. Eng. J.* **2012**, *187*, 306–310. [CrossRef]
122. Darnton, N.; Turner, L.; Breuer, K.; Berg, H.C. Moving fluid with bacterial carpets. *Biophys. J.* **2004**, *86*, 1863–1870. [CrossRef]
123. Gauger, E.M.; Downton, M.T.; Stark, H. Fluid transport at low Reynolds number with magnetically actuated artificial cilia. *Eur. Phys. J. E* **2009**, *28*, 231–242. [CrossRef]
124. Faus-Pérez, A.; Sanchis-Calvo, A.; Codoñer-Franch, P. Ciliopathies: An update. *Pediatrics Res. Int. J.* **2015**, *2015*, c1–c23. [CrossRef]
125. Chen, X.; Villa, N.S.; Zhuang, Y.F.; Chen, L.Z.; Wang, T.F.; Li, Z.D.; Kong, T.T. Stretchable supercapacitors as emergent energy storage units for health monitoring bioelectronics. *Adv. Energy Mater.* **2020**, *10*, 1902769. [CrossRef]

126. Zhang, S.Z.; Cui, Z.W.; Wang, Y.; den Toonder, J.M.J. Metachronal actuation of microscopic magnetic artificial cilia generates strong microfluidic pumping. *Lab Chip* **2020**, *20*, 3569–3581. [CrossRef] [PubMed]
127. Dong, X.G.; Lum, G.Z.; Hu, W.Q.; Zhang, R.J.; Ren, Z.Y.; Onck, P.R.; Sitti, M. Bioinspired cilia arrays with programmable nonreciprocal motion and metachronal coordination. *Sci. Adv.* **2020**, *6*, eabc9323. [CrossRef]
128. Saleem, N.; Munawar, S.; Tripathi, D. Thermal analysis of double diffusive electrokinetic thermally radiated TiO_2-Ag/blood stream triggered by synthetic cilia under buoyancy forces and activation energy. *Phys. Scr.* **2021**, *96*, 095218. [CrossRef]
129. Shaheen, S.; Beg, O.A.; Gul, F.; Maqbool, K. Electro-osmotic propulsion of jeffrey fluid in a ciliated channel under the effect of nonlinear radiation and heat source/sink. *J. Biomech. Eng.-Trans. ASME* **2021**, *143*, 051008. [CrossRef] [PubMed]
130. Saleem, N.; Munawar, S.; Mehmood, A.; Daqqa, I. Entropy production in electroosmotic cilia facilitated stream of thermally radiated nanofluid with ohmic heating. *Micromachines* **2021**, *12*, 1004. [CrossRef] [PubMed]
131. Saleem, N.; Munawar, S. Significance of synthetic cilia and Arrhenius energy on double diffusive stream of radiated hybrid nanofluid in microfluidic pump under ohmic heating: An entropic analysis. *Coatings* **2021**, *11*, 1292. [CrossRef]
132. Sreejith, M.; Chetan, S.; Khaderi, S.N. Numerical analysis of heat transfer enhancement in a micro-channel due to mechanical stirrers. *J. Therm. Sci. Eng. Appl.* **2021**, *13*, 011013. [CrossRef]
133. Rahbar, M.; Shannon, L.; Gray, B.L. Microfluidic active mixers employing ultra-high aspect-ratio rare-earth magnetic nano-composite polymer artificial cilia. *J. Micromech. Microeng.* **2014**, *24*. [CrossRef]
134. Sun, M.L.; Wang, Q.; Dai, B.Y.; Sun, W.H.; Ni, Y.R.; Lu, C.H.; Kou, J.H. Construction of a facile recyclable graphene-like C3N4 cilia array for effective visible-light-responsive photocatalytic hydrogen production. *Energy Fuels* **2020**, *34*, 10290–10298. [CrossRef]
135. Miao, J.; Zhang, T.; Li, G.; Shang, W.; Shen, Y. Magnetic artificial cilia carpets for transport, mixing, and directional diffusion. *Adv. Eng. Mater.* **2021**, *24*, 2101399. [CrossRef]
136. Chen, G.; Dai, Z.; Li, S.; Huang, Y.; Xu, Y.; She, J.; Zhou, B. Magnetically responsive film decorated with microcilia for robust and controllable manipulation of droplets. *ACS Appl. Mater. Interfaces* **2021**, *13*, 1754–1765. [CrossRef] [PubMed]
137. Verburg, T.; Schaap, A.; Zhang, S.Z.; den Toonder, J.; Wang, Y. Enhancement of microalgae growth using magnetic artificial cilia. *Biotechnol. Bioeng.* **2021**, *118*, 2472–2481. [CrossRef] [PubMed]
138. Chen, C.-Y.; Yao, C.Y.; Lin, C.Y.; Hung, S.H. Real-time remote control of artificial cilia actuation using fingertip drawing for efficient micromixing. *J. Lab. Autom.* **2014**, *19*, 492–497. [CrossRef]
139. Chen, C.-Y.; Pekkan, K. High-speed three-dimensional characterization of fluid flows induced by micro-objects in deep microchannels. *BioChip J.* **2013**, *7*, 95–103. [CrossRef]
140. Khatavkar, V.V.; Anderson, P.D.; den Toonder, J.M.J.; Meijer, H.E.H. Active micromixer based on artificial cilia. *Phys. Fluids* **2007**, *19*, 083605. [CrossRef]
141. Zhou, B.P.; Xu, W.; Syed, A.A.; Chau, Y.Y.; Chen, L.Q.; Chew, B.; Yassine, O.; Wu, X.X.; Gao, Y.B.; Zhang, J.X.; et al. Design and fabrication of magnetically functionalized flexible micropillar arrays for rapid and controllable microfluidic mixing. *Lab Chip* **2015**, *15*, 2125–2132. [CrossRef] [PubMed]
142. Alfadhel, A.; Kosel, J. Magnetic nanocomposite cilia tactile sensor. *Adv. Mater.* **2015**, *27*, 7888–7892. [CrossRef] [PubMed]
143. Alfadhel, A.; Khan, M.A.; de Freitas, S.C.; Kosel, J. Magnetic tactile sensor for braille reading. *IEEE Sens. J.* **2016**, *16*, 8700–8705. [CrossRef]
144. Asadnia, M.; Kottapalli, A.G.P.; Karavitaki, K.D.; Warkiani, M.E.; Miao, J.; Corey, D.P.; Triantafyllou, M. From biological cilia to artificial flow sensors: Biomimetic soft polymer nanosensors with high sensing performance. *Sci. Rep.* **2016**, *6*, 32955. [CrossRef] [PubMed]
145. Slinker, K.A.; Kondash, C.; Dickinson, B.T.; Baur, J.W. CNT-based artificial hair sensors for predictable boundary layer air flow sensing. *Adv. Mater. Technologies* **2016**, *1*, 1600176. [CrossRef]
146. Ribeiro, P.; Khan, M.A.; Alfadhel, A.; Kosel, J.; Franco, F.; Cardoso, S.; Bernardino, A.; Schmitz, A.; Santos-Victor, J.; Jamone, L. Bioinspired ciliary force sensor for robotic platforms. *IEEE Robotics Automat. Lett.* **2017**, *2*, 971–976. [CrossRef]
147. Kim, H.N.; Jang, K.-J.; Shin, J.-Y.; Kang, D.; Kim, S.M.; Koh, I.; Hong, Y.; Jang, S.; Kim, M.S.; Kim, B.-S. Artificial slanted nanocilia array as a mechanotransducer for controlling cell polarity. *ACS Nano* **2017**, *11*, 730–741. [CrossRef]
148. Kim, K.J.; Palmre, V.; Stalbaum, T.; Hwang, T.; Shen, Q.; Trabia, S. Promising developments in marine applications with artificial muscles: Electrodeless artificial cilia microfibers. *Mar. Technol. Soc. J.* **2016**, *50*, 24–34. [CrossRef]
149. Sarlo, R.; Leo, D. Airflow sensing with arrays of hydrogel supported artificial hair cells. In Proceedings of the ASME 2015 Conference on Smart Materials, Adaptive Structures and Intelligent Systems, Colorado Springs, CO, USA, 21–23 September 2015; p. V002T06A008.
150. Alfadhel, A.; Khan, M.A.; Cardoso, S.; Kosel, J. A single magnetic nanocomposite cilia force sensor. In Proceedings of the Sensors Applications Symposium (SAS), 2016 IEEE, Catania, Italy, 20–22 April 2016; pp. 1–4.
151. Schroeder, P.; Schotter, J.; Shoshi, A.; Eggeling, M.; Bethge, O.; Hutten, A.; Bruckl, H. Artificial cilia of magnetically tagged polymer nanowires for biomimetic mechanosensing. *Bioinspiration Biomim.* **2011**, *6*, 046007. [CrossRef] [PubMed]
152. Alfadhel, A.; Khan, M.A.; Cardoso, S.; Leitao, D.; Kosel, J. A Magnetoresistive tactile sensor for harsh environment applications. *Sensors* **2016**, *16*, 650. [CrossRef]
153. Zhang, X.; Shen, N.; Xu, Q.; Pei, Y.; Lian, Y.; Wang, W.; Zhang, G.; Zhang, W. Design and implementation of anulus-shaped ciliary structure for four-unit MEMS vector hydrophone. *Int. J. Metrol. Qual. Eng.* **2021**, *12*, 4. [CrossRef]

154. Zhang, X.Y.; Xu, Q.D.; Zhang, G.J.; Shen, N.X.; Shang, Z.Z.; Pei, Y.; Ding, J.W.; Zhang, L.S.; Wang, R.X.; Zhang, W.D. Design and analysis of a multiple sensor units vector hydrophone. *AIP Adv.* **2018**, *8*, 085124. [CrossRef]
155. Fan, Z.F.; Chen, J.; Zou, J.; Bullen, D.; Liu, C.; Delcomyn, F. Design and fabrication of artificial lateral line flow sensors. *J. Micromech. Microeng.* **2002**, *12*, 655–661. [CrossRef]
156. Qualtieri, A.; Rizzi, F.; Epifani, G.; Ernits, A.; Kruusmaa, M.; De Vittorio, M. Parylene-coated bioinspired artificial hair cell for liquid flow sensing. *Microelectron. Eng.* **2012**, *98*, 516–519. [CrossRef]
157. Alfadhel, A.; Kosel, J. Magnetic micropillar sensors for force sensing. In Proceedings of the 2015 IEEE Sensors Applications Symposium (SAS), Zadar, Croatia, 13–15 April 2015; pp. 1–4.
158. Liu, Y.F.; Fu, Y.F.; Li, Y.Q.; Huang, P.; Xu, C.H.; Hu, N.; Fu, S.Y. Bio-inspired highly flexible dual-mode electronic cilia. *J. Mater. Chem. B* **2018**, *6*, 896–902. [CrossRef] [PubMed]
159. Chen, N.N.; Tucker, C.; Engel, J.M.; Yang, Y.C.; Pandya, S.; Liu, C. Design and characterization of artificial haircell sensor for flow sensing with ultrahigh velocity and angular sensitivity. *J. Microelectromech. Syst.* **2007**, *16*, 999–1014. [CrossRef]
160. Hwang, Y.; Yoo, S.; Lim, N.; Kang, S.M.; Yoo, H.; Kim, J.; Hyun, Y.; Jung, G.Y.; Ko, H.C. Enhancement of interfacial adhesion using micro/nanoscale hierarchical cilia for randomly accessible membrane-type electronic devices. *ACS Nano* **2020**, *14*, 118–128. [CrossRef] [PubMed]
161. He, Q.; Wu, Y.F.; Feng, Z.P.; Sun, C.C.; Fan, W.J.; Zhou, Z.H.; Meng, K.Y.; Fan, E.D.; Yang, J. Triboelectric vibration sensor for a human-machine interface built on ubiquitous surfaces. *Nano Energy* **2019**, *59*, 689–696. [CrossRef]
162. Mani, K.; Chang Chien, T.C.; Panigrahi, B.; Chen, C.-Y. Manipulation of zebrafish's orientation using artificial cilia in a microchannel with actively adaptive wall design. *Sci. Rep.* **2016**, *6*, 36385. [CrossRef]
163. Lu, Y.-H.; Mani, K.; Panigrahi, B.; Hajari, S.; Chen, C.-Y. A shape memory alloy-based miniaturized actuator for catheter interventions. *Cardiovasc. Eng. Technol.* **2018**, *9*, 405–413. [CrossRef]
164. Lei, Y.; Sheng, Z.; Zhang, J.; Liu, J.; Lv, W.; Hou, X. Building magnetoresponsive composite elastomers for bionic locomotion applications. *J. Bionic Eng.* **2020**, *17*, 405–420. [CrossRef]
165. Lu, H.; Zhang, M.; Yang, Y.; Huang, Q.; Fukuda, T.; Wang, Z.; Shen, Y. A bioinspired multilegged soft millirobot that functions in both dry and wet conditions. *Nat. Commun.* **2018**, *9*, 3944. [CrossRef]
166. Li, F.; Liu, W.T.; Stefanini, C.; Fu, X.; Dario, P. A novel bioinspired PVDF micro/nano hair receptor for a robot sensing system. *Sensors* **2010**, *10*, 994–1011. [CrossRef]
167. Dayal, P.; Kuksenok, O.; Balazs, A.C. Directing the behavior of active, self-oscillating gels with light. *Macromolecules* **2014**, *47*, 3231–3242. [CrossRef]
168. Zhang, X.X.; Wang, F.Y.; Yu, Y.R.; Chen, G.P.; Shang, L.R.; Sun, L.Y.; Zhao, Y.J. Bio-inspired clamping microneedle arrays from flexible ferrofluid-configured moldings. *Sci. Bull.* **2019**, *64*, 1110–1117. [CrossRef]
169. Ben, S.; Zhou, T.; Ma, H.; Yao, J.; Ning, Y.; Tian, D.; Liu, K.; Jiang, L. Multifunctional magnetocontrollable superwettable-microcilia surface for directional droplet manipulation. *Adv. Sci.* **2019**, *6*, 1900834. [CrossRef] [PubMed]
170. Cao, M.; Jin, X.; Peng, Y.; Yu, C.; Li, K.; Liu, K.; Jiang, L. Unidirectional wetting properties on multi-bioinspired magnetocontrollable slippery microcilia. *Adv. Mater.* **2017**, *29*, 1606869. [CrossRef] [PubMed]
171. Wang, H.J.; Zhang, Z.H.; Wang, Z.K.; Liang, Y.H.; Cui, Z.Q.; Zhao, J.; Li, X.J.; Ren, L.Q. Multistimuli-responsive microstructured superamphiphobic surfaces with large-range, reversible switchable wettability for oil. *ACS Appl. Mater. Interfaces* **2019**, *11*, 28478–28486. [CrossRef]
172. Yang, C.; Wu, L.; Li, G. Magnetically responsive superhydrophobic surface: In situ reversible switching of water droplet wettability and adhesion for droplet manipulation. *ACS Appl. Mater. Interfaces* **2018**, *10*, 20150–20158. [CrossRef]
173. Al-Azawi, A.; Cenev, Z.; Tupasela, T.; Peng, B.; Ikkala, O.; Zhou, Q.; Jokinen, V.; Franssila, S.; Ras, R.H.A. Tunable and magnetic thiol-ene micropillar arrays. *MacroMol. Rapid Commun.* **2020**, *41*, 1900522. [CrossRef]
174. Khan, M.A.; Mohammed, H.; Kosel, J. Broadband magnetic composite energy harvester. *Adv. Eng. Mater.* **2018**, *20*, 1800492. [CrossRef]
175. Peng, F.P.; Xu, W.X.; Hu, Y.Y.; Fu, W.J.; Li, H.Z.; Lin, J.Y.; Xiao, Y.F.; Wu, Z.; Wang, W.; Lu, C.H. The design of an inner-motile waste-energy-driven piezoelectric catalytic system. *New J. Chem.* **2021**, *45*, 7671–7681. [CrossRef]
176. Zhang, S.; Zuo, P.; Wang, Y.; Onck, P.; Toonder, J.M.D. Anti-biofouling and self-cleaning surfaces featured with magnetic artificial cilia. *ACS Appl. Mater. Interfaces* **2020**, *12*, 27726–27736. [CrossRef]
177. Cui, Z.W.; Zhang, S.Z.; Wang, Y.; Tormey, L.; Kanies, O.S.; Spero, R.C.; Fisher, J.K.; den Toonder, J.M.J. Self-cleaning surfaces realized by biologically sized magnetic artificial cilia. *Adv. Mater. Interfaces.* **2021**, *7*, 2102016. [CrossRef]
178. Ben, S.; Tai, J.; Ma, H.; Peng, Y.; Zhang, Y.; Tian, D.L.; Liu, K.S.; Jiang, L. Cilia-inspired flexible arrays for intelligent transport of viscoelastic microspheres. *Adv. Funct. Mater.* **2018**, *28*, 1706666. [CrossRef]
179. Ma, S.S.; Lin, L.G.; Wang, Q.; Zhang, Y.H.; Zhang, H.L.; Gao, Y.X.; Xu, L.; Pan, F.S.; Zhang, Y.Z. Bioinspired EVAL membrane modified with cilia-like structures showing simultaneously enhanced permeability and antifouling properties. *Coll. Surf. B-Biointerfaces* **2019**, *181*, 134–142. [CrossRef] [PubMed]
180. Zhang, D.P.; Wang, W.; Peng, F.P.; Kou, J.H.; Ni, Y.R.; Lu, C.H.; Xu, Z.Z. A bio-inspired inner-motile photocatalyst film: A magnetically actuated artificial cilia photocatalyst. *Nanoscale* **2014**, *6*, 5516–5525. [CrossRef] [PubMed]

181. Peng, F.P.; Zhou, Q.; Zhang, D.N.; Lu, C.H.; Ni, Y.R.; Kou, J.H.; Wang, J.; Xu, Z.Z. Bio-inspired design: Inner-motile multifunctional ZnO/CdS heterostructures magnetically actuated artificial cilia film for photocatalytic hydrogen evolution. *Appl. Catal. B-Environ.* **2015**, *165*, 419–427. [CrossRef]
182. Peng, F.; Ni, Y.; Zhou, Q.; Kou, J.; Lu, C.; Xu, Z. Construction of ZnO nanosheet arrays within BiVO4 particles on a conductive magnetically driven cilia film with enhanced visible photocatalytic activity. *J. Alloy. Compd.* **2017**, *690*, 953–960. [CrossRef]
183. Wang, W.; Huang, X.; Lai, M.; Lu, C. RGO/TiO_2 nanosheets immobilized on magnetically actuated artificial cilia film: A new mode for efficient photocatalytic reaction. *RSC Adv.* **2017**, *7*, 10517–10523. [CrossRef]
184. Peng, F.; Ni, Y.; Zhou, Q.; Lu, C.; Kou, J.; Xu, Z. Design of inner-motile ZnO@ TiO2 mushroom arrays on magnetic cilia film with enhanced photocatalytic performance. *J. Photochem. Photobiol. A Chem.* **2017**, *332*, 150–157. [CrossRef]
185. Peng, F.P.; Zhou, Q.; Lu, C.H.; Ni, Y.R.; Kou, J.H.; Xu, Z.Z. Construction of (001) facets exposed ZnO nanosheets on magnetically driven cilia film for highly active photocatalysis. *Appl. Surf. Sci.* **2017**, *394*, 115–124. [CrossRef]
186. Huang, Y.; Stogin, B.B.; Sun, N.; Wang, J.; Yang, S.K.; Wong, T.S. A switchable cross-species liquid repellent surface. *Adv. Mater.* **2017**, *29*, 1604641. [CrossRef]
187. Zhang, S.Z.; Wang, Y.; Onck, P.R.; den Toonder, J.M.J. Removal of microparticles by ciliated surfaces-an experimental study. *Adv. Funct. Mater.* **2019**, *29*, 1806434. [CrossRef]
188. Pedersoli, L.; Zhang, S.Z.; Briatico-Vangosa, F.; Petrini, P.; Cardinaels, R.; den Toonder, J.; Pacheco, D.P. Engineered modular microphysiological models of the human airway clearance phenomena. *Biotechnol. Bioeng.* **2021**, *118*, 3898–3913. [CrossRef] [PubMed]
189. Al-Azawi, A.; Horenz, C.; Tupasela, T.; Ikkala, O.; Jokinen, V.; Franssila, S.; Ras, R.H.A. Slippery and magnetically responsive micropillared surfaces for manipulation of droplets and beads. *Aip Adv.* **2020**, *10*, 085021. [CrossRef]

Article

Steric and Slippage Effects on Mass Transport by Using an Oscillatory Electroosmotic Flow of Power-Law Fluids

Ruben Baños [1,†], José Arcos [1,*,†], Oscar Bautista [1,*,†] and Federico Méndez [2,†]

1 Instituto Politécnico Nacional, ESIME Azcapotzalco, Av. de las Granjas No. 682, Col. Santa Catarina, Del. Azcapotzalco, Ciudad de México 02250, Mexico; rbanosm1200@alumno.ipn.mx
2 Departamento de Termofluidos, Facultad de Ingeniería, UNAM, Ciudad de México 04510, Mexico; fmendez@unam.mx
* Correspondence: jarcos@ipn.mx (J.A.); obautista@ipn.mx (O.B.)
† These authors contributed equally to this work.

Abstract: In this paper, the combined effect of the fluid rheology, finite-sized ions, and slippage toward augmenting a non-reacting solute's mass transport due to an oscillatory electroosmotic flow (OEOF) is determined. Bikerman's model is used to include the finite-sized ions (steric effects) in the original Poisson-Boltzmann (PB) equation. The volume fraction of ions quantifies the steric effects in the modified Poisson-Boltzmann (MPB) equation to predict the electrical potential and the ion concentration close to the charged microchannel walls. The hydrodynamics is affected by slippage, in which the slip length was used as an index for wall hydrophobicity. A conventional finite difference scheme was used to solve the momentum and species transport equations in the lubrication limit together with the MPB equation. The results suggest that the combined slippage and steric effects promote the best conditions to enhance the mass transport of species in about 90% compared with no steric effect with proper choices of the Debye length, Navier length, steric factor, Womersley number, and the tidal displacement.

Keywords: steric effect; power-law fluids; boundary slip; oscillatory electroosmotic flow; mass transport rate

Citation: Baños, R; Arcos, J.; Bautista, O.; Méndez, F. Steric and Slippage Effects on Mass Transport by Using an Oscillatory Electroosmotic Flow of Power-Law Fluids. *Micromachines* **2021**, *12*, 539. https://doi.org/10.3390/mi12050539

Academic Editor: Kwang-Yong Kim

Received: 22 April 2021
Accepted: 7 May 2021
Published: 10 May 2021

1. Introduction

Lab-on-a-chip technology requires the manipulation and control of fluid flow to transport, mixing, and separation of reagents in nanoliter volumes in microfluidic devices widely used in chemical, medical, and biological applications, among others. These tasks are typically difficult to achieve because the laminar viscous flow governs electrokinetic transport phenomena (electroosmosis and electrophoresis) and due to the small mass-diffusivities of the species. In these applications, a broad kind of fluids are handled inside the microfluidic devices, from simple electrolytic solutions treated as Newtonian fluids to complex cell suspensions, biological fluids, such as blood, saliva, and DNA solutions, and polymer melts where viscosity is assumed to depend on the shear rate (i.e., the fluid is non-Newtonian). Therefore, understanding the fundamental behavior of the combined effects of fluid rheology, interfacial phenomena (steric and slippage effects), and the flow behavior in mass transport of a neutral solute through pure electroosmotic flow (EOF) or oscillatory electroosmotic flow (OEOF) is essential for the analysis and design of microfluidic components, such as microchannels, micro-mixers, and micro-pumps, that can be implemented in the design of biochips.

Taylor [1] was the first to establish that dispersion of a soluble substance (solute) driven by a Poiseuille flow in a circular cylindrical tube is controlled by the coefficient of diffusivity, which can be calculated directly from the solute profile. Aris [2] reported alternative treatments to Taylor's analysis of simultaneous convection and diffusion in dispersion. Likewise, when a solute is introduced into a pulsating flow through a circular tube, the solute has an increased mass transfer beyond molecular diffusion due to Taylor dispersion. The transport and separation phenomena of mass species are critical steps in realizing lab-on-a-chip. These steps are complicated because microfluidic devices manipulate flows with Reynolds numbers small enough for inertial effects to be irrelevant and species with small mass-diffusivity coefficients with magnitude $D \sim O(10^{-9})\mathrm{m}^2\,\mathrm{s}^{-1}$ [3]. In the case of a pure EOF, where a typical plug-like velocity profile exists, it does not influence species' transport and dispersion processes, and the mass transport is governed by pure diffusion.

Oscillatory and pulsating flows have mainly focused on enhancing the dispersion or separation of a passive solute under different conditions [4–6]. Kurzweg and Jaeger [7] performed the separation process with oscillatory flows of different species in which the lower diffuser, under a specific frequency, travels faster than the rapid diffuser to achieve the cross-over condition. Likewise, Thomas and Narayanan [8] showed how particles subjected to oscillatory and pulsating motions have a zigzag movement during mass transport, and that is why the transport of species is improved.

The dispersion of a neutral solute in an EOF was improved by oscillatory effects induced in the flow, i.e., the mass transport is caused by an oscillatory electroosmotic flow (OEOF) [9–12]. Despite the low diffusivity of the species in liquid solutions, time-periodic electroosmotic flow (AC) has a significant advantage over pure electroosmotic flows (DC) in biotechnology and physical separation methods. Hence, in this context, OEOF is a good candidate as a viable mechanism to induce efficient separation processes [13] due to the cross-over phenomenon between two solutes with two different mass-diffusivity coefficients [14].

The steric effect is an interfacial phenomenon in electrokinetics that frequently occurs in microfluidic confinements. EOF and OEOF are founded in electroosmosis, which refers to ionized liquid's motion relative to the stationary charged surface by an applied external electric field. The ionic size effects are controlled by the mean volume fraction of each ion in bulk given by $\nu = a^3 n_0$ [15], where a is the effective ion size, and n_0 is the bulk number concentration. Recently, studies have analyzed the steric effects using Bikerman's modified Poisson-Boltzmann (MPB) equation to account for the crowding of finite-sized ions in EOF by considering: non-linear biofluids, such as solutions of blood, saliva, protein, DNA, polymeric solutions, and colloidal suspensions; these fluids reveal non-linear rheology encountered in biomicrofluidic systems using the power-law viscosity model [16]; step-change in the wall temperature [17]; viscoelectric effects [18]; wall slip effects and steric interactions [19]; and using OEOF few works have been conducted [20]. The vast majority of theoretical work on colloidal electrokinetics using OEOF utilizes the Poisson–Boltzmann equation, wherein ions are treated as point charges and non-interacting. The slip condition is another interfacial phenomenon that enhances the electrokinetic effects (electrophoresis, electro-osmosis, streaming current or potential, etc.) [21]. In the literature concerning mass transport under slippage effect, Muñoz et al. [22] considered oscillatory electroosmotic flow, and they found that the dispersivity may be maximized up to two orders of magnitude compared with that obtained using the classical no-slip condition. Moreover, the mass transport and separation of species in OEOF can be improved by controlling the external electrical signal type [23].

The dispersion of solutes in physiological systems, where fluids are significantly more viscous, the effects of non-Newtonian rheologies, such as shear thinning, could also be considered, due to that the mass transport of neutral species is the result of an interaction between transverse diffusion and the structure of the flow. Hydrodynamics and dispersion phenomena of EOF were addressed in the context of non-Newtonian fluids [24–26]. For instance, recently, the unsteady solute dispersion by electrokinetic flow was studied [27] by considering wall absorption. Besides, some studies were conducted on the hydrodynamics of the OEOF for non-Newtonian fluids [28–30], and, recently, the mass transfer of an electroneutral solute in a concentric-annulus microchannel driven by an OEOF for a fluid whose behavior follows the Maxwell model was reported by Peralta et al. [31]. However, minimal effort has been devoted to understanding the fundamental physics that can serve us to improve mass transport and species separation under the influence of combined effects of rheology, finite ion size, and slippage that are present in the microfluidic devices.

In previous works, the analysis of dispersion and separation of neutral solutes has been performed on OEOF using non-Newtonian fluids. In contrast, this paper analyzes the combined effects of finite-sized ions, fluid rheology, and slippage on the mass transport of a neutral solute; these effects are controlled by the steric factor ν, the power law-index n, and the Navier slip length λ_N, respectively. A common non-linear rheological model for a fluid with shear-dependent viscosity is the power-law model since it efficiently describes, for engineering and microfluidics purposes, the rheology of many fluid substances over a wide range of shear rates. Although the main deficiency of the power-law model is the divergence of the effective viscosity in the limits of both zero and very large shear rates, it has the advantage of both applicability and simplicity to justify its use in investigations of shear-dependent flow behaviors. In general, lab-on-a-chip devices are designed to carry out the following functions: sample introduction, injection, mixing, reaction, transport, separation, and detection through a series of micrometre-scale channels [32]. In this context, to understand the fundamental physics, as well as provide useful information and criteria for designing micro-fluidic devices, the present study is, thus, aimed at the theoretical investigation of the transport and separation phenomena of mass species in an OEOF in a microchannel. Additionally, the start-up from the rest of the flow was analyzed, and, for larger times after the initial transient has died out, the flow was periodic in time, and then the concentration and mass transport rate were determined.

2. Problem Formulation

The sketch in Figure 1 represents the physical model of the OEOF through a long horizontal microchannel. It consists of two parallel plates separated by a distance (height) of $2h$. The length of the microchannel is L, and the depth in the z-direction is W, both assumed to be much larger than the height, i.e., $L, W >> 2h$; therefore, the flow field is assumed independent of the z coordinate. The origin of the cartesian coordinate system (x, y) is located at the left end and at the center of the microchannel. The microchannel is filled with a non-Newtonian liquid that obeys the power-law rheological model, which connects two reservoirs at the ends, such that the concentration $c(x, y, t)$ of the non-reacting solute at the left-reservoir is maintained at a constant concentration C_1, while the other extreme is found to a prescribed uniform concentration C_2 and assuming that $C_1 > C_2$. Besides, the effect of a certain degree of slip at the microchannel walls, quantified by the slip length λ_N, is also considered [33]. The OEOF will occur in the microchannel under the simultaneous influence of the externally imposed oscillatory electric field $E_x(t)$ and the induced electric field into the non overlapping EDLs. The microchannel walls are charged with a uniform high zeta potential ζ enough to cause crowding of ions of the dilute solution near the surface walls [15]. In this context, the zeta potential is several times higher than the thermal voltage $k_B T/ze$ in the MPB with $\nu \neq 0$ (representing ionic size effects), and z, e, k_B, and T are the valency of ions, the magnitude of the fundamental charge on an electron, the Boltzmann constant and the absolute temperature, respectively. The start-up of this OEOF from rest occurs, and, for sufficiently long times after imposition of $E_x(t)$, the velocity field is strictly periodic.

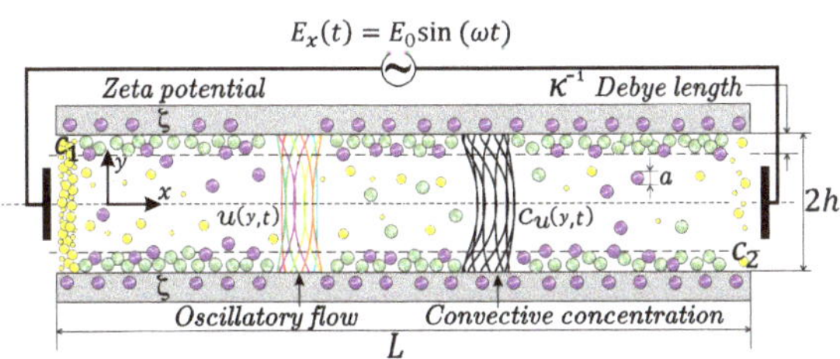

Figure 1. Schematic depiction for mass transport due to an oscillatory electroosmotic flow.

2.1. Electrical Field: Steric Effects

When an electrolyte solution is in contact with a uniformly charged surface, electrical ($ze\psi$) and chemical ($k_BT\ln n_\pm$) effects modify the electro-chemical potential $\mu_\pm$ ($\pm$denotes the sign of the charge ze); then, the electrochemical potential derives in $\mu_\pm = \pm ze\psi - k_BT\ln n_\pm$, where ψ is the electric potential, and $n_\pm$ is the ionic concentration in the diffuse electric double layer. Positive and negative ions are separated within the diffuse double layer by Boltzmann distribution $n_\pm = n_0 e^{\mp ze\psi/k_BT}$, where an equilibrium exists between external electric and osmotic forces. The electrical potential ψ is determined with the classical complete Poisson-Boltzmann equation $\epsilon\nabla^2\Phi = -\rho_e$. Here, $\Phi = \phi(x,t) + \psi(y)$ is defined by the linear superposition [34] of the local electric potential $\phi(x,t)$ and the corresponding potential $\psi(y)$ induced into the EDL, ϵ is the permittivity of the solution. The volume charge density ρ_e in the neighborhood of the surface is $\rho_e = ze(n_+ - n_-)$. However, the above treatment is valid if ions are treated as electric charges having no volume [35] and no other individual effects [36]. The present investigation involves large potentials; then the Poisson-Boltzmann equation has shortcomings because it neglects steric effects. The modified PB equations that consider the steric effects in $n_\pm$ due to finite-sized ions [15] is derived from the modified chemical potential $\mu_\pm$ using Bikerman's model [37], given by

$$\mu_\pm = \pm ze\psi - k_BT\ln n_\pm - k_BT\ln(1 - n_+a^3 - n_-a^3). \tag{1}$$

The third term on the right-hand side of Equation (1) considers the finite-sized ions. Under equilibrium conditions, the ions' electrochemical potential is constant $\nabla\mu_\pm$=0, and it derives in:

$$\frac{\nabla n^\pm}{n^\pm} - \frac{\nabla(1 - n_+a^3 - n_-a^3)}{1 - n_+a^3 - n_-a^3} = \mp\frac{ze}{k_BT}\nabla\psi. \tag{2}$$

Integrating Equation (2) from a point in the bulk solution (where $\psi = 0$ and $n_\pm = n_\infty$) leads to the modified Boltzmann distribution:

$$n_\pm = \frac{n_0\exp(\pm ze\psi/k_BT)}{1 + 2\nu\sinh^2(ze\psi/2k_BT)}. \tag{3}$$

Substituting ion distribution $n_\pm$ into the volume charge density ρ_e takes the following form:

$$\rho_e = -2zen_0\frac{\sinh(ze\psi/k_BT)}{1 + 2\nu\sinh^2(ze\psi/2k_BT)}, \tag{4}$$

where $\nu = 2a^3n_0$ is the bulk volume fraction of ions. To account for steric effects associated with the finite-sized ions and solvent molecules, Equation (3) was combined with the complete Poisson equation yielding the MPB equation:

$$\nabla^2\Phi = \frac{2zen_0}{\epsilon}\frac{\sinh(ze\psi/k_BT)}{1 + 2\nu\sinh^2(ze\psi/2k_BT)}. \tag{5}$$

Equation (5) is essentially a modified form of the PB equation considering finite-sized ions effects. The omission of the temporal term in Poisson's equation is because the characteristic time scale ($\sim10^{-12}$ s) of the electro-migration in the EDL is much less than the corresponding time-scale ($\sim10^{-2}$ s) for the viscous diffusion [38].

For a long microchannel, $L \gg h$, the term $\partial^2\Phi/\partial x^2$ in Equation (5) may be neglected [39]. One gets the simplified version of the MPB equation given by

$$\frac{d^2\psi}{dy^2} = \frac{2zen_0}{\epsilon}\frac{\sinh(ze\psi/k_BT)}{1 + 2\nu\sinh^2(ze\psi/2k_BT)}. \tag{6}$$

The boundary conditions of Equation (6) are $d\psi/dy = 0$ at $y = 0$ and $\psi = \zeta$ at $y = h$. Using the following dimensionless variables $\bar{y} = y/h$ and $\bar{\psi} = ze\psi/k_BT$ into Equation (6), it derives in:

$$\frac{d^2\bar{\psi}}{d\bar{y}^2} = \bar{\kappa}^2 \frac{\sinh\bar{\psi}}{1 + 2\nu\sinh^2(\bar{\psi}/2)}, \tag{7}$$

with the corresponding boundary conditions,

$$\bar{\psi} = \kappa_\psi \quad \text{at} \quad \bar{y} = 1, \tag{8}$$

and

$$\frac{d\bar{\psi}}{d\bar{y}} = 0 \quad \text{at} \quad \bar{y} = 0. \tag{9}$$

The parameter $\bar{\kappa} = \kappa h$ is the ratio of the microchannel height to the Debye length, defined by $\kappa^{-1} = \epsilon k_B T/2e^2z^2n_0$. The competition between the wall ζ potential and the thermal potential is given by $\kappa_\psi = ze\zeta/k_BT$. When $\kappa_\psi < 1$, it means low ζ potentials. In contrast, for $\kappa_\psi >> 1$, it represents the case of high ζ potentials. In this work, $\kappa_\psi = 2$, and high zeta potentials $\kappa_\psi = 10$ were considered since, in dilute liquids, the steric effects are visible at high zeta potentials at which high ionic concentrations at the microchannel wall cause extreme accumulation of counterions [40].

2.2. Velocity Field

By assuming that the microchannel is very long and focusing on the central region, away from the microchannel entry and exit, it is assumed that the flow is unidirectional and fully developed. Therefore, the momentum equation is given by

$$\rho_f \frac{\partial u}{\partial t} = \frac{\partial \tau_{xy}}{\partial y} + \rho_e E_x(t). \tag{10}$$

Here, $u(y,t)$ is the velocity component in the x direction, ρ_f is the mass density and τ_{xy} represents the shear stress. Equation (10) is subject to the symmetry boundary condition of the velocity ($\partial u/\partial y = 0$) at the center of the microchannel. The Navier slip boundary condition at the interface between the fluid and the microchannel wall is considered, given by $\mathbf{u}_s = \lambda_N\{\mathbf{D}\cdot\mathbf{n} - [(\mathbf{D}\cdot\mathbf{n})\cdot\mathbf{n}]\mathbf{n}\}$ [41]. Here, $\mathbf{u}_s$ is the fluid velocity at the microchannel wall, $\mathbf{n}$ represents the unit vector normal to the microchannel surface pointing toward the fluid, the rate of strain tensor is defined as $\mathbf{D} = \nabla\mathbf{u} + (\nabla\mathbf{u})^{tr}$, $\mathbf{u}$ is the velocity field, and tr denotes the transpose of $\nabla\mathbf{u}$. For the present problem, the slip boundary condition is simplified to

$$u = -\lambda_N \frac{\partial u}{\partial y} \quad \text{at} \quad y = h. \tag{11}$$

Besides, it is assumed that the fluid is at rest for $t = 0$, that is,

$$u = 0 \quad \text{at} \quad t = 0, \quad \text{for} \quad -h \leqslant y \leqslant h. \tag{12}$$

The shear stress τ_{xy} for non-Newtonian fluids where the dynamic viscosity $\eta(\dot{\gamma})$ is a function of the strain rate $\dot{\gamma}$, is defined as $\tau_{xy} = \eta(\dot{\gamma})\dot{\gamma}$. For the unidirectional and fully developed OEOF, $\dot{\gamma} = \partial u/\partial y$, and the dynamic viscosity $\eta(\dot{\gamma})$ is a function of the velocity gradient, according to the power-law model as follows [42]:

$$\eta(\dot{\gamma}) = m\left(\frac{\partial u}{\partial y}\right)^{(n-1)}, \tag{13}$$

where m denotes the consistency index, and n is the power-law index. Thus, substituting ρ_e defined in Equation (4) and $E_x(t)$ into (10), it derives in

$$\rho_f \frac{\partial u}{\partial t} = \frac{\partial}{\partial y}\left\{\eta \frac{\partial u}{\partial y}\right\} - \frac{2zen_\infty \sinh(ze\psi/k_B T)}{1 + 2\nu \sinh^2(ze\psi/2k_B T)} E_0 \sin(\omega t). \tag{14}$$

Using the following dimensionless variables, $\bar{u} = u/U_{HS}, \bar{y} = y/h, \tau = \omega t/2\pi$, where $U_{HS} = -\epsilon\zeta_0 E_0/\mu$ is the Helmholtz-Smoluchowski velocity. Therefore, the dimensionless version of the momentum Equation (14) is as follows:

$$\frac{Wo^2}{2\pi}\frac{\partial \bar{u}}{\partial \tau} = \bar{m}\left[\bar{\eta}\frac{\partial^2 \bar{u}}{\partial \bar{y}^2}\right] - \frac{\bar{\kappa}^2}{\kappa_\psi}\frac{\sinh \bar{\psi}}{1 + 2\nu \sinh^2(\bar{\psi}/2)} \sin(2\pi\tau), \tag{15}$$

where the dimensionless dynamic viscosity is defined as

$$\bar{\eta} = \left(\frac{\partial \bar{u}}{\partial \bar{y}}\right)^{(n-1)}. \tag{16}$$

The parameter $\bar{m} = n(m/\mu)(U_{HS}/h)^{n-1}$ is the dimensionless consistency index and relates the characteristic shear stresses for Non-Newtonian and Newtonian fluids. The Womersley number is defined as $Wo = h\sqrt{\omega/\nu_0}$ and relates the ratio of the viscous diffusion time-scale to the oscillation time-scale, where $\nu_0 = \mu/\rho_f$. Here, it is important to note that the Womersley is based on physical properties of a Newtonian fluid. This is because in the process of nondimensionalizing the mathematical problem, the characteristic scale for the velocity, U_{HS}, corresponds to the classical Helmholtz-Smoluchowsky velocity [39], where the viscosity μ of a Newtonian fluid is considered. However, the non-Newtonian parameters of the Power-Law fluid are taken into account in the definitions of the dimensionless parameter $\bar{m}$. Equation (15) is subject to the following dimensionless boundary and initial conditions:

$$\frac{\partial \bar{u}}{\partial \bar{y}} = 0 \quad \text{at} \quad \bar{y} = 0, \tag{17a}$$

$$\bar{u} = -\delta\frac{\partial \bar{u}}{\partial \bar{y}} \quad \text{at} \quad \bar{y} = 1, \tag{17b}$$

$$\bar{u} = 0 \quad \text{at} \quad \tau = 0 \quad \text{for} \quad -1 \leqslant \bar{y} \leqslant 1. \tag{17c}$$

Here, $\delta = \lambda_N/h$ denotes the ratio between the Navier length and the microchannel height.

2.3. Concentration Field

For the fully developed OEOF described in Section 2.2, diffusion and convection mechanisms govern the mass transport of passive species. Assuming that the transport phenomenon is not affected by any of the electrical potentials, and the particles do not interact with each other, the concentration field of the solute $c(x, y, t)$ with constant molecular diffusion coefficient D can be found by solving the species conservation equation, given by [39]

$$\frac{\partial c}{\partial t} + u\frac{\partial c}{\partial x} = D\left(\frac{\partial^2 c}{\partial x^2} + \frac{\partial^2 c}{\partial y^2}\right). \tag{18}$$

The boundary and initial conditions associated to Equation (18) are

$$c(y,t) = C_1 \quad \text{at} \quad x = 0, \tag{19a}$$
$$c(y,t) = C_2 \quad \text{at} \quad x = L, \tag{19b}$$
$$\frac{\partial c(x,t)}{\partial y} = 0 \quad \text{at} \quad y = h, \tag{19c}$$
$$\frac{\partial c(x,t)}{\partial y} = 0 \quad \text{at} \quad y = 0, \tag{19d}$$
$$c(x) = C_1 + (C_2 - C_1)(x/L) \quad \text{at} \quad t = 0. \tag{19e}$$

In Equations (19a) and (19b), C_1 and C_2 are fixed values of $c(y,t)$ at the left and right reservoirs, respectively. As depicted in Figure 1, $C_1 > C_2$. Equations (19c) and (19d) denote the impermeability and symmetry conditions, respectively. Here, a linear profile of the concentration distribution is imposed at $t = 0$, prescribed in Equation (19e). Because of the linearity of Equation (18), it is often convenient to use the Chatwin approximation [4,43], given by the superposition of a linear distribution for the concentration and other corresponding with convective effects by the oscillatory motion of the fluid, which is given by

$$c(x,y,t) = C_1 + \frac{C_2 - C_1}{L}x + c_u(y,t). \tag{20}$$

This approximate solution is invalid near the microchannel ends due to that Equation (20) does not satisfy the boundary conditions at both ends taking into account that $c_u(y,t)$ is different to zero. However, it is valid for long microchannels ($L \gg h$) where any end effects are neglected [13,43]. Substituting Equation (20) into Equation (18) yields

$$\frac{\partial c_u}{\partial t} + u\left(\frac{C_2 - C_1}{L}\right) = D\frac{\partial^2 c_u}{\partial y^2} \tag{21}$$

with the following boundary and initial conditions,

$$\frac{\partial c_u(t)}{\partial y} = 0 \quad \text{at} \quad y = h, \tag{22a}$$
$$\frac{\partial c_u(t)}{\partial y} = 0 \quad \text{at} \quad y = 0, \tag{22b}$$
$$c_u(y) = 0 \quad \text{at} \quad t = t_\omega. \tag{22c}$$

In Equation (22c), t_ω refers to the value of t when the initial transient step has died out. Introducing the dimensionless concentration of species as $\bar{c}_u = c_u/(C_2 - C_1)$ and the dimensionless variables defined before, Equation (21) can be rewritten in the following form:

$$\frac{Wo^2 Sc}{2\pi}\frac{\partial \bar{c}_u}{\partial \tau} + Pe_D \alpha \bar{u} = \frac{\partial^2 \bar{c}_u}{\partial \bar{y}^2}, \tag{23}$$

where $Sc = \nu_0/D$ is the Schmidt number that measures the competition between the momentum and the mass diffusivities. $Pe_D = U_{HS}h/D$ is the diffusive Péclet number, measuring the ratio of the convective to the diffusive transport rate. Here, in a similar manner as the definition of the Womersley number, the Schmidt and Péclet numbers are based on physical properties of a Newtonian fluid. However, in the convection-diffusion Equation (23), the influence of the fluid's reology is taken into account in the velocity profile $\bar{u}(\bar{y},\bar{\tau})$. Hence, the dimensionless form of the boundary and initial conditions (22a)–(22c) are:

$$\left.\frac{\partial \bar{c}_u}{\partial \bar{y}}\right|_{\bar{y}=1} = 0, \tag{24a}$$

$$\left.\frac{\partial \bar{c}_u}{\partial \bar{y}}\right|_{\bar{y}=0} = 0, \tag{24b}$$

$$\bar{c}_u(\bar{y},\tau = \tau_\omega) = 0. \tag{24c}$$

In Equation (24c), τ_ω denotes the value of the dimensionless time τ from which the flow becomes periodic, i.e., the transient stage has died out. In physical units, such condition is achieved when the time t assumes a value of the characteristic scale of the viscous diffusion time-scale ($t \sim h^2/\nu_0$). On the other hand, for the periodic condition of the flow, $1/\omega \sim h^2/v$, and, taking into account the definition of the dimensionless time, it yields that $\tau_\omega = t_\omega \omega/2\pi \sim O(10^1)$. This value of τ_ω is used for all the numerical calculations.

2.4. Mass Transport Rate

The time-averaged mass transfer in the system Q_x was evaluated during one period of oscillation, defined by [43]:

$$Q_x = \frac{1}{4h}\frac{\omega}{\pi}\int_{-h}^{h}\int_{t_\omega}^{t_\omega+\frac{2\pi}{\omega}} J_x dt dy, \tag{25}$$

where J_x represents the total flux density defined as the sum of convective $j_{x,conv}$ and diffusive $j_{x,diff}$ flux densities in the x-direction, as follows:

$$J_x = j_{x,conv} + j_{x,diff} = u(y,t)c(x,y,t) - D\frac{\partial c(x,y,t)}{\partial x}. \tag{26}$$

Substituting Equation (26) into Equation (25), the dimensionless rate of mass transport $\bar{Q}_x$ in terms of dimensionless variables is as follows

$$\bar{Q}_x = 1 - \frac{Pe_D}{2\alpha}\int_{-1}^{1}\int_{\tau_\omega}^{\tau_\omega+1} \bar{u}\bar{c}_u d\bar{y} d\tau. \tag{27}$$

Here, $\bar{Q}_x = Q_x L/(C_1 - C_2)D$. Due to the flow's oscillatory character, the Helmholtz-Smoluchwosky velocity U_{HS}, and the corresponding Péclet number, Pe_D, will no longer remain constant for particular values of the angular frequency ω, or consequently Wo. To show the frequency dependence of $\bar{Q}_x$, it is necessary to use the concept of tidal displacement $\triangle z$. This quantity is defined as the cross-stream averaged maximum axial distance for which the fluid elements travel during the one-half period of the oscillation [13],

$$\triangle z = \frac{1}{2h\pi}\left|\int_{-h}^{h}\int_{t_\omega}^{t_\omega+\frac{\pi}{\omega}} u \quad dt \quad dy\right|. \tag{28}$$

After introducing the dimensionless variables, $\bar{u} = u/U_{HS}$, $\bar{y} = y/h$, and $\tau = \omega t/2\pi$ into Equation (28), can be simplified in the following form:

$$\triangle z = \frac{U_{HS}}{\omega}\left|\int_{-1}^{1}\int_{\tau_\omega}^{\tau_\omega+1/2} \bar{u} \quad d\tau \quad d\bar{y}\right|. \tag{29}$$

From Equation (29), the Helmholtz-Smoluchwosky velocity U_{HS} is obtained as:

$$U_{HS} = \frac{\omega \Delta z}{\left|\int_{-1}^{1}\int_{\tau_\omega}^{\tau_\omega+1/2} \bar{u} \quad d\tau \quad d\bar{y}\right|}, \tag{30}$$

where it is evident the frequency dependence of U_{HS}. Substituting Equation (30) into the Péclet number definition, it provides a relationship as a function of Sc, Wo, and ΔZ, as follows:

$$Pe_D = \frac{Wo^2 Sc \triangle Z}{\left| \int_{-1}^{1} \int_{\tau_\omega}^{\tau_\omega + 1/2} \bar{u} \quad d\tau \quad d\bar{y} \right|}. \tag{31}$$

For different fixed values of the dimensionless tidal displacement $\triangle Z = \triangle z/h$, significant changes in Pe_D are obtained for each flow situation. In this context, the concentration field and the overall mass transfer will be affected by the frequency.

3. Numerical Scheme

Figure 2 shows a flowchart of the methodology used to solve the formulated problem. The algorithm was developed in Fortran PowerStation version 4.0 with Microsoft Developer Studio software; the process is described more precisely as indicated below.

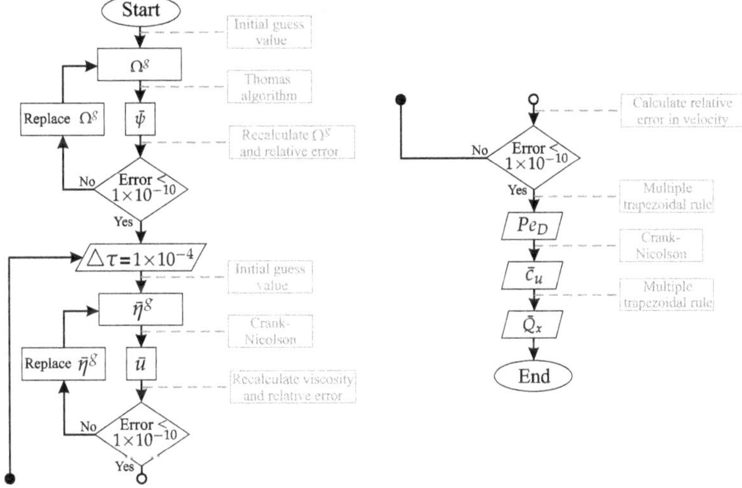

Figure 2. Schematic diagram of the numerical algorithm.

3.1. Electric Potential Field

The MPB equation (7) was approximated by the second-order centered-space difference, it derives in

$$\bar{\psi}_{i+1} - (2 + \Delta\bar{y}^2\bar{\kappa}^2\Omega^g)\bar{\psi}_i + \bar{\psi}_{i-1} = 0, \tag{32}$$

where

$$\Omega^g = \sinh(\bar{\psi}_i)/\bar{\psi}_i\left[1 + 2\nu\sinh^2(\bar{\psi}_i/2)\right]. \tag{33}$$

For an initial guess value in Ω^g, the equation system in (32) can be solved simultaneously by applying the Thomas algorithm [44]. Using the solution obtained for ψ_i, the value of Ω is recalculated according to (33), and the new value replaces the previous one. This process is repeated until a numerical error value of 1×10^{-10} is reached.

3.2. Velocity Field

Equation (15) was solved using the Crank-Nicolson method, applying a central difference scheme [44]. The resulting discretization of Equation (15) is as follows:

$$-\theta_1\bar{u}_{i+1}^{l+1} + \theta_2\bar{u}_i^{l+1} - \theta_1\bar{u}_{i-1}^{l+1} = \theta_3, \tag{34}$$

where θ_1, θ_2, and θ_3 are defined as:

$$\theta_1 = \frac{\bar{\eta}^g \gamma}{2\triangle\bar{y}^2}, \tag{35a}$$

$$\theta_2 = 2\theta_1 + \frac{1}{\triangle\tau}, \tag{35b}$$

$$\theta_3 = \frac{\bar{u}_i^l}{\triangle\tau} + \theta_1(\bar{u}_{i+1}^l - 2\bar{u}_i^l + \bar{u}_{i-1}^l) + \Lambda, \tag{35c}$$

and

$$\bar{\eta}^g = \left[\left(\frac{\bar{u}_{i+1}^l - \bar{u}_i^l}{\triangle\bar{y}}\right)^2\right]^{\left(\frac{n-1}{2}\right)}. \tag{36}$$

Here, $\triangle\tau$ and $\triangle\bar{y}$ are the time step and the size step in the $\bar{y}$ direction, respectively. In this work, a $\triangle\bar{y}$-step of 2×10^{-3} and $\triangle\tau$-step of 1×10^{-4} have been used in all numerical essays. The parameters γ and Λ in Equations (35a) and (35c) are defined as follows:

$$\gamma = \frac{2\pi\bar{m}}{Wo^2}, \tag{37}$$

and

$$\Lambda = \frac{2\pi\bar{\kappa}^2}{Wo^2\kappa_\psi}\frac{\sinh\bar{\psi}_i}{1 + 2\nu\sinh^2(\bar{\psi}_i/2)}\sin(2\pi\tau^l). \tag{38}$$

The numerical value of ψ_i is provided in (38) by the iterative procedure described in (32) and (33). To solve the velocity field a similar procedure to that explained in the previous section was applied. The solution begins by providing an initial guess value to $\bar{\eta}^g$. To solve the non-linear equation system generated by (34), a tridiagonal matrix algorithm (TDMA) was used. With the values obtained for the velocity field $\bar{u}_{i+1}^{l+1}$, the term $\bar{\eta}^g$ is recalculated at the next iteration and the process is repeated until the required relative error is achieved. Given that Equation (36) is a function of $\bar{y}$, it was recalculated for each node in the space and time. This solution procedure is useful because, when $\partial\bar{u}/\partial\bar{y} = 0$ with an index n smaller than unity, the value of $\bar{\eta}^g$ is undetermined; thereby, a numerical value very close to zero is assigned to $\bar{\eta}^g$ when $\partial\bar{u}/\partial\bar{y} \to 0$ to avoid a singularity.

3.3. Concentration Field and Mass Transport Rate

After solving the electrical potential and the velocity field, the Péclet number has to be determined. Average velocity is required as it is indicated in Equation (31), this average was calculated using the multiple-trapezoid rule. The concentration field in Equation (23) was approximated by using the second-order central-difference formula for the second derivative diffusion term and the forward difference formula for the first-order time derivative. Then, the concentration equation can be discretized as follows:

$$-\beta_2\bar{c}_{u,i+1}^{n+1} + (1 + 2\beta_2)\bar{c}_{u,i}^{n+1} - \beta_2\bar{c}_{u,i-1}^{n+1} = \beta_1, \tag{39}$$

where

$$\beta_1 = -2\beta_2\Delta\bar{y}^2 Pe_\omega\alpha\bar{u}_i^n + \bar{c}_{u,i}^n + \beta_2(\bar{c}_{u,i+1}^n - 2\bar{c}_{u,i}^n + \bar{c}_{u,i-1}^n), \tag{40a}$$

$$\beta_2 = \frac{\pi\Delta\tau}{Wo^2 Sc\Delta\bar{y}^2}. \tag{40b}$$

Finally, with the results obtained for $\bar{u}$ and $\bar{c}_u$, the dimensionless rate of mass transport $\bar{Q}_x$ defined in Equation (27) is calculated by applying the multiple-trapezoidal rule. The dimensionless initial condition $\overline{T}_\omega$ for the concentration field can be any value of the non-dimensional time in such a way that the transient stage for the velocity field has diet out.

A description of the validation process between the numerical solution and the analytical solution of the velocity field reported by Huang and Lai [13] is presented. The main differences between both studies are the following: The flow configuration considered by Huang and Lai [13] is a two-dimensional rectangular microchannel of length L and width h filled with a Newtonian liquid which is an electrolyte. The coordinate system (x, y) is located on the lower wall at the microchannel inlet. The associated boundary conditions for the velocity are the no-slip condition at the channel walls.

Conversely, in the present study, the microchannel consists of two parallel plates with a length L and distant by $2\,h$. The origin of the cartesian coordinate system (x, y) is located to the left end of the microchannel center. A non-Newtonian liquid that obeys the power-law rheological model flows through the microchannel. For the hydrodynamics, the problem formulation includes the effects of slip, which enter the problem through a Navier-slip model, and the symmetry boundary condition at the microchannel's midplane. The dimensionless velocity distribution across the channel-width with a Womersley number $W = 5$ and a dimensionless Debye length $\lambda^* = 70$ reported by Huang and Lai corresponds to the velocity profile with $Wo = 2.5$ and $\bar{\kappa} = 35$ in the present study assuming the slippage is absent, $\delta = 0$, with a power-law index $n = 1$. An excellent agreement between both solutions is observed in Figure 3.

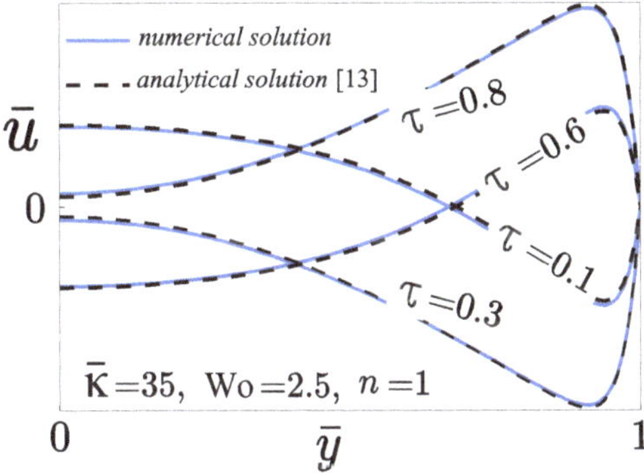

Figure 3. Comparison of the analytical solution [13] and the numerical solution (present work) for the dimensionless velocity $\bar{u}$ across the microchannel in a Newtonian fluid.

4. Results and Discussion

This section highlights the numerical solution of the MPB, momentum, and conservation species equations, and mass transport rate provided in a microchannel due to an OEOF of power law fluids by considering the steric and slippage effects. For pertinent results, the appropriate dimensionless parameters are highlighted in Table 1, by considering the relevant geometrical and physicochemical properties, as shown in Table 2.

In Figure 4, the absence and the presence of the steric effect and the slippage or the combined effects on the evolution over time of the velocity $\bar{u}$ for Newtonian and non-Newtonian fluids is presented. The blue, black, and green lines represent the flow solution for Shear-thinning (with $n = 0.8$), Shear-thickening (with $n = 1.4$), and Newtonian (with $n = 1$) behavior of the fluid, respectively. Once the electric field $E_x(t)$ is applied, the flow initiates the transient state where the oscillatory effects have no relevance; that is, the velocity exhibits a gradual increase in its magnitude, and the periodic flow behavior starts when $\tau \sim O(10^{-1})$, as shown in Figure 4a–d.

Table 1. Order of magnitude of the dimensionless quantities.

Dimensionless Quantities	Definition	Order of Magnitude
Sc	Schmidt number, (ν_0/D)	$\sim O(10^2–10^3)$
Wo	Womersley number, $(h\sqrt{\omega/\nu_0})$	$\sim O(1)$
α	Aspect ratio, (h/L)	$\sim O(10^{-3})$
$\bar{m}$	Consistency index, $n(\frac{m}{\mu})(\frac{U_{HS}}{h})^{n-1}$	$\sim O(10^{-1}–10^0)$
ΔZ	Tidal displacement, $(\triangle z/h)$	$\sim O(1)$
δ	Slip lenght, (λ_N/h)	~ 0.05
κ_ψ	Potential ratio, $(ze\zeta/k_B T)$	$\sim 2, 10$
$\bar{\kappa}$	Electrokinetic parameter, (κh)	$\sim 10, 20$
ν	Steric factor, $(2a^3 n_0)$	$\sim O(0–0.4)$

Table 2. Physical properties and geometrical parameters used for estimating the dimensionless parameters from the present analysis.

Parameter	Definition	Value
a	Ion size	$\sim$2 nm [15]
c_0	molar concentration	$\sim$(50–100) mol m^{-3} [15]
D	Diffusion coefficient	$\sim(10^{-9}–10^{-8})$ m^2 s^{-1} [45]
e	Electron charge	$\sim 1.602 \times 10^{-19}$C *
E_0	Electric field	$\sim 10^3$ V/m [46]
h	Microchannel half-height	$\sim$(5–100) μm *
k_B	Boltzmann constant	$\sim 1.38 \times 10^{-23}$ J K^{-1} *
L	Micro-channel length	$\sim 10^{-2}$ m
m	Consistency index	$\sim(10^{-3}–10^{-4})$ Pa s^n [47]
n	Power-law index	(0.8, 1, 1.4) [42]
n_0	Ionic concentration	$\sim 10^{25}$ m^{-3} [15]
N_A	Avogadro number	$\sim 6.022 \times 10^{23}$ mol^{-1} *
T	Absolute temperature	$\sim$298 K *
ϵ	Permittivity of the solution	6.95×10^{-10} C^{2}N^{-1}m^{-2} *
ζ	Zeta potential	$\sim$(50–260) mV [48]
κ^{-1}	Debye length	$\sim$(15–300) nm *
λ_N	Navier length	$\sim(10^{-9}–10^{-6})$ m [49]
μ	Newtonian viscosity	$\sim 10^{-3}$ Pa s *
ρ_f	Fluid density	$\sim 10^3$ kg m^{-3} *
ω	Angular frequency	$\sim$400 Hz–5 kHz [50]

* Values taken from Reference [34].

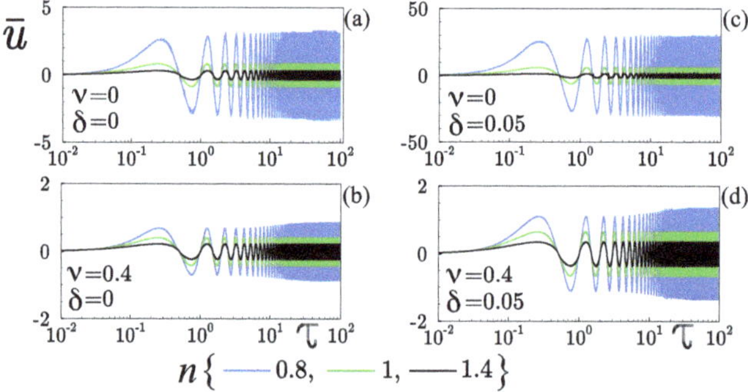

Figure 4. Evolution over time of the dimensionless velocity $\bar{u}$ at $\bar{y} = 0.9$, affected by the slippage (δ) and the finite-sized ions (ν). Here, $Wo = 0.5$, $\bar{\kappa} = 20$, and $\kappa_\psi = 10$. (**a**) $\nu = 0, \delta = 0$; (**b**) $\nu = 0.4, \delta = 0$; (**c**) $\nu = 0, \delta = 0.05$ and (**d**) $\nu = 0.4, \delta = 0.05$.

As the dimensionless time τ goes on, the flow begins a state of damping; the amplitude begins by growing in this way but eventually settles down to a constant value, acquiring an oscillatory and periodic state. Figure 4a,b show the cases when the slippage is absence ($\delta = 0$), the velocity decreases with the finite-sized ions (ν) in Newtonian and non-Newtonian flows. In Figure 4c,d shows the effect of the slippage ($\delta = 0.05$), and the velocity is reduced up to one order of magnitude by considering the finite-sized ions ($\nu = 0.4$). That reduction in the velocity is attributed to the fact that the finite ionic volumes give rise to the reduction of the ionic concentration inside EDL, therefore decreasing the oscillatory electroosmotic body force. Hydrophobic condition promotes the increment in the flow velocity affected by the steric effect, as shown in Figure 4b,d. It is illustrated in Figure 4d, with $\delta = 0.05$, the dimensionless velocity at $\tau \sim O(10^1)$ when the periodic state is reached, $\bar{u} \approx 1.4$, representing an increase of approximately 75% with respect to the no-slippage case, where $\bar{u} \approx 0.8$, as is shown in Figure 4b. In shear-thinning fluids (Figure 4a,c), due to the shear-rate dependent viscosity, the start-up transient motion will die out in a faster dimensionless time $\tau \sim O(10^0)$ because of the viscosity decreases (see Figure 5a, blue curve) when the shear rate increases. That is due to the ions are treated as electric charges having no volume into the EDL, increasing the electroosmotic body force and consequently modify the shear rate of the fluid. Conversely, the finite-sized ions in shear-thinning fluids delayed the transient state to tend to its periodic-state in a dimensional time of order $\sim O(10^1)$, as shown in Figure 4b,d. In Newtonian and shear-thickening fluids, the transient state will tend to its periodic-state much faster than shear-thinning fluids in a dimensional time of order $\sim O(10^{-1})$, as shown in all the panels of Figure 4. Therefore, this paper is mainly focused on the transport of neutral solutes in the periodic -state. A dimensionless time of order $\tau = 40$ was selected to ensure all initial transients have died out in the flow. The hydrodynamic and the concentration of species presented in Figures 5–8, and the rate of mass transport shown in Figures 9–12, are all referred to the periodic state of the OEOF.

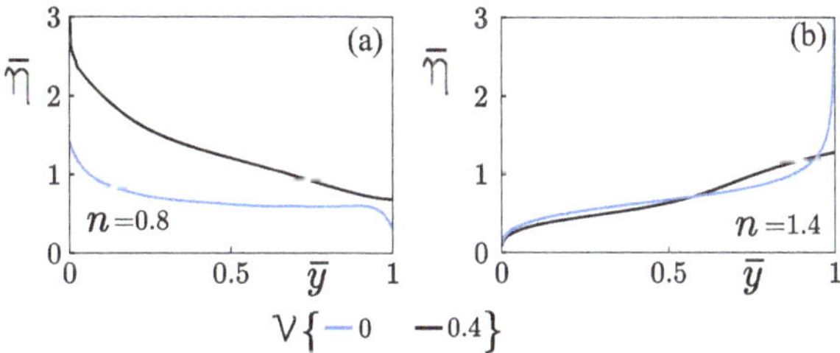

Figure 5. Variation of dynamic viscosity across the microchannel with and without the influence of steric effects. (**a**) Shear-thinning and (**b**) shear-thickening fluids.

Figure 5 shows the influence of the finite-sized ions in the fluid's rheology by considering the hydrophobic condition. Both Figure 5a,b were determined with $\delta = 0.05$, $Wo = 1$, $\bar{\kappa} = 10$, $\kappa_\psi = 10$, and $\tau = 49.4$. In Figure 5a (shear-thinning fluids), the absence of the steric effect gives rise to a non-linear variation of the dynamic viscosity across the microchannel given in Equation (16), and it decreases from the centerline up to the wall. The steric effect's presence promotes higher values of dynamic viscosity than when the steric effect is absent. That increment in the dynamic viscosity is due to the finite-sized ions reducing the ion-concentration into the EDL and, consequently, the electric body force, offering the fluid greater resistance to deformed. In shear-thickening fluids, dynamic viscosity increases from the centerline up to the wall with and without steric effects, as is shown in Figure 5b. The dynamic viscosity near walls by considering $\nu = 0$ is higher than that obtained with $\nu = 0.4$; this occurs due to velocity gradients strongly affected with finite-sized ions be discussed later in Figure 6b,d. Interestingly, it is observed that, in specified values of $\delta = 0.05$, $Wo = 1$, $\bar{\kappa} = 10$, $\kappa_\psi = 10$, $\tau = 49.4$, the fluid acts as an inviscid flow at the centerline for the special case of $n = 1.4$.

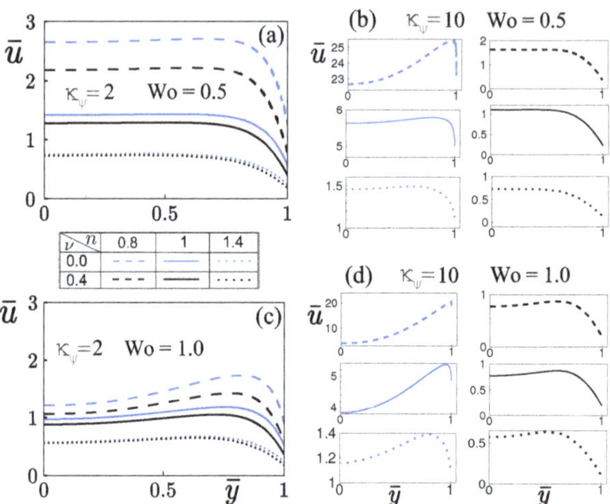

Figure 6. The velocity profiles are shown for shear-thinning and shear-thickening fluids. (**a**): $\kappa_\psi = 2$, and $Wo = 0.5$. (**b**): $\kappa_\psi = 10$, and $Wo = 0.5$. (**c**): $\kappa_\psi = 2$, and $Wo = 1.0$. (**d**): $\kappa_\psi = 10$, and $Wo = 1.0$.

In Figure 6, the influence of $\kappa_\psi = 2$, and high zeta potentials ($\kappa_\psi = 10$), with two different values of the Womersley number on the velocity profiles $\bar{u}$, as a function of the coordinate $\bar{y}$ for shear-thinning ($n = 0.8$), Newtonian ($n = 1$), and shear-thickening fluids ($n = 1.2$) are plotted. All panels were determined with $\delta = 0.05$, $\bar{\kappa} = 10$, and $\tau = 49.2$. Dashed, solid, and dotted lines correspond to shear-thinning, Newtonian and shear-thickening fluids, respectively. Curves in blue do not consider the finite-sized ions (steric effects), and black curves consider the steric effects. In Figure 6a, the steric effect promotes a decrease in non-Newtonian and Newtonian fluids' velocity; that behavior is more representative in shear-thinning fluids. This reduction in hydrodynamics can be avoided and can even be exceeded by applying high zeta potentials of the order of $\kappa_\psi \sim O(10^1)$, keeping fixed Wo, as shown in Figure 6b. It is evident from Figure 6a,b that the velocity increases in one order of magnitude by increasing κ_ψ from 2 to 10; however, with $\kappa_\psi = 10$ (Figure 6b), the velocity's reduction by steric effects in Newtonian and Non-Newtonian fluids is more significant than that obtained by considering $\kappa_\psi = 2$. That occurs because the body electric forced is reduced by the finite-sized ions into the EDL; this causes a significant decrement in the slippage velocity at the walls changing the velocity profiles from concave to convex shape (Figure 6b,d). In addition, steric effects give rise to a strong reduction in the velocity gradients near the microchannel walls offering the fluid a high resistance to be deformed. The rheology of the power-law fluids plays an important role in reducing velocity gradients due to the dynamic viscosity increases with the steric effect being more pronounced this behavior in shear-thinning fluids, as shown in Figure 5a.

According to Equation (16) and confirmed in Figure 5, in shear-thinning fluids ($n < 1$), the dynamic viscosity is infinite at channel center due to the absence of velocity gradient, as shown in Figure 6a–d, and decreases at channel wall where higher velocity gradients are reached, while the opposite is true for shear-thickenings ($n > 1$). In this context, the shear-thinning behavior is more remarkable when Wo increases from 0.5 to 1, as depicted in Figure 6a,c, maintaining fixed the zeta potential $\kappa_\psi = 2$. For shear-thickening fluids (Figure 6c), the steric effect does not modify the velocity profiles, becoming a little concave in the center of the channel, and the important changes in velocity occur in the neighborhood of the sidewalls, as shown in Figure 6a,c. In Figure 6d, steric effects in Newtonian and power-law fluids provoke a representative decrease in the velocity on the microchannel cross-section. That reduction in velocity is more significant in shear-thinning fluids in one order of magnitude in comparison when the steric effects are absent. In addition, steric effects modified the shape of the velocity profiles from concave to convex due to the rheology of power-law fluids explained in Figure 5.

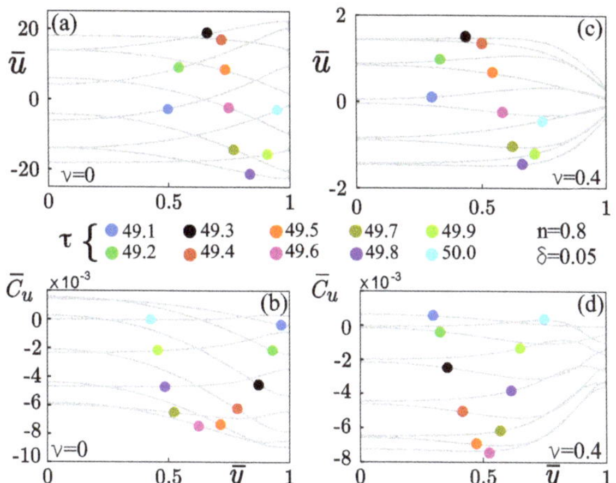

Figure 7. (**a**,**c**): Velocity profiles $\bar{u}$ across the microchannel at one period of oscillation. (**b**,**d**): Distribution of the convective concentration $\bar{c}_u$ across the microchannel due to the hydrodynamic in (**a**,**c**), respectively.

The combined effects of the hydrophobic surface condition and the finite-sized ions on the velocity and species concentration fields are shown in Figure 7. All panels were determined with $n = 0.8$, $Wo = 1$, $\bar{\kappa} = 10$, $\kappa_\psi = 10$, $\delta = 0.05$, $Sc = 500$, $\alpha = 0.001$. Because a pure AC electric field drives the electroosmotic flow, the movement of the flow acquires an oscillatory behavior, as shown in Figure 7a,c, and, consequently, the corresponding concentration field, as depicted in Figure 7b,d. The velocity $\bar{u}$ is plotted over one period of oscillation τ (=49.1–50) as a function of the transversal coordinate $\bar{y}$, for shear-thinning fluids with $n = 0.8$, $Wo = 1$, $\bar{\kappa} = 10$, $\kappa_\psi = 10$, $Sc = 500$, and $\alpha = 0.001$, by considering a dimensionless Navier slip length, $\delta = 0.05$. When the ions are treated as point charges, i.e., $\nu = 0$, the high zeta-potential $\kappa_\psi = 10$ magnifies the slippage effect on the hydrodynamics of the OEOF, as indicated in Figure 7a. That is because, in the absence of the ionic size effects, shear-thinning fluids $(n < 1)$ exhibit smaller viscosity at the wall than that with steric effects, as pointed in Figure 5a. That explains why the fluid has a low resistance to be deformed, causing high-velocity gradients in the walls' vicinity. According to the oscillatory electroosmotic body force included in Equation (15),

$$\bar{\rho}_e \bar{E}_x(\tau) = \frac{\bar{\kappa}^2}{\kappa_\psi} \frac{\sinh \bar{\psi}}{1 + 2\nu \sinh^2(\bar{\psi}/2)} \sin(2\pi\tau), \tag{41}$$

when the steric effect is absent ($\nu = 0$) in Equation (41), and considering fixed values in $\bar{\kappa}$ and $\bar{\psi} = \kappa_\psi$, the electric body force at the wall, where A is a constant, $\bar{\rho}_e \bar{E}_x \sim A \sin(2\pi\tau)$ promotes the highest velocity gradients, namely for half-period τ (=49.1–49.4) the flow moves in the positive axial direction and the body force $A \sin(2\pi\tau) > 0$, while, for τ (=49.6–49.9), the flow moves in the opposite direction and the function $A \sin(2\pi\tau) < 0$; in both scenarios described before, the velocity profiles acquire concave shape in the flow direction. At specific times τ (=49.5 and 50), the electric body force $A \sin(2\pi\tau) = 0$, causing a significant decrease in velocity and velocity gradients. Then, velocity profiles acquired a convex shape, and the velocity is close to zero in the neighborhood of the sidewalls, as is shown in Figure 7a. On the other hand, in Figure 7c, the finite-sized ions ($\nu \neq 0$) affect the slippage condition at the wall, giving rise to a reduction in the velocity up to one order of magnitude, and consequently, the velocity gradients are smaller than that obtained with $\nu = 0$. That reduction in velocity gradients is attributed to reducing the ionic concentration inside EDL as stated in Equation (3), decreasing the oscillatory electroosmotic body force in Equation (41). For example, an estimation of $\bar{\rho}_e \bar{E}_x(\tau)|_{\bar{y}=1}$ near the microchannel-wall, where $\bar{\kappa} \sim O(10^1)$, $\kappa_\psi \sim O(10^1)$, and $\bar{\psi} \sim \kappa_\psi$; for a specific $\tau = 49.2$, when $\nu = 0$ the electric body force $\bar{\rho}_e \bar{E}_x(\tau)|_{\bar{y}=1} \sim O(10^4)$, and when steric effects are considered with $\nu = 0.4$, the electric force $\bar{\rho}_e \bar{E}_x(\tau)|_{\bar{y}=1} \sim O(10^1)$.

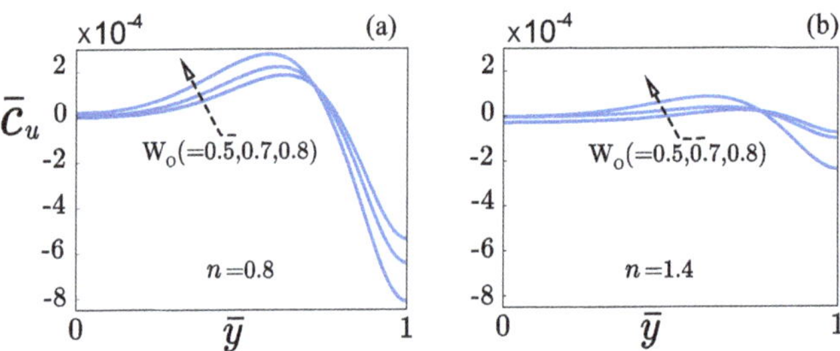

Figure 8. Distribution of dimensionless convective concentration $\bar{c}_u$ across the microchanne at different Womersley numbers *Wo*. (**a**) Shear-thinning fluids. (**b**) Shear-thickening fluids.

The dimensionless concentration $\bar{c}_u$ plotted in Figure 7b,d corresponds to the velocity field $\bar{u}$ in igure 7a,c over the same period of oscillation. Figure 7b shows the concentration, $\bar{c}_u$, without steric effects. In the first half period of oscillation τ (=49.1–49.6), the concave flow profile causes transversal concentration gradients, and the neutral dilute solute diffuses from the centerline of the microchannel to lateral walls. During the second half interval of time τ (=49.7–50), the flow profile is reversed, and the solute moves from the boundary walls to the centerline, where the solute's concentration is small compared to its value at the neighborhood of the walls. Figure 7d shows the concentration distribution with the steric effect. Due to a strong decrease in the slip-velocity at the walls originated by steric effects, the values of $\bar{c}_u$ near the walls are small compared to its value at the centerline; therefore, neutral dilute solute diffuses from the lateral walls to the centerline of the microchannel.

The rheology plays an essential role on the distribution of the concentration of the solute $\bar{c}_u$, as shown in Figure 8. For instance, higher concentration gradients across the microchannel-width appear with shear-thinning fluids (Figure 8a) in comparison against those present in shear-thickening fluids (Figure 8b). It occurs when *Wo* increases from 0.5 to 0.8, keeping fixed $\nu = 0.4$, $\kappa_\psi = 10$, $\bar{\kappa} = 10$, $Sc = 500$, $\triangle Z = 1$, $\delta = 0.05$, and $\alpha = 0.001$. In Figure 8a, the distribution in $\bar{c}_u$ is more dissimilar as *Wo* increases. This consequence arises from the non-uniformity in the velocity flow becoming stronger in shear-thinning fluids, as shown in Figure 6d. The concentration $\bar{c}_u$ in shear-thickening fluids is less affected by the Womersley number *Wo* as presented in Figure 8b in comparison with shear-thinning fluids, due to a significant resistance offering by the fluid to be deformed, as shown in Figure 6d, where the velocity is reduced by about 30% concerning that obtained in shear-thinning fluids under the influence of steric effects.

Oscillatory flows become a subject of considerable interest due to their potential to provide significant advantages in the manipulation of the transport of species, causing that those with low diffusivities can be transported faster than species with higher diffusivities, and vice versa. In this context, the rate of mass transport $\bar{Q}_x$ is shown in Figure 9a,b as a function of the Womersley number *Wo* with $\bar{\kappa}_\psi = 10$, $\bar{\kappa} = 10$, $\delta = 0.05$, $\alpha = 0.001$ and $\Delta Z = 1$. Dashed, solid, and dotted lines correspond to shear-thinning, Newtonian and shear-thickening fluids, respectively. Curves in blue do not consider the finite-sized ions (steric effects), and black curves consider the steric effects. Figure 9a analyzes a fast diffuser ($Sc = 500$), while Figure 9b analyzes a slow diffuser ($Sc = 2000$).

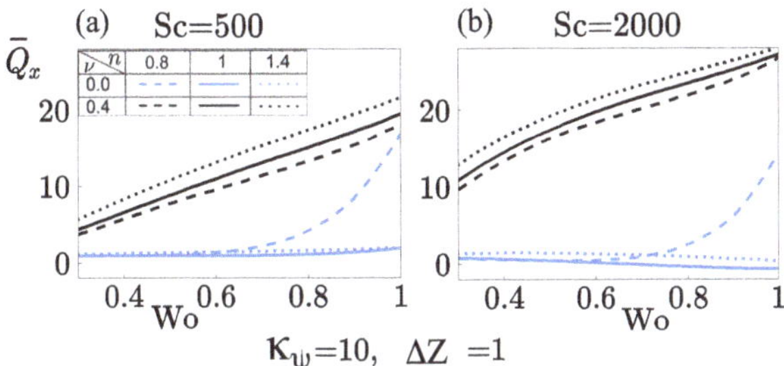

Figure 9. Rate of mass transport $\bar{Q}_x$ as a function of Wo at different flow behavior indices n and the steric factor ν. (**a**) $Sc = 500$ and (**b**) $Sc = 2000$.

The mass transport $\bar{Q}_x$, in a fast diffuser in the absence of steric effects, Wo does not influence on $\bar{Q}_x$ in the interval $0.3 \lesssim Wo \lesssim 0.6$ in fluids with n (=0.8, 1.0, 1.4), as indicated in Figure 9a. Shear-thinning fluids with $Wo \gtrsim 0.6$ promote a rapid increase in $\bar{Q}_x$ while Newtonian and shear-thickening fluids remain unaltered with Wo. That occurs in fluids with $n < 1$ because of high slippage velocity at the walls and the concave shape of velocity profiles in the flow direction with one order of magnitude higher than fluids with $n = (1, 1.4)$ where the velocity profiles acquire flattened shape, as shown in blue curves in Figure 6b. Conversely, in the presence of steric effects, Wo enhances $\bar{Q}_x$ in fluids with n (=0.8, 1.0, 1.4), as indicated by black-curves in Figure 9a. It is observed that independently of the value that Wo takes, the convex-shape (black lines, in Figure 6b,d) of the velocity profiles in fluids with n (=0.8, 1.0, 1.4) remain in the same order of magnitude. However, the shear-thickening fluid is a candidate to transport more species due to shear-thinning fluids offer greater resistance to being deformed in the presence of steric effects, as it is demonstrated in Figure 5a. Figure 9b shows how the importance of the hydrodynamic behavior mentioned above in benefiting and enhance the mass transport of the slow diffuser ($Sc = 2000$) with a shear-thickening as a carrier fluid in the presence of steric effects. That is, species with low diffusivity ($Sc = 2000$) can be transported up to 64.6% faster than species with high diffusivity ($Sc = 500$) when the steric effect is considered. Contrarily, when the steric effect is neglected, the more diffusive species travel slightly faster up to 21.5%.

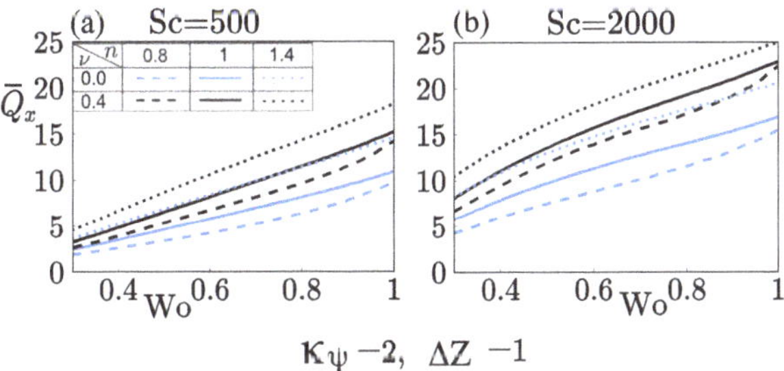

Figure 10. Rate of mass transport $\bar{Q}_x$ as a function of Wo at different flow behavior indices n and the steric factor ν. (**a**) $Sc = 500$ and (**b**) $Sc = 2000$.

Figure 10 highlights the effect of the zeta potential $\bar{\kappa}_\psi = 2$ on $\bar{Q}_x$, with $\bar{\kappa} = 10$, $\delta = 0.05$, $\alpha = 0.001$, and $\Delta Z = 1$. As expected, the zeta potential attenuates the mass transport of fast and slow diffusers with shear-thickening fluid as carrier compared to high-zeta potential, as described in Figure 9. However, the zeta potential improves $\bar{Q}_x$ with shear-thinning fluids, as shown in Figure 10. In addition, the steric effect enhances the mass transport in fast and slow diffusers when Wo increases. The reason is that independently the value Wo takes, the reduction in velocity by steric effects remains in the same order of magnitude, except in shear-thickening fluids where steric effects are tiny, as shown in Figure 6a,b.

Furthermore, Figures 10a and 11a were compared to analyze the influence of tidal displacement ΔZ on the mass transport rate $\bar{Q}_x$. A tidal displacement $\Delta Z = 2$ causes an increment in the mass transport rate about two times with $\bar{\kappa}_\psi = 2$ compared to that obtained with $\Delta Z = 1$, and that increment was estimated up to four times higher when high zeta potentials ($\bar{\kappa}_\psi = 2$) were considered (see Figures 9a and 11b). From Equation (31) it is evident that the tidal displacement influences the concentration $\bar{c}_u$ through Pe_ω; as ca be deduced from Equation (39), $\bar{c}_u$ will affect the mass transport rate, $\bar{Q}_x$. Then, the tid displacement increases the convective effects in the concentration field and enhances th mass transport rate in comparison when a smaller tidal displacement is used.

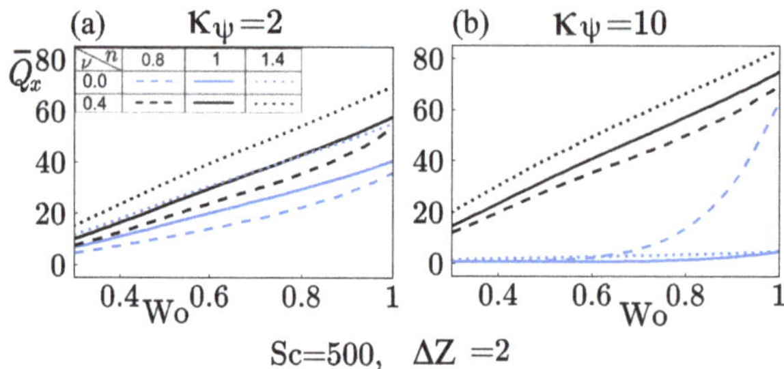

Figure 11. Rate of mass transport $\bar{Q}_x$ as a function of Wo at different flow behavior indices n and the steric factor ν. (**a**) $\kappa_\psi = 2$ and (**b**) $\kappa_\psi = 10$.

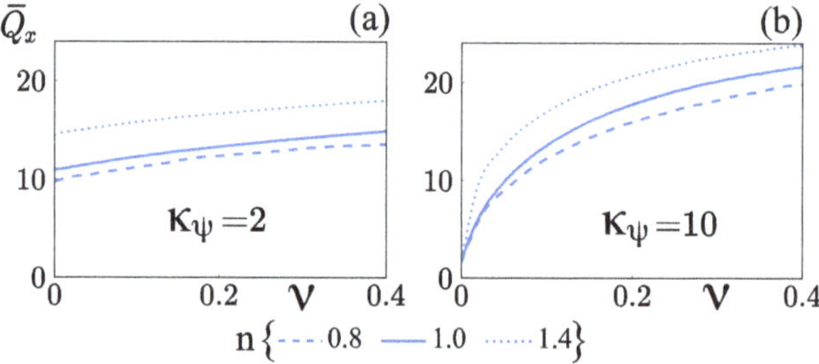

Figure 12. Rate of mass transport $\bar{Q}_x$ as a function of the steric effect factor ν, at different flo behavior indices n. (**a**) $\kappa_\psi = 2$ and (**b**) $\kappa_\psi = 10$.

Finally, the dependency of the rate of mass transport $\bar{Q}_x$ on the steric factor is studie and Figure 12 displays the results with $\bar{\kappa} = 10$, $\delta = 0.05$, $\alpha = 0.001$, $Sc = 500$, $\Delta Z =$ and $Wo = 1$. When the ions are assumed to have finite volumes, the mass transport in power-law and Newtonian fluids is an ascending function of the steric factor, ν, for $\bar{\kappa}_\psi = 2$ and high ($\bar{\kappa}_\psi = 10$) zeta potentials, as it is shown in Figure 12a,b. In Figure 12a, the values of $\bar{Q}_x$ for $\nu = 0$ and $\nu = 0.4$ with the power-law index n (=0.8, 1.0, 1.4) correspond to the values of $\bar{Q}_x$ in Figure 10a with $Wo = 1$. In Figure 12a, the mass transport $\bar{Q}_x$ in shear-thickening fluids is higher than in Newtonian and shear-thinning fluids. This is due to in shear-thickening fluids there is a little influence of the steric factor on the velocity profiles, while, for Newtonian and shear-thinning fluids the velocity profiles are reduced by steric effects, as shown in Figure 6c. This occurs because, fluids with $n > 1$ offer more resistance to be deformed in comparison with Newtonian and shear-thinning fluids. This is confirmed in Figure 5, the steric effect in fluids with $n = 1.4$ has a little impact on the apparent viscosity (see Figure 5b), while, in fluids with $n = 0.8$, an important increment in the viscosity distribution occurs due to steric effects, as shown in Figure 5a.

In Figure 12b, the values of $\bar{Q}_x$ as a function of ν with the power-law index n (=0.8, 1.0, 1.4) and $\kappa_\psi = 10$ are plotted. The increase of $\bar{Q}_x$ with ν is more pronounced with $\kappa_\psi = 10$ than with $\kappa_\psi = 2$. That is due to near the wall, the reduction of the body force due to steric effects is significant; consequently, a higher zeta potential gives rise to a smaller velocity, as it is shown in Figure 6d, where for fluids with $n = 0.8$, the velocity is reduced in one order of magnitude, and fluids with $n = 1.0$ and $n = 1.4$, the reduction in the velocity are maintained in the same order of magnitude.

5. Conclusions

The effects of non-Newtonian rheology, steric effect, and slippage on an OEOF were analyzed. The start-up condition for the flow was considered, and after all initial transients have died out and the flow is periodic, the mass concentration and transport of species were determined numerically. The influence of the main parameters that describe the dynamic behavior of the flow is the steric factor ν, the normalized slip length, δ, an electrokinetic parameter $\bar{\kappa}$, the normalized zeta potential κ_ψ, the Womersley (Wo) and Schmidt (Sc) numbers, and the power-law index n. The ionic size effects were controlled by the mean volume fraction of each ion in bulk given by the factor $\nu = 2a^3 n_0$. The numerical results are compared against an OEOF solution for a Newtonian fluid without slippage, and steric effects, previously reported by H. Huang and C. Lai [13]. The following conclusions from this study were drawn:

1. In shear-thinning fluids, the steric effect under hydrophobic conditions has a noticeable impact on the rheology of the fluid, causing higher values in the dynamic viscosity, $\bar{\eta}$, compared with the absence of finite-size ions. In shear-thickening fluids, since steric effects reduce the oscillatory electroosmotic body force up to three orders of magnitude compared with no steric case, the dynamic viscosity decreases near the microchannel wall.
2. Finite-size ions reduce oscillatory electroosmotic body force by preventing EDL from being highly concentrated. As a result, steric effects result in a decrease in velocity in shear-thinning fluids up to one order of magnitude compared with no steric case, with $\kappa_\psi = 10$. However, in shear-thickening fluids, the steric effects are negligible on the velocity when $\kappa_\psi = 2$.
3. The suggested values of $Wo \sim O(10^{-1})$, $\kappa_\psi = 10$ and $\nu = 0.4$ promote the best conditions for the mass transport $\bar{Q}_x$ for any Sc number values. A value of $\nu = 0.4$ with $\kappa_\psi = 10$ increases the value of $\bar{Q}_x$ in about 90 % compared with no steric effect. In a similar way, in about 20 % the value of $\bar{Q}_x$ was increased with $\kappa_\psi = 2$.
4. The steric effect enhances the mass transport in fast and slow diffusers when Wo increases by using $\bar{\kappa}_\psi = 2$ or high zeta potentials ($\bar{\kappa}_\psi = 10$). However, at high zeta potentials ($\bar{\kappa}_\psi = 10$), $\bar{Q}_x$ increases up to one order of magnitude compared with that obtained with $\bar{\kappa}_\psi = 2$. Additionally, steric effect promotes that slow diffusers ($Sc = 2000$) can travel faster than fast diffusers ($Sc = 500$) at $Wo < 1$, as shown in Figure 9 (black lines). The opposite behavior occurs in the absence of steric effects.
5. A wide variety of different physical and chemical phenomena could also be included in the model to examine their effects on mass transport with non-Newtonian fluids. Possibilities include the depletion of macromolecules near the microchannel walls, the presence of a reversible reaction or mass exchange between the microchannel wall and the fluid. In physiological systems, which can be significantly more important viscoelastic effects than higher purity for typical aqueous solutions, could be considered.

Author Contributions: Conceptualization, R.B., J.A., O.B., and F.M.; Data curation, R.B., J.A., O.B., and F.M.; Formal analysis, R.B., J.A., O.B., and F.M.; Funding acquisition, J.A. and O.B.; Investigation, R.B., J.A., and O.B.; Methodology, R.B., J.A., and O.B.; Project administration, R.B., O.B., and F.M.; Resources, R.B. and O.B.; Software, R.B.; Supervision, O.B. and F.M.; Validation, R.B.; Visualization,

R.B., J.A., O.B., and F.M.; Writing—original draft, R.B., J.A., O.B. and F.M.; Writing—review & editing, R.B., J.A., O.B., and F.M. All authors have read and agreed to the published version of the manuscript.

Funding: This research was supported by the Instituto Politécnico Nacional in México, grant numbers SIP-20211004 and SIP-20210909.

Acknowledgments: R.B. acknowledges the support from Instituto Politécnico Nacional program for the Ph.D. Fellowship at SEPI-ESIME-IPN. In addition, R.B. thanks the BEIFI program from the IPN.

Conflicts of Interest: The authors declare no conflict of interest. The funders had no role in the design of the study; in the collection, analyses, or interpretation of data; in the writing of the manuscript, or in the decision to publish the results.

Nomenclature

a	Ion size, m
c	Concentration field of the solute, mol m^{-3}
c_0	Molar ion concentration of the electrolyte solution, mol m^{-3}
C_1, C_2	Fixed concentration values at the ends of the microchannel, mol m^{-3}
c_u	Convective species concentration, mol m^{-3}
$\bar{c}_u$	Dimensionless convective species concentration
D	Diffusion coefficient, m^2 s^{-1}
$\mathbf{D}$	Rate of strain tensor
e	Electron charge, C
$E_x(t)$	External electric field, Vm^{-1}
E_0	Intensity of the applied electric field, Vm^{-1}
h	Microchannel Half-Height, m
$j_{x,conv}$	Convective flux density, mol m^2 s^{-1}
$j_{x,diff}$	Diffusive flux density x, mol m^2 s^{-1}
J_x	Total flux density, mol m^2 s^{-1}
K_B	Boltzmann constant, J K^{-1}
L	Microchannel length, m
m	Fluid consistency index, Pa s^n
$\bar{m}$	Dimensionless consistency index
n	Power-law index
$\mathbf{n}$	Unit vector normal to the microchannel surface
n_0	Ionic concentration, m^{-3}
Pe_D	Diffusive Péclet number
Q_x	Flow rate, mol m^{-2} s^{-1}
$\bar{Q}_x$	Dimensionless flow rate
Sc	Schmidt number
t	Time, s
t_ω	Periodic time, s
T	Absolute temperature, K
u	Longitudinal velocity component, m s^{-1}
$\mathbf{u}_s$	Fluid velocity at the microchannel wall, m s^{-1}
U_{HS}	Helmholtz-Smoluchowsky velocity
$\bar{u}$	Dimensionless longitudinal velocity component
W_o	Womersley number
x, y	Space coordinates, m
$\bar{x}, \bar{y}$	Dimensionless space coordinates
z	valency of both the ions
Greek symbols	
α	Aspect ratio
$\dot{\gamma}$	Strain rate
Δz	Tidal displacement, m
ΔZ	Dimensionless tidal displacement
δ	Dimensionless slip lenght

ϵ	Permittivity of the solution, C V^{-1} m^{-1}
ζ	Zeta potential, V
η	Apparent viscosity of the fluid, Pa s
κ^{-1}	Debye length, m
κ_{ψ}	Relation between the wall ζ potential and the thermal potential
$\bar{\kappa}$	Ratio of the microchannel height to the Debye length
λ_N	Navier length, m
μ	Viscosity of a Newtonian fluid, Pa s
ν	Bulk volume fraction of ions
v_0	Kinematic viscosity, m^2 s^{-1}
ρ_f	Fluid density, kg m^{-3}
ρ_e	Electric charge density, C m^{-3}
τ	Dimensionless time
τ_{ω}	Dimensionless periodic time
τ_{xy}	Unidirectional shear stress
Φ	Total electric potential, V
ϕ	External electric potential, V
ψ	Electric potential, V
ω	Angular frequency, rad s^{-1}
Subscripts and superscripts	
i	Nodal position in $\bar{x}$ coordinate
j	Nodal position in $\bar{y}$ coordinate
l	Nodal position in time
tr	Matrix transpose
Abbreviations	
AC	Alternating current
DC	Direct current
DNA	Deoxyribonucleic acid
EDL	Electrical double layer
EOF	Electroosmotic flow
MPB	Modified Poisson-Boltzmann
OEOF	Oscillatory electroosmotic flow
PB	Poisson-Boltzmann
TDMA	Tridiagonal matrix algorithm

References

1. Taylor, G.I. Dispersion of soluble matter in solvent flowing slowly through a tube. *Proc. R. Soc. Lond. Ser. A* **1953**, *219*, 186–203.
2. Aris, R. On the dispersion of a solute in pulsating flow through a tube. *Proc. R. Soc. Lond. Ser. A* **1960**, *259*, 370–376.
3. Atkins, P.W. *Physical Chemistry*; Oxford University Press: Oxford, UK, 1994.
4. Chatwin, P.C. On the longitudinal dispersion of passive contaminant in oscillatory flows in tubes. *J. Fluid Mech.* **1975**, *71*, 513–527. [CrossRef]
5. Watson, E.J. Diffusion in oscillatory pipe flow. *J. Fluid Mech.* **1983**, *133*, 233–244. [CrossRef]
6. Kurzweg, U.H.; Howell, G.; Jaeger, M.J. Enhanced dispersion in oscillatory flows. *Phys. Fluids* **1984**, *27*, 1046–1048. [CrossRef]
7. Kurzweg, U.; Jaeger, M. Diffusional separation of gases by sinusoidal oscillations. *Phys. Fluids* **1987**, *30*, 1023–1025. [CrossRef]
8. Thomas, A.M.; Narayanan, R. The use of pulsatile flow to separate species. *Ann. N. Y. Acad. Sci.* **2002**, *974*, 42–56. [CrossRef]
9. Ramon, G.; Agnon, Y.; Dosoretz, C. Solute dispersion in oscillating electro-osmotic flow with boundary mass exchange. *Microfluid. Nanofluid.* **2011**, *10*, 97–106. [CrossRef]
10. Peralta, M.; Arcos, J.; Méndez, F.; Bautista, O. Oscillatory electroosmotic flow in a parallel-plate microchannel under asymmetric zeta potentials. *Fluid Dyn. Res.* **2017**, *49*, 035514. [CrossRef]
11. Rojas, G.; Arcos, J.; Peralta, M.; Méndez, F.; Bautista, O. Pulsatile electroosmotic flow in a microcapillary with the slip boundary condition. *Colloid Surf. A-Physicochem. Eng. Asp.* **2017**, *513*, 57–65. [CrossRef]
12. Medina, I.; Toledo, M.; Méndez, F.; Bautista, O. Pulsatile electroosmotic flow in a microchannel with asymmetric wall zeta potentials and its effect on mass transport enhancement and mixing. *Chem. Eng. Sci.* **2018**, *184*, 259–272. [CrossRef]
13. Huang, H.; Lai, C. Enhancement of mass transport and separation of species by oscillatory electroosmotic flows. *Proc. R. Soc. A* **2016**, *462*, 2017–2038. [CrossRef]
14. Thomas, A.; Narayanan, R. Physics of oscillatory flow and its effect on the mass transfer and separation of species. *Phys. Fluids* **2001**, *13*, 859–866. [CrossRef]
15. Kilic, M.S.; Bazan, M.Z.; Ajdari, A. Steric effects in the dynamics of electrolytes at large applied voltages. I. Double-layer charging. *Phys. Rev. E* **2007**, *75*, 021502. [CrossRef]

16. Yazdi, A.A.; Sadeghi, A.; Saidi, M.H. Steric effects on electrokinetic flow of non-linear biofluids. *Colloid Surf. A-Physicochem. Eng. Asp.* **2015**, *484*, 394–401. [CrossRef]
17. Dey, R.; Ghonge, T.; Chakraborty, S. Steric-effect-induced alteration of thermal transport phenomenon for mixed electroosmotic and pressure driven flows through narrow confinements. *Int. J. Heat Mass Transf.* **2013**, *56*, 251–262. [CrossRef]
18. Marroquin-Desentis, J.; Méndez, F.; Bautista, O. Viscoelectric effect on electroosmotic flow in a cylindrical microcapillary. *Fluid Dyn. Res.* **2016**, *48*, 035503. [CrossRef]
19. Garai, A.; Chakraborty, S. Steric effect and slip-modulated energy transfer in narrow fluidic channels with finite aspect ratios. *Electrophoresis* **2010**, *31*, 843–849. [CrossRef] [PubMed]
20. Storey, B.D.; Edwards, L.R.; Kilic, M.S.; Bazant, M.Z. Steric effects on ac electro-osmosis in dilute electrolytes. *Phys. Rev. E* **2008**, *77*, 036317. [CrossRef]
21. Joly, L.; Ybert, C.; Trizac, E.; Bocquet, L. Hydrodynamics within the Electric Double Layer on Slipping Surfaces. *Phys. Rev. Lett.* **2004**, *93*, 257805. [CrossRef]
22. Muñoz, J.; Arcos, J.; Bautista, O.; Méndez, F. Slippage effect on the dispersion coefficient of a passive solute in a pulsatile electro-osmotic flow in a microcapillary. *Phys. Rev. Fluids* **2018**, *3*, 084503. [CrossRef]
23. Teodoro, C.; Bautista, O.; Méndez, F. Mass transport and separation of species in an oscillatory electroosmotic flow caused by distinct periodic electric fields. *Phys. Scr.* **2019**, *94*, 115012. [CrossRef]
24. Babaie, A.; Sadeghi, A.; Saidi, M.H. Combined electroosmotically and pressure driven flow of power-law fluids in a slit microchannel. *J. Non-Newton. Fluid Mech.* **2011**, *166*, 792–798. [CrossRef]
25. Sánchez, S.; Arcos, J.; Bautista, O.; Méndez, F. Joule heating effect on a purely electroosmotic flow of non-Newtonian fluids in a slit microchannel. *J. Non-Newton. Fluid Mech.* **2013**, *192*, 1–9. [CrossRef]
26. Ng, C.; Qi, C. Electroosmotic flow of a power-law fluid in a non-uniform microchannel. *J. Non-Newton. Fluid Mech.* **2014**, *208–209*, 118–125. [CrossRef]
27. Sadegui, M.; Saidi, M.H.; Moosavi, A.; Sadegui, A. Unsteady solute dispersion by electrokinetic flow in a polyelectrolyte layer-grafted rectangular microchannel with wall absorption. *J. Fluid Mech.* **2020**, *887*, A13. [CrossRef]
28. Peralta, M.; Bautista, O.; Méndez, F.; Bautista, E. Pulsatile electroosmotic flow of a Maxwell fluid in a parallel flat plate microchannel with asymmetric zeta potentials. *Appl. Math. Mech.* **2018**, *39*, 667–684. [CrossRef]
29. Mederos, G.; Arcos, J.; Bautista, O.; Méndez, F. Hydrodynamics rheological impact of an oscillatory electroosmotic flow on a mass transfer process in a microcapillary with a reversible wall reaction. *Phys. Fluids* **2020**, *32*, 122003. [CrossRef]
30. Baños, R.; Arcos, J.C.; Bautista, O.; Méndez, F.; Merchan-Cruz, E.A. Mass transport by an oscillatory electroosmotic flow of power-law fluids in hydrophobic slit microchannels. *J. Braz. Soc. Mech. Sci.* **2021**, *43*, 1–15.
31. Peralta, M.; Arcos, J.; Méndez, F.; Bautista, O. Mass transfer through a concentric-annulus microchannel driven by an oscillatory electroosmotic flow of a Maxwell fluid. *J. Non-Newton. Fluid Mech.* **2020**, *279*, 104281. [CrossRef]
32. Chang, C.C.; Yang, R.J. Electrokinetic mixing in microfluidic systems. *Microfluid. Nanofluidics* **2007**, *3*, 501–525. [CrossRef]
33. Navier, C.L.M.H. Mémoire sur les lois du mouvement des fluides. *Mémoires de l'Académie Royale des Sciences de l'Institut de France VI*. 1823. pp. 389–440. Available online: https://cdarve.web.cern.ch/publications_cd/navier_darve.pdf (accessed on 21 April 2021).
34. Masliyah, J.H.; Bhattacharjee, S. *Electrokinetic and Colloid Transport Phenomena*; John Wiley & Sons: Hoboken, NJ, USA, 2006.
35. Torrie, G.M.; Valleau, J.P. Electrical double layers. I. Monte Carlo study of a uniformly charged surface. *J. Chem. Phys.* **1980**, *73*, 5807–5816. [CrossRef]
36. Lyklema, J. *Fundamentals of Interface and Colloid Science. Volumen II: Solid-Liquid Interfaces*; Academic Press: Cambridge, MA, USA, 1995.
37. Bikerman, J.J. Structure and capacity of electrical double layer. *Philos. Mag. Ser.* **1942**, *7*, 384–397. [CrossRef]
38. Hsu, J.P.; Kuo, Y.C.; Tseng, S. Dynamic interactions of two electrical double layers. *J. Colloid Interface Sci.* **1997**, *195*, 388–394. [CrossRef]
39. Probstein, R.F. *Physicochemical Hydrodynamics: An Introduction*; John Wiley & Sons: Hoboken, NJ, USA, 2010.
40. Kilic, M.S.; Bazan, M.Z.; Ajdari, A. Steric effects in the dynamics of electrolytes at large applied voltages. II. Modified Poisson-Nernst-Planck equations. *Phys. Rev. E* **2007**, *75*, 021503. [CrossRef]
41. Leal, L.G. *Advanced Transport Phenomena: Fluid Mechanics and Convective Transport Processes*; Cambridge University Press: Cambridge, UK, 2007.
42. Osswald, T.; Hernández-Ortiz, J.P. *Polymer Processing: Modeling and Simulation*; Carl Hanser Verlag: Munich, Germany, 2006.
43. Kurzweg, U.H. Enhanced diffusional separation in liquids by sinusoidal oscillatory. *Sep. Sci. Technol.* **1988**, *23*, 105–117. [CrossRef]
44. Anderson, J.D.; Wendt, J. *Computational Fluid Dynamics*; Springer: Berlin/Heidelberg, Germany, 1995.
45. Hacioglu, A.; Narayanan, R. Oscillating flow and separation of species in rectangular channels. *Phys. Fluids* **2016**, *28*, 073602. [CrossRef]
46. Morgan, H.; Green, N.G. *AC Electrokinetics: Colloids and Nanoparticles*; Research Studies Press: Oxford, UK, 2006.
47. Zhao, C.; Zhang, W.; Yang, C. Dynamic Electroosmotic Flows of Power-Law Fluids in Rectangular Microchannels. *Micromachines* **2017**, *8*, 34. [CrossRef]
48. Griffiths, S.k.; Nilson, R.H. Electroosmotic fluid motion and late-time solute transport for large zeta potentials. *Anal. Chem.* **2000**, *72*, 4767–4777. [CrossRef] [PubMed]

49. Hervet, H.; Léger, L. Flow with slip at the wall: From simple to complex fluids. *C. R. Physique* **2003**, *4*, 241–249. [CrossRef]
50. Green, N.G.; Ramos, A.; Gonzalez, A.; Morgan, H.; Castellanos, A. Fluid flow induced by nonuniform ac electric fields in electrolytes on microelectrodes. I. Experimental measurements. *Phys. Rev. E* **2000**, *61*, 4011–4018. [CrossRef] [PubMed]

 micromachines

MDPI

Article

Mixing Mechanism of Microfluidic Mixer with Staggered Virtual Electrode Based on Light-Actuated AC Electroosmosis

Liuyong Shi [1], Hanghang Ding [1], Xiangtao Zhong [1], Binfeng Yin [2], Zhenyu Liu [3] and Teng Zhou [1,*]

1 Mechanical and Electrical Engineering College, Hainan University, Haikou 570228, China; shiliuyong@hainanu.edu.cn (L.S.); dinghanghang@hainanu.edu.cn (H.D.); zhongxiangtao97@foxmail.com (X.Z.)
2 School of Mechanical Engineering, Yangzhou University, Yangzhou 225127, China; binfengyin@yzu.edu.cn
3 Changchun Institute of Optics, Fine Mechanics and Physics (CIOMP), Chinese Academy of Sciences, Changchun 130033, China; liuzy@ciomp.ac.cn
* Correspondence: zhouteng@hainanu.edu.cn; Tel.: +86-186-8963-7366

Abstract: In this paper, we present a novel microfluidic mixer with staggered virtual electrode based on light-actuated AC electroosmosis (LACE). We solve the coupled system of the flow field described by Navier–Stokes equations, the described electric field by a Laplace equation, and the concentration field described by a convection–diffusion equation via a finite-element method (FEM). Moreover, we study the distribution of the flow, electric, and concentration fields in the microchannel, and reveal the generating mechanism of the rotating vortex on the cross-section of the microchannel and the mixing mechanism of the fluid sample. We also explore the influence of several key geometric parameters such as the length, width, and spacing of the virtual electrode, and the height of the microchannel on mixing performance; the relatively optimal mixer structure is thus obtained. The current micromixer provides a favorable fluid-mixing method based on an optical virtual electrode, and could promote the comprehensive integration of functions in modern microfluidic-analysis systems.

Keywords: light-actuated AC electroosmosis (LACE); microfluidic mixer; optical virtual electrode; electrokinetics; computational fluid dynamics (CFD)

Citation: Shi, L.; Ding, H.; Zhong, X.; Yin, B.; Liu, Z.; Zhou, T. Mixing Mechanism of Microfluidic Mixer with Staggered Virtual Electrode Based on Light-Actuated AC Electroosmosis. *Micromachines* **2021**, *12*, 744. https://doi.org/10.3390/mi12070744

Academic Editors: Kwang-Yong Kim and Nam-Trung Nguyen

Received: 25 April 2021
Accepted: 23 June 2021
Published: 24 June 2021

1. Introduction

Micromixing [1] is one of the key technologies of micro-total analysis systems (µTAS) [2] and lab-on-a-chip (LOC) [3] devices, and it is widely used in the fields of biology [4], chemistry [5], medicine [6], and engineering, both in academia and industry, because of its low cost, high speed, and low sample consumption [7–9]. However, compared with macrofluids, microfluidic flow is highly ordered laminar flow due to the low Reynolds numbers, and molecular diffusion is the main mechanism of microfluidic mixing [10]. Small molecules (rapidly diffusing species) can achieve diffusion mixing within tens of microns during a few seconds, but the equilibrium time required for mixing large molecules (peptides, proteins, and high-molecular-weight nucleic acids) within the same distance is from several minutes to several hours. For many chemical analyses, this means that longer channel lengths and mixing times are required to achieve the homogenization of the species concentration, and such delays are impractical. Therefore, to improve the mixing efficiency of the micromixer by optimizing the size of the device and the time of sample analysis is the main object of the design and development of micromixers today.

According to mixing mechanism, micromixers can be basically divided into active [11] and passive [12,13] micromixers. Passive micromixers rely on the geometry of microchannels to generate complex flow fields, such as a flow split [14], flow recombination [15], flow separation, chaotic advection [16], and hydrodynamic focus to realize the efficient mixing of fluids [17,18]. Active micromixers, via external energy or stimuli, improve the mixing of fluids either with moving parts [19] or with external force fields such as acoustic [20,21], electric [22–24], magnetic [25,26], thermal [27–30], and pressure [31] fields. The rotating electroosmotic vortex generated under the action of applied potential is widely used in active micromixers to manipulate and control different fluids to achieve mixing. In the past few years, a large number of electroosmotic microfluidic mixers were designed and developed [32]. In general, when a fixed microelectrode [33] is embedded on the sidewall of the microchannel, a pair of electroosmotic rotating vortices can be generated near the microelectrode to achieve the mixing of fluids in the microchannel [34]. The implantation of a heterogeneous surface charge [35–38] and a nonuniform surface potential charge [39] on the side and/or bottom walls of the microchannel to induce localized nonaxial flow structure or vortices can achieve better mixing efficiency [40]. However, the method of embedding fixed microelectrodes or performing heterogeneous processing on the inner wall of the microchannel both increases the difficulty and cost of processing the microchannel, and reduces the flexibility of microchip functionality.

With the development of light-induced technology [41–43], a rotating vortex generated by LACE can both manipulate microparticles in the fluids [44,45] and be used for the mixing of different fluids [46]. However, the mixer structure and mixing efficiency in [46] could still be further optimized and improved. On this basis, we improved the inlet layout of the mixer and designed a new type of staggered electrode, further revealing the fluid-mixing mechanism under the combined action of the staggered electrode and the applied electric field by multiphysics field coupling numerical simulation. Moreover, the influence of geometric parameters, including the length, width, and spacing of the staggered virtual electrodes, and the height of the microchannel, on mixing efficiency was studied (this is rarely mentioned in previous studies, and the circular spot or ring virtual electrode was generally employed before); thus, a relatively optimal mixer structure was obtained.

2. Theory and Methods

2.1. Micromixer Structure

Figure 1a shows the schematic diagram of a three-dimensional structure of the microfluidic mixer with staggered virtual electrode based on LACE. In this model, two kinds of fluids with the same solute but different solute concentrations (c_1 = 0 mol/m^3, c_2 = 1 mol/m^3) were driven by pressure from Inlets 1 and 2 into the straight mixing microchannel. The length of the straight microchannel was L_c = 200 μm, and width and height were W_c = H_c = 20 μm. The center line on the inlet boundary perpendicular to the bottom wall of the microchannel was taken as the dividing line between Inlets 1 and 2. As shown in Figure 1b, the top and bottom walls of the microchannel were transparent indium tin oxide (ITO) glass. The photoconductive layer was coated at the surface of the bottom wall of the microchannel by plasma-enhanced chemical vapor deposition (PECVD). From top to bottom, the photoconductive layer was a multilayer film structure composed of SiC_x film with a thickness of 25 nm, intrinsic a-Si:H with a thickness of 2 μm, and n^+ a-Si:H with a thickness of 50 nm [47].

In general, the conductivity of the photoconductive layer was approximately σ_D = 6.7 × 10^{-5} S/m, and when light was projected onto the photoconductive layer, the conductivity of the illuminated area increased sharply to σ_L = 0.2 S/m [48]. This means that a specific nonuniform electric field was generated in the microchannel when a specific optical pattern was projected onto the photoconductive layer. Since this optical pattern could produce a nonuniform electric field similar to the fixed metal microelectrode, it could called the "optical virtual electrode". The generation and adjustment of the virtual electrode could be realized by a series of optical lenses and plane mirrors based on a digital micromirror device (DMD). The position of the virtual electrodes staggered along the axial direction of the microchannel is shown in Figure 1c, which clearly shows the characteristic size (L_c, W_c) of the microchannel, the characteristic size (L_o = 10 μm, W_o = 10 μm, S_o = 10 μm, D_o = 90 μm) of the staggered light spot projected onto the photoconductive layer, and the position of each section (A–A, B–B, C–C, D–D). In this study, the number of staggered virtual electrodes is 8.

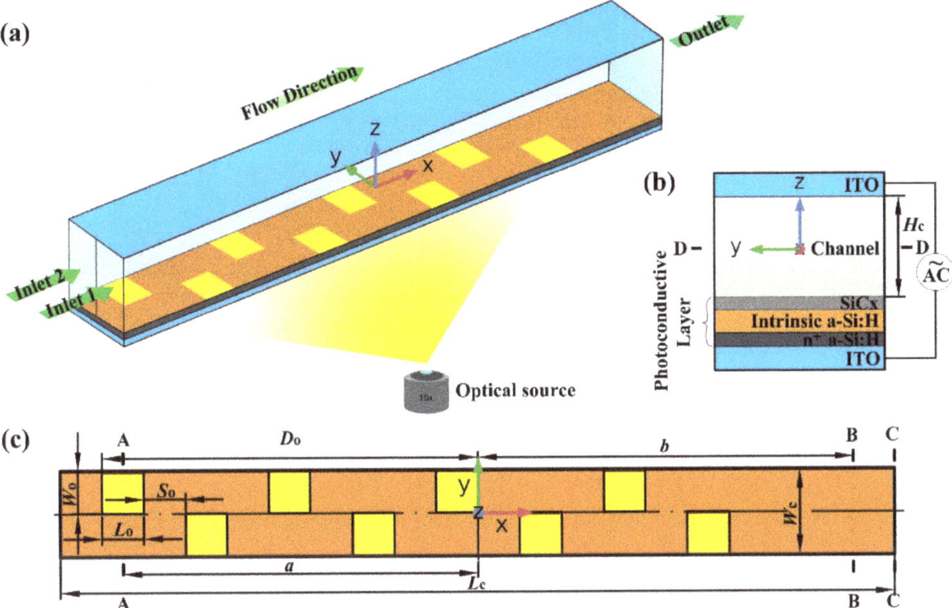

Figure 1. Microfluidic mixer with staggered virtual electrode based on LACE. (**a**) Three-dimensional structure of micromixer; green arrows indicate flow direction. (**b**) Cross-sectional (yz section) view of micromixer and photoconductive layer. (**c**) Top view of micromixer. Orange and yellow areas indicate photoconductive layer and the staggered light spots projected onto photoconductive layer, respectively. A–A, B–B, and C–C represent cross-section (yz section) at a distance of −85, 90, and 100 μm (exit position) from microchannel center, respectively. D–D represents section (xy section) at the microchannel center. Origin of the Cartesian coordinate system is located at the microchannel center.

2.2. Governing Equations

The transient Navier–Stokes equations for incompressible fluids describe the flow in the microchannel:

$$\rho\frac{\partial u}{\partial t} - \nabla\cdot\mu\left(\nabla u + (\nabla u)^T\right) + \rho u\cdot\nabla u + \nabla p = 0 \quad (1)$$

$$\nabla\cdot u = 0 \quad (2)$$

where ρ is density, μ is dynamic viscosity, u is velocity, σ is conductivity, and p is pressure. ρ = 1000 kg/m^3, μ = 0.001 kg/(m·s), and σ = 0.11845 S/m when we use an aqueous solution as fluid medium.

Uniform fluid velocity was applied at inlet boundaries 1 and 2:

$$u_1 = U_1 \boldsymbol{n} \quad (3)$$

$$u_2 = U_2 \boldsymbol{n} \quad (4)$$

Here, U_1 and U_2 are the fluid velocities at Inlets 1 and 2, respectively; $\boldsymbol{n}$ is the normal vector of the inlet boundary; $U_m = U_1 = U_2 = 2\times10^{-4}$ m/s is the mean inlet velocity.

The mixed fluid freely flowed out from the outlet boundary of the microchannel. The total stress component was perpendicular to boundary

$$\left(-p\boldsymbol{I} + \mu\left(\nabla u + (\nabla u)^T\right)\right)\cdot\boldsymbol{n} = 0, \quad (5)$$

$$p = 0 \quad (6)$$

where $\boldsymbol{I}$ is the unit tensor.

Compared with the characteristic size of the microchannel, the thickness (Debye length) of the electrical double layer on the inner wall surface of the microchannel was very small, so the Helmholtz–Smoluchowski slip-boundary condition between electroosmotic velocity and the tangential component of the applied potential could be applied to the entire top and bottom walls of the microchannel:

$$u_{\text{slip}} = -\frac{\varepsilon_0 \varepsilon_f \zeta}{\mu} \boldsymbol{E}_\text{t} \tag{7}$$

where $\zeta = -0.1$ V is the zeta potential on the top and bottom walls of the microchannel; $\boldsymbol{E}_\text{t}$ is the tangential component of the applied potential vector; ε_0 is the vacuum dielectric constant; and $\varepsilon_f = 80.2$ is the relative permittivity of the fluid.

Symmetrical boundary conditions applied to the two side walls of the microchannel described no penetration and vanishing shear stresses.

Assuming that there was no concentration gradient in the ions carrying current, a Laplace equation was used to describe the applied potential:

$$\nabla^2 \varphi = 0 \tag{8}$$

$$\boldsymbol{E} = -\nabla \varphi \tag{9}$$

$\boldsymbol{E}$ is the applied electric-field intensity vector, and φ is the applied potential.

$$\varphi = V_0 \sin(2\pi f t) \tag{10}$$

$V_0 = 1$ V is the peak-to-peak value of the applied potential, and $f = 5$ Hz is the electric-field frequency. The duration of one electric-field cycle was $T = 1/f = 0.2$ s. The top wall of the microchannel was grounded.

All other boundaries were electrically insulated:

$$\boldsymbol{n} \cdot \nabla \varphi = 0 \tag{11}$$

In the microchannel, the convection–diffusion equation is used to describe the concentration of dissolved species in the fluid:

$$\frac{\partial c}{\partial t} + \nabla \cdot (-D \nabla c) = -\boldsymbol{u} \cdot \nabla c \tag{12}$$

where c is the species concentration, and $D = 10^{-11}$ m^2/s [11] is the diffusion coefficient of the solute.

At Inlets 1 and 2, the solute gave concentrations of $c_1 = 0$ mol/m^3 and $c_2 = 1$ mol/m^3.

The mixed fluid flowed out from the outlet by convection, and the boundary conditions along the outlet were as follows:

$$(D \nabla c) \cdot \boldsymbol{n} = 0 \tag{13}$$

All other boundaries were set to no flux:

$$(c\boldsymbol{u} - D \nabla c) \cdot \boldsymbol{n} = 0 \tag{14}$$

The mixing efficiency of the two fluids near the outlet of the microchannel was calculated by using the following formula [49]:

$$M = 1 - \sqrt{\frac{\iint_\Gamma (c - \bar{c})^2}{S\bar{c}^2}} \tag{15}$$

where Γ is the cross-section used to measure the concentration. To avoid the outlet effect, the B–B cross-section was selected here. c and $\bar{c}$ are the concentration and average concentration on Γ, respectively. S is the area of the Γ. $M = 0$ and $M = 1$ mean in which the two fluids were completely separated and mixed. Therefore, the greater the value of M was, the better the mixing effect of the two fluids was.

In this paper, commercial finite-element software package COMSOL Multiphysics (Version 5.3a, COMSOL Group, Stockholm, Sweden) was used to solve the coupling system of laminar flow, electric currents, and transport of diluted species. To ensure the credibility of the simulation results, a grid-dependency test was carried out to determine the optimal number of grid elements. We calculated mixing-efficiency index M of the B-B cross-section when $t = 4$ T for 9 different numbers of grid elements from 750 to 3,510,190, and results are shown in Figure 2. The mixing-efficiency index remained unchanged when the number of grid elements exceeded 612,909. Therefore, considering the accuracy and efficiency of simulation, 612,909 was selected as the optimal number of grid system elements. According to the data statistics of COMSOL 5.3a, the numbers of hexahedral, quads, edge, and vertex elements for the required calculation in the simulation model were 612,909, 80,127, 3652, and 76, respectively. Maximal and minimal element sizes were 0.544 and 0.0355 μm, respectively. The minimal element quality and average element quality could reach 0.5766 and 0.9812, respectively. This proves that the solution and analysis of the simulation model in this grid system were relatively accurate and reliable.

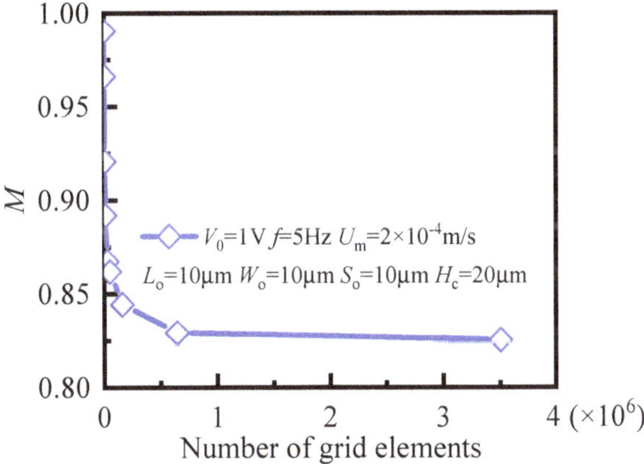

Figure 2. Grid-dependency test for mixing-efficiency index M at B-B cross-section.

3. Results and Discussion

3.1. Mixing Process and Mechanism

The famous Helmholtz–Smoluchowski slip-boundary condition (Equation (7)) shows that the magnitude of electroosmotic slip velocity u_{slip} is closely related to tangential component E_t of the applied potential. In addition, tangential component E_t of the total electric field on the top and bottom walls of the microchannel included E_x and E_y, considering that the slip velocity resulting from E_x had little perturbance on the main flow direction, so the generation of electroosmotic flow on the cross-section mainly depended on E_y. Thus, by giving the electric-field, flow-field, and concentration-distribution evolution (as shown in Figure 3) of the A–A cross-section in the microchannel at five different moments during one electric-field cycle, we elaborate the effect of the staggered virtual electrodes on the fluid mixing and reveal its mixing mechanism. Here, the selected A–A cross-section was located at the center of the virtual electrode adjacent to the inlet of the microchannel, as shown in Figure 1c. Since a sinusoidal AC signal was applied in this study, the five different taken moments were $t = 1/4$ T, $t = 3/8$ T, $t = 1/2$ T, $t = 5/8$ T and $t = 3/4$ T.

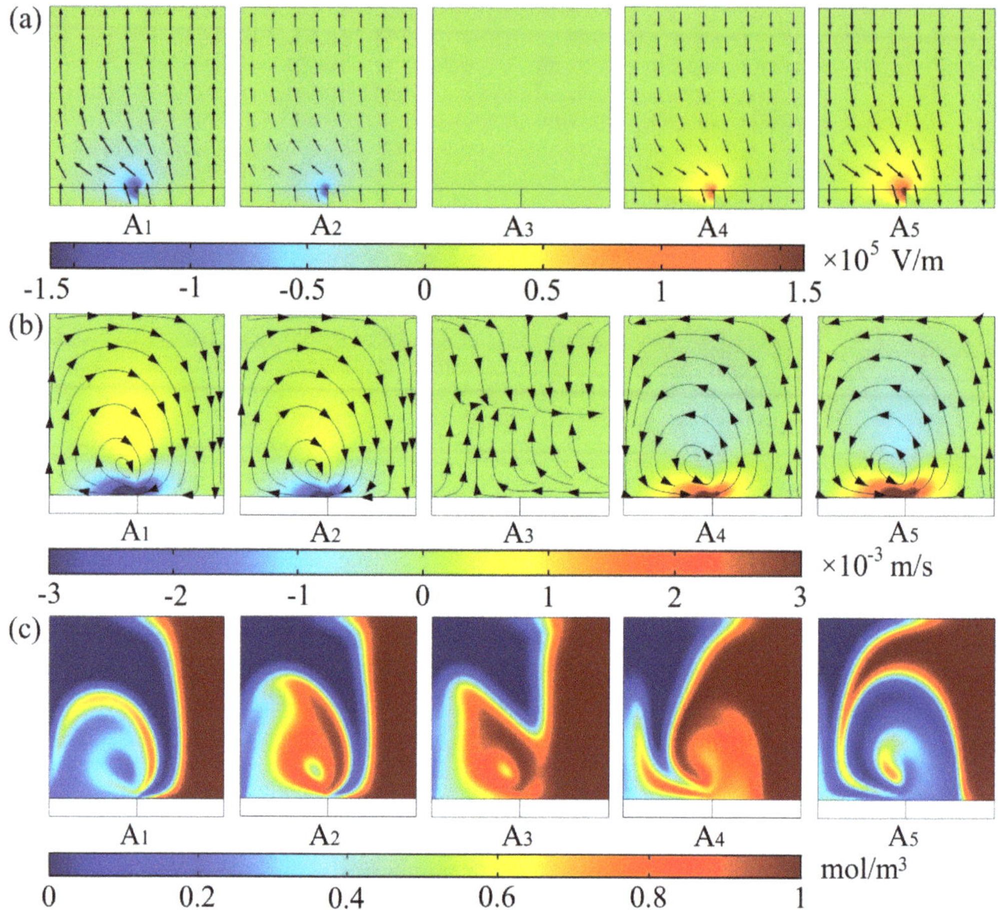

Figure 3. (**a**) Y component of applied potential, (**b**) flow field, and (**c**) concentration-distribution evolution of A–A cross-section in the microchannel at five different moments during one cycle. A, cross-section in the microchannel, and subscript i represents different moments (A_1) $t = 1/4$ T, (A_2) $t = 3/8$ T, (A_3) $t = 1/2$ T, (A_4) $t = 5/8$ T and (A_5) $t = 3/4$ T. Peak-to-peak value of applied potential was $V_0 = 1$ V, electric-field frequency was $f = 5$ Hz, and inlet mean velocity was $U_m = 2 \times 10^{-4}$ m/s. Length was $L_o = 10$ μm, width was $W_o = 10$ μm and spacing was $S_o = 10$ μm of the virtual electrode. Aspect ratio of the microchannel was $W_c/H_c = 1$. (**a**) Black arrows, electric-field intensity vectors; and color legend, magnitude of E_y. (**b**) Black lines, streamlines; black arrows, flow direction; color legend, magnitude of velocity component in y direction. (**c**) Color legend, concentration distribution from 0 to 1 mol/m^3.

A_1, A_2, and A_3 in Figure 3a show that, at the middle moment of the first half cycle ($t = 1/4$ T), the y component of the electric field in the A–A section became the largest, and it was weakened when $t = 3/8$ T, until it completely decreased to 0 V/m at the end of the first half cycle ($t = 1/2$ T). A_4 and A_5 show that the magnitude of E_y gradually increased when $t = 5/8$ T, and reaches the maximal value again at the middle moment of the second half cycle ($t = 3/4$ T), which was basically equal to that of the first half cycle. Different from the first half cycle, the direction of E_y in the second half cycle was the opposite. Due to the different voltage drop between the illuminated and dark areas in the photoconductive layer, the electric field on the A–A section exhibited nonuniform distribution, and it was larger near the boundary between the illuminated and dark areas. In addition, since the virtual electrodes on the bottom wall of the microchannel were arranged in a staggered manner, electric-field distribution on the cross-sections of the two staggered electrodes adjacent to each other at the same time was basically equal in magnitude, and exactly the opposite in direction.

As shown in A_1 in Figure 3b, when $t = 1/4$ T, a clockwise rotating vortex with an influence range of almost the entire microchannel was generated on the A–A section. The center of the rotating vortex was located near the boundary between the illuminated and dark areas with large flow velocity. A_2 shows that the size and rotation direction of the rotating vortex were basically unchanged when $t = 3/8$ T, while the magnitude of the velocity near the bottom wall of the microchannel was smaller than that when $t = 1/4$ T. A_3 shows that, when $t = 1/2$ T, the fluid in the microchannel returned to highly ordered laminar flow, that is, there was no electroosmotic flow generated on A–A section. A_4 shows that the rotation direction of the rotating vortex was counterclockwise when $t = 5/8$ T, which is opposite to that of the first half cycle. When $t = 3/4$ T from A_5, the magnitude of the velocity of the rotating vortex reached the maximum again, and the rotation direction was still opposite to that of the first half cycle. The reason for the above changes in the flow field is that the magnitude of the electroosmotic slip velocity was proportional to the tangential component of the applied electric-field intensity. From the perspective of the entire cycle, the size and direction of the rotating vortex generated in the microchannel were closely related to the magnitude and direction of E_y, which is consistent with the description in Figure 3a.

As shown in A_1 in Figure 3c, the rotating vortex generated on the A–A section strongly disturbed the interface between the two fluids when $t = 1/4$ T. The fluid elements were folded, and stretched widely and repeatedly in the direction perpendicular to the main flow, and the two fluids were wrapped around each other A_2 and A_3 show that the mixing area of the two fluids was still expanding from $t = 3/8$ T to $t = 1/2$ T, although the rotating vortex gradually decayed until it disappears. A_4 and A_5 show that the fluid elements were folded and stretched in opposite directions due to the change in the rotational direction of the rotating vortex; when $t = 5/8$ T, the two fluids were wrapped layer by layer until $t = 3/4$ T. On the whole, the alternating rotating vortex on A–A section strengthened the convection between the two fluids and increased the turbulence of fluid motion, but the two fluids on A–A section were still well-separated. This shows that only one virtual electrode could not achieve a high level of mixing performance in the whole microchannel.

To explore the effect of the rotating vortex generated on the cross-section on the axial mainstream of the microchannel and further reveal its mixing mechanism, we studied the flow fields of the D–D section in the microchannel at five different moments during one cycle (as shown in Figure 4). Here, the selected five different moments were consistent with those in Figure 3. As shown in Figure 1b, the D–D section was the xy section when $z = 0$.

Figure 4a shows that, under the action of the rotating vortex generated on the cross-section of the microchannel, the direction of the main flow was significantly deflected when $t = 1/4$ T, and the maximal deflection angle was almost close to 90 degrees. In addition, due to the staggered arrangement of the virtual electrodes on the bottom wall of the microchannel, the main flow deflected to the opposite direction every time it passed through a virtual electrode, which greatly enhanced the convection between two fluids. Since there was no spot irradiation near the outlet of the microchannel, the fluid near the outlet returned to a highly ordered laminar flow. Figure 4b shows that the main flow still deflected when passing through the electrode ($t = 3/8$ T), but the degree of deflection became smaller, which is related to the decrease in flow velocity of the rotating vortex generated on the cross-section. The flow field when $t = 1/2$ T in Figure 4c shows that the main flow in the whole microchannel was restored to highly orderly laminar flow since the electric field was zero at this time. The effect of fluid-relaxation time was not considered here because of the low electric-field frequency. Figure 4d shows that, under the action of the rotating vortex with opposite rotation direction compared with that in the first half cycle, the main flow direction deflected again when $t = 5/8$ T, and the deflection direction of the fluid at the same position was opposite to that of the first half cycle. As shown in Figure 4e, the deflection angle of the mainstream again reached the maximum when $t = 3/4$ T. Meanwhile, under the pumping action of E_x, the fluid near the outlet could quickly pass through the outlet of the microchannel.

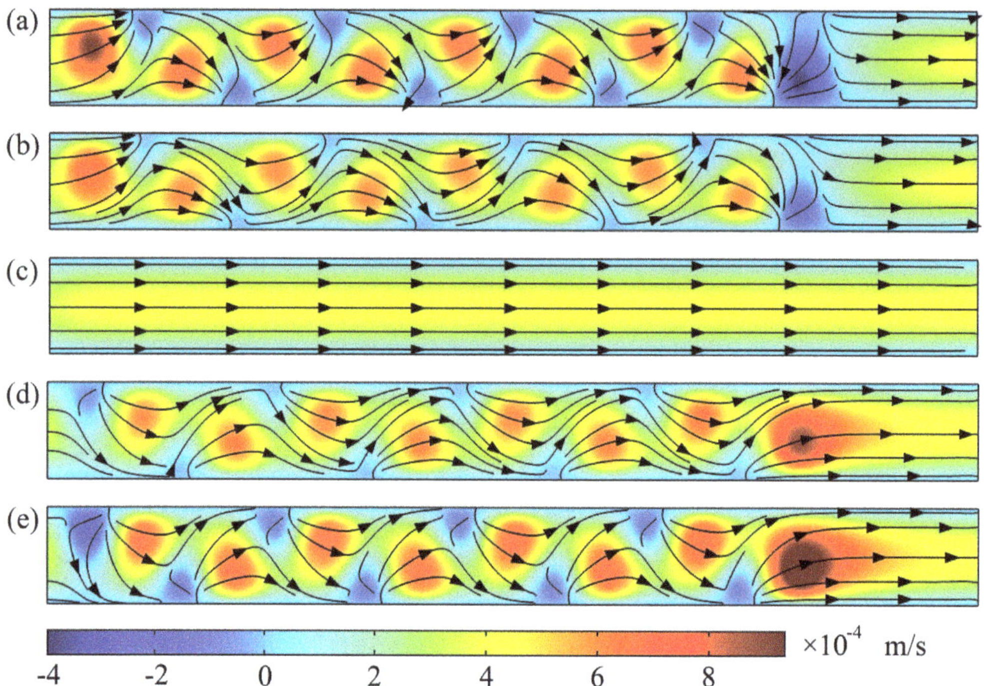

Figure 4. Flow fields of D–D section in the microchannel at five different moments during one cycle: (**a**) $t = 1/4$ T, (**b**) $t = 3/8$ T, (**c**) $t = 1/2$ T, (**d**) $t = 5/8$ T and (**e**) $t = 3/4$ T. Peak-to-peak value of applied potential was $V_0 = 1$ V, electric-field frequency was $f = 5$ Hz, and inlet mean velocity was $U_m = 2 \times 10^{-4}$ m/s. Length was $L_o = 10$ μm, width was $W_o = 10$ μm, and spacing was $S_o = 10$ μm of the virtual electrode. Aspect ratio of the microchannel was $W_c/H_c = 1$. Black lines, streamlines; black arrows, flow direction; color legend, magnitude of the velocity component in the x direction.

Under the coupling action of an alternating electric field and a series of staggered virtual electrodes, the mainstream both flowed along the S-shaped path in the microchannel, and also periodically and alternately changed with time, which greatly improved the convection between the two fluids and positively affected fluid mixing.

Figure 5 shows the concentration-distribution evolution of the D–D section in the microchannel at 10 different moments during four cycles. The 10 different selected moments were (a) $t = 0$ T, (b) $t = 1/4$ T, (c) $t = 3/8$ T, (d) $t = 1/2$ T, (e) $t = 5/8$ T, (f) $t = 3/4$ T, (g) $t = 1$ T, (h) $t = 2$ T, (i) $t = 3$ T, and (j) $t = 4$ T.

Figure 5a shows that, at the initial moment ($t = 0$ T), molecular diffusion at the interface between the two fluids in the microchannel was the main mixing mechanism, so the two fluids were well-separated. As shown in Figure 5b, in the vicinity of the virtual electrode, the rotating vortex generated on the cross-section disturbed the interface between the two fluids when $t = 1/4$ T, which made the fluid elements in the D–D section fold, and widely and repeatedly stretch in the direction perpendicular to the mainstream, and the two fluids were wrapped around each other. Figure 5c shows that, although the velocity of the rotating vortex on the cross-section decreased due to the attenuation of the electric field when t = 3/8 T, the interface between the two fluids was serrated. The convection between the fluids was gradually strengthened, and the mixing area dramatically expanded. Figure 5d shows that the rotating vortex on the cross-section disappeared when $t = 1/2$ T, but the mixing area was still expanding under the action of the mainstream. Figure 5e shows that, when $t = 5/8$ T, due to the opposite rotation vortex in the cross-section compared with that in the first half cycle, the main flow stretched and folded in the opposite direction, and the interface between the two fluids gradually presented an arc shape. Figure 5f shows that, under the action of the gradually increasing applied potential, the length of the fluid interface increased when $t = 3/4$ T, and the two fluids in the microchannel were wrapped around each other.

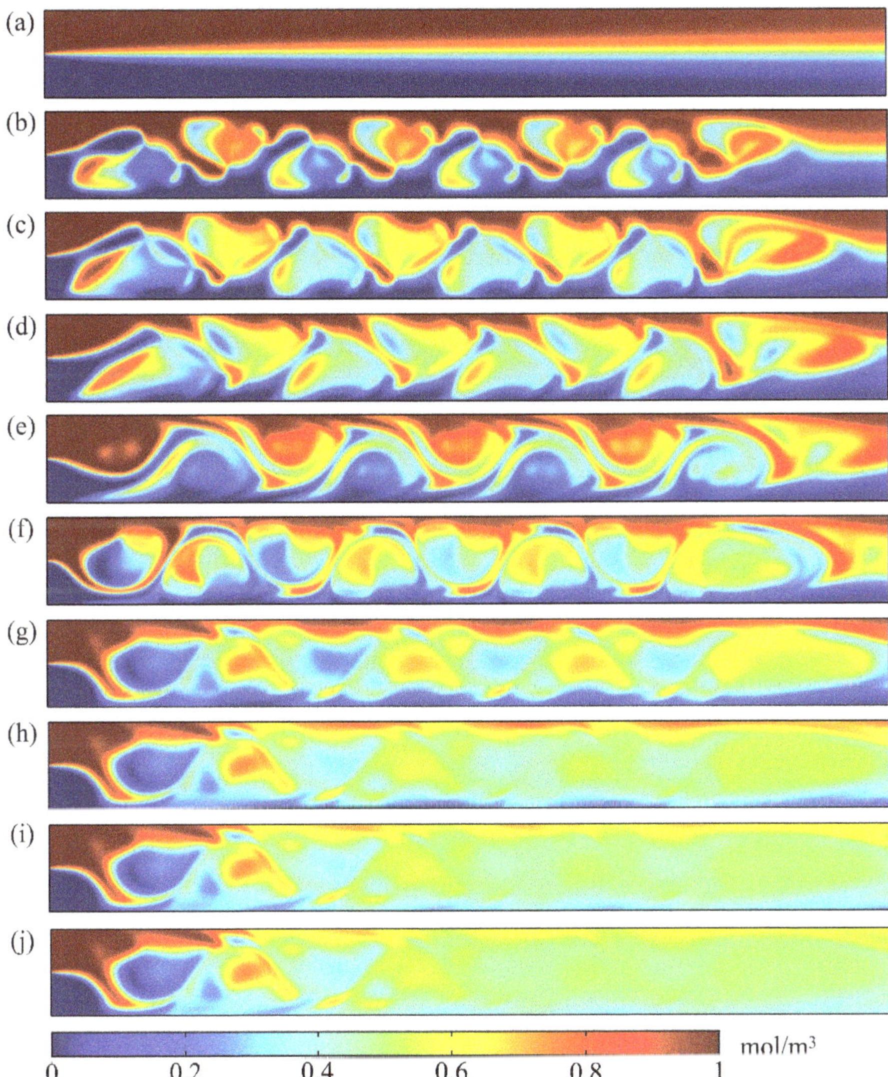

Figure 5. Concentration-distribution evolution of D–D section in microchannel at 10 different moments during 4 cycles: (**a**) $t = 0$ T, (**b**) $t = 1/4$ T, (**c**) $t = 3/8$ T, (**d**) $t = 1/2$ T, (**e**) $t = 5/8$ T, (**f**) $t = 3/4$ T, (**g**) $t = 1$ T, (**h**) $t = 2$ T, (**i**) $t = 3$ T, and (**j**) $t = 4$ T. Peak-to-peak value of applied potential was $V_0 = 1$ V, electric-field frequency was $f = 5$ Hz, and inlet mean velocity was $U_m = 2 \times 10^{-4}$ m/s. Length was $L_o = 10$ μm, width was $W_o = 10$ μm, and spacing was $S_o = 10$ μm of the virtual electrode. Aspect ratio of the microchannel was $W_c/H_c = 1$. Color legend, concentration distribution from 0 to 1 mol/m^3.

Although the two fluids on the D–D section were alternately folded and stretched under the action of the rotating vortex on the cross-section, the two fluids still had a clear dividing line during one cycle, as shown in Figure 5g. Figure 5h shows that, when t = 2 T, the fluids were basically mixed except those near the inlet and side wall of the microchannel. When t = 3 T, as shown in Figure 5i, the fluids near the side wall of the microchannel gradually realized mixing. Figure 5j shows that, when t = 4 T, the fluids after passing through the virtual electrode were basically completely mixed. Figure 5 shows that the mixing effect of the fluids gradually improved as the applied potential was prolonged. In addition, the two fluids near the inlet are always well separated. As the number of virtual electrodes through which the fluids flowed increased, the mixing effect of the fluids became more favorable, and the two fluids basically achieved complete mixing near the outlet.

3.2. *Influence of Key Geometric Parameters on Mixing Performance*

In this section, we evaluate the mixing performance of the micromixer by studying the concentration-distribution evolution near the outlet of the microchannel. To avoid the outlet effect, we chose the B–B section within the microchannel, which was 10 μm away from the outlet, as shown in Figure 1c.

The concentration-distribution evolution of the B–B cross-section in the microchannel at ten different moments during four cycles is shown in Figure 6. Here, the 10 different selected moments were consistent with those in Figure 5. As shown in B_1, the two fluids near the outlet presented highly ordered laminar flow at the initial moment (t = 0 T). B_2 indicated that, although electric-field intensity in the microchannel reached the maximum when t = 1/4 T, the two fluids were still well-separated because of the short duration of the rotating vortex in the initial stage. B_3 shows that, when t = 3/8 T, although the electric field gradually weakened, slight distortion occurred at the interface between the two fluids under the action of the continuous rotating vortex. B_4 indicates that there was no electric field in the microchannel when t = 1/2 T, but the interface between the two fluids was significantly distorted. B_5 shows that the direction of all rotating vortices in the cross-section of the microchannel changed when t = 5/8 T, and the interface between the two fluids was distorted in opposite directions. B_6 shows that, under the action of continuously changing applied potential, with the increase in electric-field intensity in the microchannel, the two fluids could wrap around each other when t = 3/4 T. B_7 shows that, under the action of applied potential during one cycle (t = 1 T), the two fluids realized mutual wrapping. B_8 indicates that, with the periodic change in applied potential, except those near the side wall of the microchannel, the two fluids achieved a better mixing effect at the end of the second cycle (t = 2 T). B_9 and B_{10} show that the mixing degree of the two fluids in the microchannel became increasingly uniform by increasing the duration time of the applied potential, and the two fluids were basically completely mixed at the end of the fourth cycle (t = 4 T).

To obtain the optimal mixing performance, we continued to study the influence of several key geometric parameters on mixing performance, namely, length L_o, width W_o, and spacing S_o of the virtual electrode, and height H_c of the microchannel. Here, we apply mixing-efficiency index M (Equation (15)) calculated on the B–B section at the end of the fourth cycle (t = 4 T) as the evaluation criterion of mixing performance. The dependence of mixing-efficiency index M on length L_o of the virtual electrode is shown in Figure 7a. Mixing-efficiency index M increased with the increase in length L_o of the virtual electrode. When the length of the virtual electrode was L_o = 0 μm (no virtual electrode), due to the absence of an electroosmotic rotating vortex, the fluid hardly had any additional mixing effect (M = 0.290). However, with the increase in the length of the virtual electrode, the axial length of the electroosmotic rotating vortex generated on the cross-section of the microchannel became larger. The rotating vortex with larger axial length had stronger stability and greater distortion to the main flow in the microchannel. For instance, when L_o was equal to 13 μm, the efficiency reached the maximal value (M = 0.889).

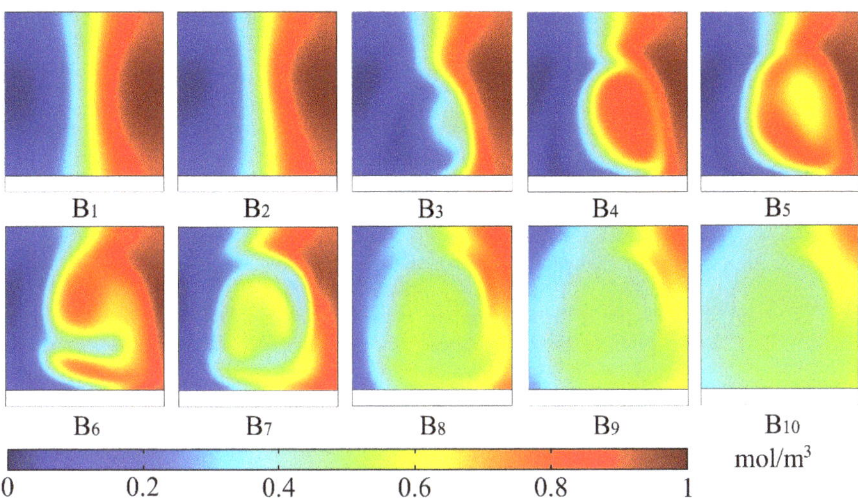

Figure 6. Concentration-distribution evolution of B–B cross-section (near the outlet) at 10 different moments in the microchannel during four cycles. B, cross-section; subscript i, different moments: (**B$_1$**) t = 0 T, (**B$_2$**) t = 1/4 T, (**B$_3$**) t = 3/8 T, (**B$_4$**) t = 1/2 T, (**B$_5$**) t = 5/8 T, (**B$_6$**) t = 3/4 T, (**B$_7$**) t = 1 T, (**B$_8$**) t = 2 T, (**B$_9$**) t = 3 T and (**B$_{10}$**) t = 4 T. Peak-to-peak value of applied potential was V_0 = 1 V, electric-field frequency was f = 5 Hz, and inlet mean velocity was $U_m = 2 \times 10^{-4}$ m/s. Length was L_o = 10 μm, width was W_o = 10 μm, and spacing was S_o = 10 μm of the virtual electrode. Aspect ratio of the microchannel was W_c/H_c = 1. Color legend, concentration distribution from 0 to 1 mol/m^3.

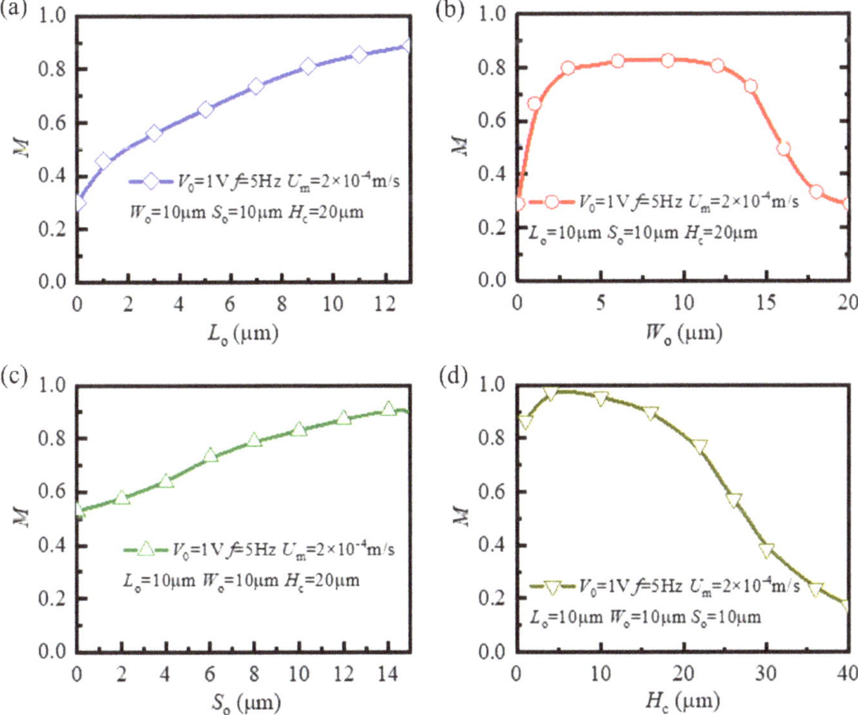

Figure 7. Mixing-efficiency index M near outlet depending on different geometric parameters. (**a**) Length L_o, (**b**) width W_o, (**c**) spacing S_o of virtual electrode; (**d**) height H_c of the microchannel.

Figure 7b shows the variation of mixing-efficiency index M with width W_o of the virtual electrode. As the width of the virtual electrode increased, mixing-efficiency index M generally first increased, reached the maximum when width W_o was about half of the width of the channel, and then decreased. According to the study of the electric field on the cross-section of the microchannel and the resulting electroosmotic rotating vortex in Figure 3, the center of the rotating vortex was located approximately at the boundary between the illuminated and dark areas. This means that only when the center of the rotating vortex was close to the center of the microchannel, was the disturbance degree to the fluid interface the maximum, and this achieved the best mixing effect. When width W_o was excessively large or small, the center of the rotating vortex was located inside the fluid on one side, which was not conducive to mixing the two fluid samples.

Figure 7c shows variation in mixing-efficiency index M with spacing S_o of the virtual electrode. As the distance between two adjacent virtual electrodes increased, the mixing-efficiency index basically linearly also increased. Figure 3 shows that the rotating vortex was generated on the cross-section where each virtual electrode was located at the microchannel. At the same time, the rotation direction of the rotating vortex on the adjacent cross-section was the opposite because the virtual electrodes were staggered in this study. When the fluids in the mainstream direction flowed through two adjacent virtual electrodes in turn, under the action of two rotating vortices with opposite rotating directions perpendicular to the mainstream direction, the interface between the two fluids at the two adjacent electrodes was distorted in opposite directions (as shown in Figure 4). However, when the two adjacent cross-sections were very close, such as $S_o = 0$ μm, the overall distortion of the two fluids in the main flow direction in the microchannel was weakened, which was not conducive to the mixing of the two fluids. This is consistent with the minimal value ($M = 0.529$) of the calculation results of the mixing-efficiency index in Figure 7c. With the increase in the space between the virtual electrodes, the overall distortion degree of the two fluids in the mainstream direction in the microchannel gradually increased. For example, when $S_o = 15$ μm, the two fluids in the mainstream direction were mixed to the greatest extent, and the mixing-efficiency index reached the maximal value ($M = 0.904$).

Mixing-efficiency index M changing with height H_c of the microchannel is shown in Figure 7d. Mixing-efficiency index M decreased with the increase in microchannel height. Figure 3 shows that the center of the rotating vortex on the cross-section of the microchannel was closer to the bottom wall of the microchannel, and the higher flow velocity was also concentrated near the boundary between the illuminated and dark areas. In addition, when the applied potential remained unchanged, the larger channel height resulted in a weaker electric field in the microchannel, which reduced electroosmotic slip velocity. Therefore, when the microchannel height increased, the disturbance of the rotating vortex to the interface between two fluids gradually weakened, which led to the deterioration of mixing performance. For example, when $H_c = 40$ μm, there was little mixing effect between the two fluids ($M = 0.173$). When the height of the microchannel was excessively small compared with its width ($H_c < 5$ μm, $W_c = 20$ μm), although electric-field intensity in the microchannel was large enough, the circular rotating vortex in the cross-section may have been compressed, resulting in a poor mixing effect of the fluids. Only when the height of microchannel was appropriately comparable with its width, such as $H_c = 6$ μm, could the rotating vortex that formed on the cross-section of the microchannel disturb the interface between the two fluids to the maximal extent, achieving the best mixing effect ($M = 0.978$).

4. Conclusions

In this paper, we presented a novel microfluidic mixer with staggered virtual electrode based on light-actuated AC electroosmosis. The coupling systems of Navier–Stokes equations, a Laplace equation, and a convection–diffusion equation were solved by a finite-element method. The flow field, electric field, and concentration distribution on the section of microchannel were studied. Simulation results showed that the electroosmotic rotating vortex generated on the cross-section of the microchannel enhanced convection and improved the mixing effect between fluids. In addition, we studied the effect of several key geometric parameters such as the length, width, and spacing of the staggered virtual electrode, and the height of the microchannel on mixing performance; the relatively optimal mixer structure was thus obtained (L_o = 13 μm, W_o = 10 μm, S_o = 15 μm, H_c = 6 μm). On the basis of its functional flexibility of optical virtual electrodes, the current fluid micromixer could promote the comprehensive integration of functions in modern microfluidic analysis systems. For example, reagent mixing, reaction, and detection steps may be concentrated in a single microfluidic channel.

Author Contributions: H.D. and X.Z., mathematical-model construction and validation; L.S. and T.Z., formal result analysis the and manuscript writing; B.Y. and Z.L., paper revision. All authors have read and agreed to the published version of the manuscript.

Funding: This research was funded by the Hainan Provincial Natural Science Foundation of China (grant numbers 519MS021 and 2019RC032), the National Natural Science Foundation of China (grants 61964006 and 52075138), and the Natural Science Foundation of Jiangsu (grant number GBK20190872).

Conflicts of Interest: The authors declare no conflict of interest.

References

1. Jeong, G.S.; Chung, S.; Kim, C.B.; Lee, S.H. Applications of micromixing technology. *Analyst* **2010**, *135*, 460–473. [CrossRef]
2. Manz, A.; Graber, N.; Widmer, H.M. Miniaturized total chemical analysis systems a novel concept for chemical sensing. *Sens. Actuators B1* **1990**, *1*, 244–248. [CrossRef]
3. Harrison, D.J.; Manz, A.; Fan, Z.; Luedi, H.; Widmer, H.M.; Manz, J.A.; Fan, Z.; Ludi, H.; Widmer, H.M. Capillary electrophoresis and sample injection systems integrated on a planar glass chip. *Anal. Chem.* **1992**, *64*, 1926–1932. [CrossRef]
4. Alam, M.K.; Koomson, E.; Zou, H.; Yi, C.; Li, C.W.; Xu, T.; Yang, M. Recent advances in microfluidic technology for manipulation and analysis of biological cells (2007–2017). *Anal. Chim. Acta* **2018**, *1044*, 29–65. [CrossRef]
5. Sasaki, N.; Kitamori, T.; Kim, H.B. AC electroosmotic micromixer for chemical processing in a microchannel. *Lab Chip* **2006**, *6*, 550–554. [CrossRef] [PubMed]
6. Jung, J.H.; Kim, G.Y.; Seo, T.S. An integrated passive micromixer-magnetic separation-capillary electrophoresis microdevice for rapid and multiplex pathogen detection at the single-cell level. *Lab Chip* **2011**, *11*, 3465–3470. [CrossRef]
7. Lee, C.Y.; Chang, C.L.; Wang, Y.N.; Fu, L.M. Microfluidic mixing: A review. *Int. J. Mol. Sci.* **2011**, *12*, 3263–3287. [CrossRef] [PubMed]
8. Haeberle, S.; Zengerle, R. Microfluidic platforms for lab-on-a-chip applications. *Lab Chip* **2007**, *7*, 1094–1110. [CrossRef] [PubMed]
9. Mark, D.; Haeberle, S.; Roth, G.; von Stetten, F.; Zengerle, R. Microfluidic lab-on-a-chip platforms: Requirements, characteristics and applications. *Chem. Soc. Rev.* **2010**, *39*, 1153–1182. [CrossRef] [PubMed]
10. Wang, Y.; Lin, Q.; Mukherjee, T. A model for laminar diffusion-based complex electrokinetic passive micromixers. *Lab Chip* **2005**, *5*, 877–887. [CrossRef]
11. Zhou, T.; Wang, H.; Shi, L.; Liu, Z.; Joo, S.W. An enhanced electroosmotic micromixer with an efficient asymmetric lateral structure. *Micromachines* **2016**, *7*, 218. [CrossRef] [PubMed]
12. Bayareh, M.; Ashani, M.N.; Usefian, A. Active and passive micromixers: A comprehensive review. *Chem. Eng. Process.-Process Intensif.* **2020**, *147*, 107771. [CrossRef]
13. Zhou, T.; Xu, Y.; Liu, Z.; Joo, S.W. An enhanced one-layer passive microfluidic mixer with an optimized lateral structure with the dean effect. *J. Fluids Eng.* **2015**, *137*. [CrossRef]
14. Hossain, S.; Kim, K.-Y. Mixing analysis in a three-dimensional serpentine split and recombine micromixer. *Chem. Eng. Res. Des.* **2015**, *100*, 95–103. [CrossRef]
15. Sheu, T.S.; Chen, S.J.; Chen, J.J. Mixing of a split and recombine micromixer with tapered curved microchannels. *Chem. Eng. Sci.* **2012**, *71*, 321–332. [CrossRef]
16. Chew, Y.T.; Xia, H.M.; Shu, C.; Wan, S.Y.M. Techniques to enhance fluid micro-mixing and chaotic micromixers. *Mod. Phys. Lett. B* **2005**, *19*, 1567–1570. [CrossRef]
17. Wang, L.; Liu, D.; Wang, X.; Han, X. Mixing enhancement of novel passive microfluidic mixers with cylindrical grooves. *Chem. Eng. Sci.* **2012**, *81*, 157–163. [CrossRef]
18. Liu, A.L.; He, F.Y.; Wang, K.; Zhou, T.; Lu, Y.; Xia, X.H. Rapid method for design and fabrication of passive micromixers in microfluidic devices using a direct-printing process. *Lab Chip* **2005**, *5*, 974–978. [CrossRef]
19. Kim, Y.; Lee, J.; Kwon, S. A novel micro-mixer with a quasi-active rotor: Fabrication and design improvement. *J. Micromech. Microeng.* **2009**, *19*, 105028. [CrossRef]
20. Ahmed, D.; Mao, X.; Shi, J.; Juluri, B.K.; Huang, T.J. A millisecond micromixer via single-bubble-based acoustic streaming. *Lab Chip* **2009**, *9*, 2738–2741. [CrossRef]

21. Chen, H.; Chen, C.; Bai, S.; Gao, Y.; Metcalfe, G.; Cheng, W.; Zhu, Y. Multiplexed detection of cancer biomarkers using a microfluidic platform integrating single bead trapping and acoustic mixing techniques. *Nanoscale* **2018**, *10*, 20196–20206. [CrossRef]
22. Cartier, C.A.; Drews, A.M.; Bishop, K.J. Microfluidic mixing of nonpolar liquids by contact charge electrophoresis. *Lab Chip* **2014**, *14*, 4230–4236. [CrossRef]
23. Harnett, C.K.; Templeton, J.; Dunphy-Guzman, K.A.; Senousy, Y.M.; Kanouff, M.P. Model based design of a microfluidic mixer driven by induced charge electroosmosis. *Lab Chip* **2008**, *8*, 565–572. [CrossRef]
24. Wu, Y.; Ren, Y.; Tao, Y.; Hou, L.; Hu, Q.; Jiang, H. A novel micromixer based on the alternating current-flow field effect transistor. *Lab Chip* **2016**, *17*, 186–197. [CrossRef]
25. Yap, L.W.; Chen, H.; Gao, Y.; Petkovic, K.; Liang, Y.; Si, K.J.; Wang, H.; Tang, Z.; Zhu, Y.; Cheng, W. Bifunctional plasmonic-magnetic particles for an enhanced microfluidic sers immunoassay. *Nanoscale* **2017**, *9*, 7822–7829. [CrossRef]
26. Wen, C.Y.; Liang, K.P.; Chen, H.; Fu, L.M. Numerical analysis of a rapid magnetic microfluidic mixer. *Electrophoresis* **2011**, *32*, 3268–3276. [CrossRef]
27. Park, S.; Chuang, H.-S.; Kwon, J.-S. Numerical study and taguchi optimization of fluid mixing by a microheater-modulated alternating current electrothermal flow in a y-shape microchannel. *Sens. Actuators B Chem.* **2021**, *329*, 129242. [CrossRef]
28. Khakpour, A.; Ramiar, A. Numerical investigation of the effect of electrode arrangement and geometry on electrothermal fluid flow pumping and mixing in microchannel. *Chem. Eng. Process.-Process Intensif.* **2020**, *150*, 107864. [CrossRef]
29. Kunti, G.; Bhattacharya, A.; Chakraborty, S. Analysis of micromixing of non-newtonian fluids driven by alternating current electrothermal flow. *J. Non-Newton. Fluid Mech.* **2017**, *247*, 123–131. [CrossRef]
30. Ng, W.Y.; Goh, S.; Lam, Y.C.; Yang, C.; Rodriguez, I. Dc-biased ac-electroosmotic and ac-electrothermal flow mixing in microchannels. *Lab Chip* **2009**, *9*, 802–809. [CrossRef] [PubMed]
31. Li, Z.; Kim, S.J. Pulsatile micromixing using water-head-driven microfluidic oscillators. *Chem. Eng. J.* **2017**, *313*, 1364–1369. [CrossRef]
32. Rashidi, S.; Bafekr, H.; Valipour, M.S.; Esfahani, J.A. A review on the application, simulation, and experiment of the electrokinetic mixers. *Chem. Eng. Process.-Process Intensif.* **2018**, *126*, 108–122. [CrossRef]
33. Zambrano, H.A.; Vasquez, N.; Wagemann, E. Wall embedded electrodes to modify electroosmotic flow in silica nanoslits. *Phys. Chem. Chem. Phys.* **2016**, *18*, 1202–1211. [CrossRef]
34. Song, H.; Cai, Z.; Noh, H.M.; Bennett, D.J. Chaotic mixing in microchannels via low frequency switching transverse electroosmotic flow generated on integrated microelectrodes. *Lab Chip* **2010**, *10*, 734–740. [CrossRef]
35. Bag, N.; Bhattacharyya, S. Electroosmotic flow of a non-newtonian fluid in a microchannel with heterogeneous surface potential. *J. Non-Newton. Fluid Mech.* **2018**, *259*, 48–60. [CrossRef]
36. Bhattacharyya, S.; Bera, S. Combined electroosmosis-pressure driven flow and mixing in a microchannel with surface heterogeneity. *Appl. Math. Model.* **2015**, *39*, 4337–4350. [CrossRef]
37. Kateb, M.; Kolahdouz, M.; Fathipour, M. Modulation of heterogeneous surface charge and flow pattern in electrically gated converging-diverging nanochannel. *Int. Commun. Heat Mass Transf.* **2018**, *91*, 103–108. [CrossRef]
38. Nayak, A.K. Analysis of mixing for electroosmotic flow in micro/nano channels with heterogeneous surface potential. *Int. J. Heat Mass Transf.* **2014**, *75*, 135–144. [CrossRef]
39. Mahapatra, B.; Bandopadhyay, A. Electroosmosis of a viscoelastic fluid over non-uniformly charged surfaces: Effect of fluid relaxation and retardation time. *Phys. Fluids* **2020**, *32*, 032005. [CrossRef]
40. Ahmed, F.; Kim, K.Y. Parametric study of an electroosmotic micromixer with heterogeneous charged surface patches. *Micromachines* **2017**, *8*, 199. [CrossRef] [PubMed]
41. Hwang, H.; Park, J.K. Optoelectrofluidic platforms for chemistry and biology. *Lab Chip* **2011**, *11*, 33–47. [CrossRef]
42. Baigl, D. Photo-actuation of liquids for light-driven microfluidics: State of the art and perspectives. *Lab Chip* **2012**, *12*, 3637–3653. [CrossRef]
43. Han, D.; Park, J.K. Optoelectrofluidic enhanced immunoreaction based on optically-induced dynamic ac electroosmosis. *Lab Chip* **2016**, *16*, 1189–1196. [CrossRef] [PubMed]
44. Pei-Yu, C.; Ohta, A.T.; Jamshidi, A.; Hsin-Yi, H.; Wu, M.C. Light-actuated ac electroosmosis for nanoparticle manipulation. *J. Microelectromech. Syst.* **2008**, *17*, 525–531. [CrossRef]
45. Chiou, P.Y.; Ohta, A.T.; Wu, M.C. Massively parallel manipulation of single cells and microparticles using optical images. *Nature* **2005**, *436*, 370–372. [CrossRef] [PubMed]
46. Ding, H.H.; Zhong, X.T.; Liu, B.; Shi, L.Y.; Zhou, T.; Zhu, Y.G. Mixing mechanism of a straight channel micromixer based on light-actuated oscillating electroosmosis in low-frequency sinusoidal ac electric field. *Microfluid. Nanofluidics* **2021**, *25*, 1–15. [CrossRef]
47. Zhao, Y.; Hu, S.; Wang, Q. Simulation and analysis of particle trajectory caused by the optical-induced dielectrophoresis force. *Microfluid. Nanofluidics* **2013**, *16*, 533–540. [CrossRef]
48. Zhu, X.; Yin, Z.; Ni, Z. Dynamics simulation of positioning and assembling multi-microparticles utilizing optoelectronic tweezers. *Microfluid. Nanofluidics* **2011**, *12*, 529–544. [CrossRef]
49. Fu, H.; Liu, X.; Li, S. Mixing indexes considering the combination of mean and dispersion information from intensity images for the performance estimation of micromixing. *RSC Adv.* **2017**, *7*, 10906–10914. [CrossRef]

micromachines

MDPI

Article

Rapid AC Electrokinetic Micromixer with Electrically Conductive Sidewalls

Fang Yang [1,*], Wei Zhao [2], Cuifang Kuang [3] and Guiren Wang [2,4,*]

1 Key Laboratory for Molecular Enzymology and Engineering of Ministry of Education, School of Life Sciences, Jilin University, Changchun 130012, China
2 State Key Laboratory of Photon-Technology in Western China Energy, International Scientific and Technological Cooperation Base of Photoelectric Technology and Functional Materials and Application, Institute of Photonics and Photon-Technology, Northwest University, Xi'an 710127, China; zwbayern@nwu.edu.cn
3 State Key Laboratory of Modern Optical Instrumentation, Zhejiang University, Hangzhou 310027, China; cfkuang@zju.edu.cn
4 Department of Mechanical Engineering and Biomedical Engineering Program, University of South Carolina, Columbia, SC 29208, USA
* Correspondence: fangyang@jlu.edu.cn (F.Y.); guirenwang@sc.edu (G.W.)

Abstract: We report a quasi T-channel electrokinetics-based micromixer with electrically conductive sidewalls, where the electric field is in the transverse direction of the flow and parallel to the conductivity gradient at the interface between two fluids to be mixed. Mixing results are first compared with another widely studied micromixer configuration, where electrodes are located at the inlet and outlet of the channel with electric field parallel to bulk flow direction but orthogonal to the conductivity gradient at the interface between the two fluids to be mixed. Faster mixing is achieved in the micromixer with conductive sidewalls. Effects of Re numbers, applied AC voltage and frequency, and conductivity ratio of the two fluids to be mixed on mixing results were investigated. The results reveal that the mixing length becomes shorter with low Re number and mixing with increased voltage and decreased frequency. Higher conductivity ratio leads to stronger mixing result. It was also found that, under low conductivity ratio, compared with the case where electrodes are located at the end of the channel, the conductive sidewalls can generate fast mixing at much lower voltage, higher frequency, and lower conductivity ratio. The study of this micromixer could broaden our understanding of electrokinetic phenomena and provide new tools for sample preparation in applications such as organ-on-a-chip where fast mixing is required.

Keywords: microfluidics; electrokinetics; mixing; micromixer

Citation: Yang, F.; Zhao, W.; Kuang, C.; Wang, G. Rapid AC Electrokinetic Micromixer with Electrically Conductive Sidewalls. *Micromachines* **2022**, *13*, 34. https://doi.org/10.3390/mi13010034

Academic Editor: Kwang-Yong Kim

Received: 11 November 2021
Accepted: 23 December 2021
Published: 27 December 2021

1. Introduction

Mixing of two or more fluids is always crucial in the application of microfluidics in chemical engineering, environmental engineering, and even biomedical and biochemical analysis such as enzyme reaction, protein folding, DNA purification, etc. [1]. Fast mixing can generate stronger signals to increase the sensitivity and enable more accurate measurement of chemical reaction kinetics. However, since the flows are mainly laminar in microfluidics, mixing is carried out by molecular diffusion and fast mixing is not easily achieved. Highly efficient and fast mixing of two fluids inside microchannels could be highly challenging. Therefore, developing new techniques and methodologies to increase the interfacial surface area for enhancing the mixing processes is crucial to improve the corresponding performance of 'lab-on-a-chip' devices [2].

Many new micromixer techniques have been developed in last two decades [3]. Generally, the 'micromixer' can be categorized into two groups: passive and active mixers [4]. Passive mixers do not require external energy. They enhance mixing processes through the augmentation of diffusion through fluid folding, stretching, and tilting by special design of microchannel geometry [5]. In contrast, active mixers usually employ external energy to introduce disturbances on the flow to enhance fluid mixing. Several types of active micromixers with flow disturbances generated in terms of temperature [6], pressure [7], electrohydrodynamics [8,9], acoustics [10], as well as magnetics [11], have already been reported to effectively enhance fluid mixing in microchannels.

Among the aforementioned methods, liquid or particle motion can be effectively manipulated by electrokinetic mechanisms, including electro-osmosis, electrophoresis, dielectrophoresis, and electrowetting, etc., since electric body force (EBF) is more effective on small scales and interfaces [12,13]. In 2001, Oddy et al. [14] presented an active micromixer in which an AC electric field induces a chaotic flow field to enhance the mixing of two pressure-driven flow streams. Moreover, it was demonstrated by Shin et al. [15] that more chaotic trajectories can be generated in a cross-shaped microchannel by a time-dependent electric field. Recently, many works on electrokinetic instability (EKI) were accomplished and attracted much attention, it is a phenomenon described by charge accumulation at perturbed interfaces due to electrical conductivity gradients, which exist in the bulk flow [16–20]. Although many significant results have already been obtained through those previous works [21–25], much effort is still needed to improve our understanding of electrokinetic mixing under AC electric field, to make the electrokinetic micromixers more efficient and flexible for "lab-on-a-chip" applications.

In 2014, we demonstrated that turbulence [26] and its corresponding ultrafast mixing [27] of two pressure-driven flows can be realized electrokinetically in a microchannel with slightly divergent sidewalls (fabricated with electrodes) at low bulk-flow Reynolds number. However, parallel microchannels are mostly used in microfluidics and the mixing enhancement in the electrically conductive sidewalls in parallel has not been studied. In this paper, we present a parametric study of the rapid fluid mixing inside a T-shaped microchannel, where two streams of pressure-driven flows are disturbed by an externally time-dependent electric field, which is orthogonal to the conductivity gradient at the interface between the two fluids to be mixed. The parameters, such as electrode positions and voltage phase shift between two electrodes, were investigated.

2. Materials and Methods

The schematic of the micromixers is given in Figure 1. Both of the micromixers are T-shaped with parallel sidewalls. Two cases have been considered in this investigation: one has electrically conductive sidewalls, the other has insulated sidewalls with electrodes placed at inlet and outlet. In the former, the sidewalls of the channel are made of gold sheet (as shown in Figure 1a). Here, x and y denote the streamwise and transverse directions in the main channel, respectively. In the latter, the sidewalls of the micromixer are fabricated with acrylic, as shown in Figure 1b. Platinum electrodes are placed at the inlets and out of the microchannel. The micromixers both have rectangular cross sections of 120 μm in width and 230 μm in height, with the length of 5 mm. Two inlets and one outlet with the diameter of 1 mm were drilled at the ends of the channel.

Two fluids with different electrical conductivity and permittivity are used for the study. Each fluid enters the micro-fluidic chamber through its own inlet channel. As soon as they contact, a jump in electrical conductivity and/or permittivity is generated at the interface between the two fluids. The flow of an incompressible and Newtonian fluid in presence of an electric field is governed by the Navier–Stokes equations:

$$\rho\left(\frac{\partial \vec{V}}{\partial t}+\vec{V}\cdot\nabla\vec{V}\right)=-\nabla P+\eta\nabla^{2}\vec{V}+\vec{f}_{e} \tag{1}$$

where ρ is the fluid density, $\vec{V}$ denotes the velocity field, P refers to the pressure, and η is the (constant) dynamic viscosity. $\vec{f}_e$ is the EBF attributed to Coulombic force, as:

$$\vec{f}_e = \rho_f \vec{E} \tag{2}$$

where ρ_f is net charge density which can be expressed as [18]:

$$\rho_f = -\frac{\varepsilon \vec{E} \cdot \vec{\nabla} \sigma}{\sigma} \tag{3}$$

where ε is the permittivity of the electrolyte and $\vec{E}$ is the electric field. Due to the presence of electrical conductivity gradients, $\vec{\nabla}\sigma$ at interfaces between two streams with different electrical conductivity, which exist in the bulk flow [18,28], non-zero net charge will be accumulated at interfaces when an electric field is applied. Then, an EBF $\vec{f}_e$ results which distorts the interface of the two fluid streams. If the magnitude of the disturbance is sufficiently large, a transversal convection (secondary flow) can be induced on the interface, destabilize the interface through electrokinetic instability (EKI), and promote the mixing of the two fluids. If the there is no conductivity gradient, which means $\vec{\nabla}\sigma = 0$, then no net charge will be induced ($\rho_f = 0$), consequently, no body force on liquid will be generated ($\rho_f \vec{E} = 0$), and then no EKI occurs. In this investigation, electric conductivity is not a passive scalar [27] since the EBF can significantly manipulate the flow and accordingly affect the field of electric conductivity.

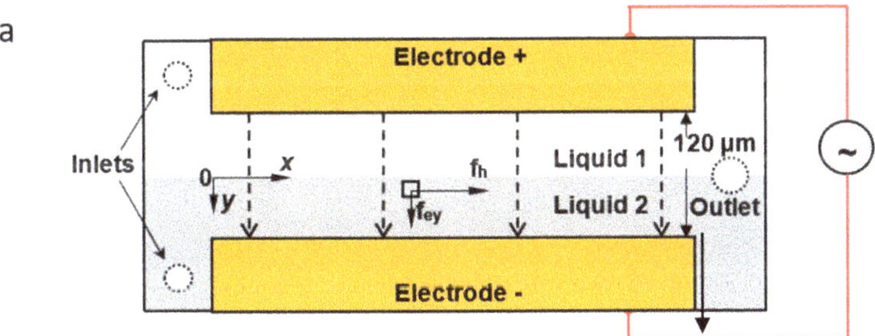

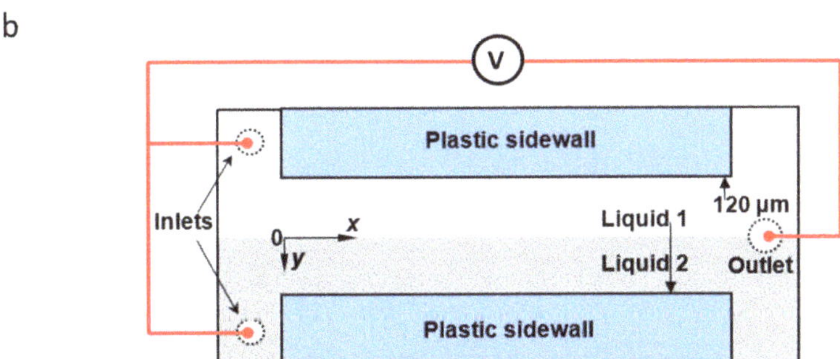

Figure 1. Flow, electrodes configuration, and force descriptions in micromixer. (**a**) Conductive sidewalls micromixer; (**b**) micromixer with electrodes located at the ends of channel.

From Equation (3), it is important to notice that indicated by the term of $\vec{E} \cdot \vec{\nabla}\sigma$, the charge density could be minimum when the external electric field is perpendicular to the electrical conductivity gradients, which is the case in that electrodes are placed at the inlets and outlets of the mixing channel. In contrast, when the external electric field is parallel to the electrical conductivity gradients, the charge density can be maximized. In our micromixer, the electrodes are directly used to form the sidewall; therefore, the external electric field is parallel to the electrical conductivity gradients (Figure 1) which will result in a maximum EBF, and strongest distortion between the interface of the two-liquid flow, initially. In the present study, AC voltage signals are used instead of DC voltage due to the fact that bubbles are more easily generated in highly conductive buffer under DC voltage due to electrolysis, which can block the microchannel and thus be detrimental to the performance of microfluidic devices [29].

A syringe pump (Harvard, Model PHD2000 Programmable, Holliston, MA, USA) was used to pump fluorescent dye solution and DI water from the inlets respectively through the micromixer toward the outlet. Flow visualization were applied to study fluid mixing. Fluorescein sodium salt ($C_{20}H_{10}Na_2O_5$) was used as the fluorescent dye trace for characterizing the mixing results. Electrically neutral dye rhodamine B (Sigma-Aldrich, Corp., Burlington, MA, USA) was also used as the scalar marker to study conductivity ratio influence on fluid mixing. Phosphate buffer (VWR VW3345-1 pH 7.2) was diluted into DI water as one of the mixing streams to control the conductivity ratio between the two streams. Figure 2 shows the schematic of the experimental setup. The microchip was placed on an inverted fluorescent microscope (Olympus-IX70, Tokyo, Japan) for fluorescence measurements. A function generator (Tektronix, Model AFG3102, Beaverton, OR, USA) was used to apply AC electric signal between these two electrodes.

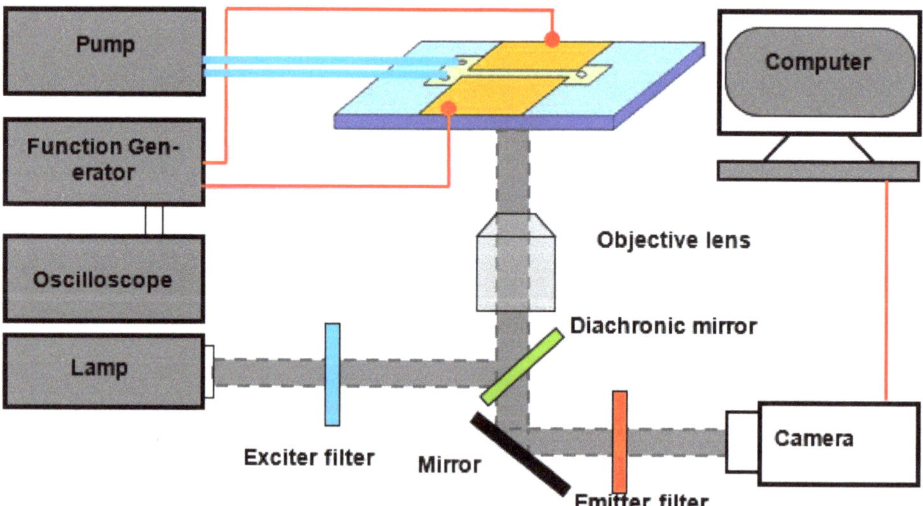

Figure 2. Experimental setup.

A mercury lamp was used as the illumination source in the present study. The excitation beam is 488 nm. Upon excitation, the fluorescent solution would emit fluorescence. A 10× objective lens (NA = 0.25) was used for the fluorescence imaging. The fluorescence signal was captured by a sensitive and high-resolution CCD camera (SensiCam QE, PCO, Bavaria, Germany), with an exposure time of 0.1 s. All concentration was quantitatively determined by measuring the fluorescence intensity within each pixel of the camera using MATLAB (MathWorks Inc., Natick, MA, USA). Mixing enhancement results were compared based on concentration profiles of the fluorescent dye along a transverse line that is perpendicular to flow direction of the microchannel at a given streamwise position.

3. Results and Discussion

3.1. Effect of External Electric Field Direction

To evaluate the influence of directions of external electric field, the mixing results in two cases were compared, e.g., electrodes are placed at the sidewalls (case A) and at the ends of the channel (case B), respectively. The latter has been studied widely as electrokinetic micromixers [18,30]. However, a direct comparison of the two arrangements of the electrodes on mixing has not been carried out before.

In the experiment, we kept flow rate at 5 μL/min and conductivity ratio of the two streams at 10:1, unless otherwise specified. The external electric fields have strength (E_A) of both 200 V/cm for the two cases. In this part, a low AC frequency ($f_{AC} = 1$ Hz) was used.

As shown in Figure 3, it is clearly illustrated that under the same E_A, the mixing is much stronger in case A than case B. For the plastic sidewalls, to achieve $E_A = 200$ V/cm, we had to use a voltage amplifier accompanied with function generator to apply AC voltage on the case B. In this case, it is difficult to apply high AC voltage and high frequency signal simultaneously. However, for the case A, no power amplifier is required to achieve $E_A = 200$ V/cm. The function generator can provide sufficient high E_A and AC frequency on the electrodes simultaneously. Figure 3 clearly indicates, as a result of the parallel of electric conductivity gradient and external electric field, that $\vec{E} \cdot \vec{\nabla} \sigma$ reaches maximum. A much larger $\vec{f}_e$ can be predicted according to Equation (3), relative to that in the case B. In the following section, the electrokinetics micromixer with conductive sidewalls will be characterized.

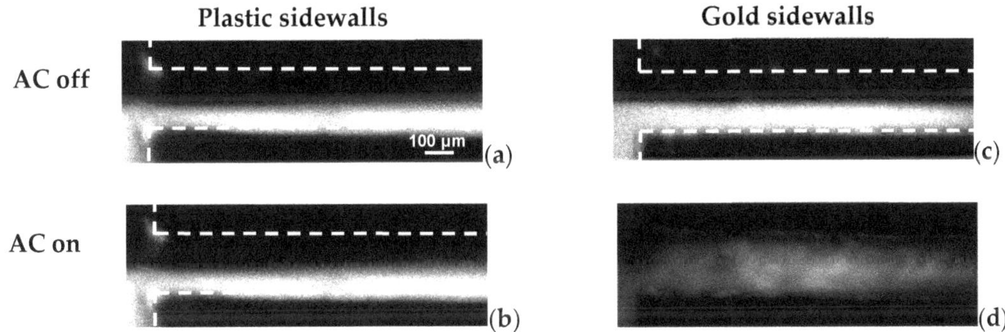

Figure 3. Comparison of mixing results under different electrode positions and corresponding electric fields. In (**a**,**b**), electrodes are located at the ends of the channel (plastic sidewalls) and the electric field is orthogonal to the initial conductivity gradient between the two streams. In (**c**,**d**) electrodes (gold sidewalls) are placed at the sidewalls and the electric field is in transverse direction and in parallel to the initial conductivity gradient between the two streams.

3.2. Effect of Electric Conductivity Ratio

According to Equations (2) and (3), two mixing streams with conductivity gradient were subject to an external electric field; mixing was directly influenced by the $\vec{f}_e$ [31].

In order to conduct a parametric study to quantify the effect of the conductivity ratio of the two streams on the mixing performance, we kept the AC signal of $f_{AC} = 10$ kHz and $E_A = 833$ V/cm (corresponding to AC amplitude of 10 $V_{p\text{-}p}$.) Three conductivity ratios ($\gamma = \sigma_1/\sigma_2$, with $\sigma_1 \geq \sigma_2$) between the two streams were investigated, and they are 1, 2, and 10, respectively.

Mixing performances under different conductivity ratio are shown in Figure 4. Figure 4a,b indicates that the mixing is stronger at $\gamma = 2$ than that at $\gamma = 1$. When $\gamma = 10$ (Figure 4c), the mixing is the strongest among these three cases. In Figure 4d, the corresponding concentration distribution (evaluated by fluorescence intensity) in the transverse direction is displayed. As we know, the stronger the mixing, the more uniform the concentration in the transverse direction at a given streamwise position. The curve should approach flat when the fluids are well mixed in the microchannel. According to Figure 4d, when the conductivity ratio γ is 10, concentration distribution reached a relatively uniform profile at $x/w = 3$ (w is the width of the microchannel) from the entrance. While γ are 1 and 2, at the same streamwise position, C distributions were far away from a flat profile. The mixing result is evaluated by a mixing index κ, which is similar (but different) to the mixing criterion used in Arockiam et al.'s work [32] and is defined as:

$$\kappa = 1 - \frac{\sqrt{\left\langle (C - \langle C \rangle)^2 \right\rangle}}{\langle C \rangle} \tag{4}$$

where $\langle \cdot \rangle$ denotes ensemble averaging. Here, $0 \leq \kappa \leq 1$. The higher κ, the stronger the mixing. It can be seen from Figure 4e, when $\gamma = 1$, κ is nearly flat which indicates the mixing is not enhanced under the external electric field. When $\gamma = 2$, κ increases gradually along streamwise direction. At $x/w = 4$, κ is about 0.65, which is approximately three times larger than that of $\gamma = 1$. When $\gamma = 10$, κ increases rapidly and reaches 0.84 at $x/w = 1.5$. It is twice larger than that of $\gamma = 2$ and the time cost is only 30 ms. Note there is a little fluctuation of κ along streamwise direction, which is because of the non-uniform excitation light distribution of the microscope.

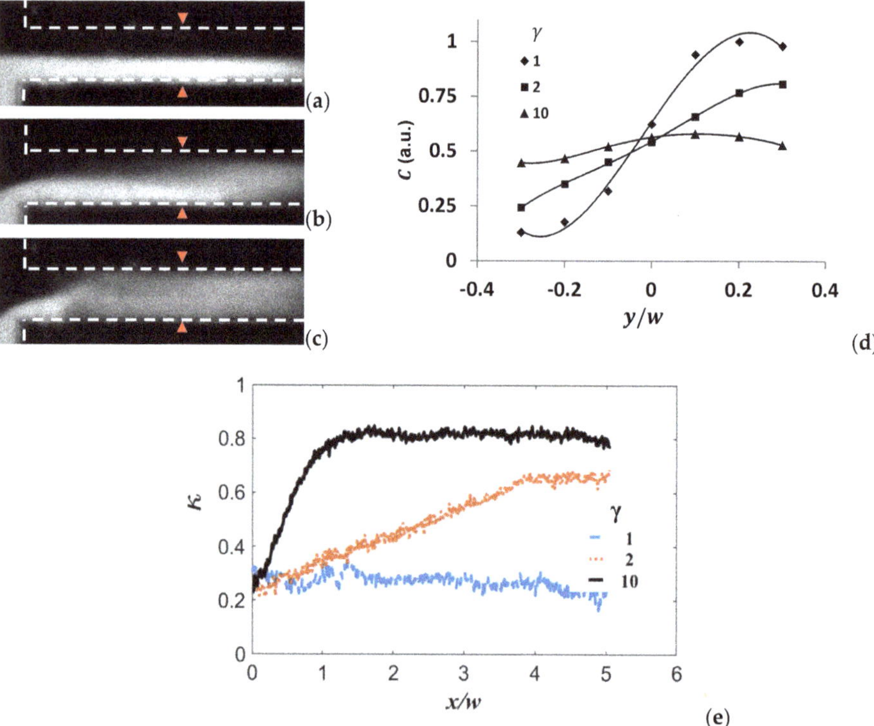

Figure 4. Visualization and comparison of mixing results in the micromixer with different conductivity ratios γ at E_A = 833 V/cm. (**a**) $\gamma = 1$, (**b**) $\gamma = 2$, (**c**) $\gamma = 10$, and (**d**) comparison of concentration profile in transverse direction at $x/w = 3$ from the channel entrance (marked by red line) with different conductivity. (**e**) Mixing index varying along streamwise direction at different conductivity ratios.

Specially, mixing results under the same low $\gamma = 2$ in both conductive sidewalls micromixer and plastic sidewalls micromixer with electrodes located at the ends of the channel were measured and compared, as shown in Figure 5. In this part, the applied electric fields are kept constant, i.e., $E_A = 200$ V/cm for the two mixers. For the case of Figure 5c, a periodic electric field was added to the applied static electric field, to enhance mixing, which is $E_A = 1667$ V/cm (with 20 $V_{p\text{-}p}$ voltage) in amplitude and $f_{AC} = 10$ Hz.

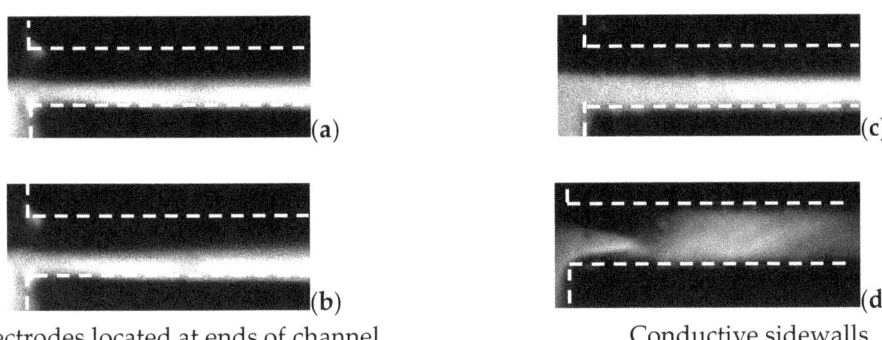

Figure 5. Visualization and comparison of mixing result in micromixer with electrodes located at ends of channel (**a**,**b**) and micromixer with conductive sidewalls (**c**,**d**), (**a**) without voltage, (**c**) with voltage; sidewall, (**b**) without voltage, and (**d**) with voltage.

Figure 5 clearly shows that, under electric field $E_A = 200$ V/cm, obviously stronger mixing has been achieved in the micromixer with conductive sidewalls, as shown in Figure 5d. However, in the micromixer with the nonconductive sidewalls, where the electrodes are located at ends of the channel, no obvious mixing enhancement was observed (Figure 5b).

3.3. Effect of AC Frequency

Influence of AC frequency on mixing has also been investigated in a wide range from 100 Hz to 2 MHz. Fluid mixing under DC voltage was also presented as a comparison. To study the effect of AC frequency on mixing, the electric field was kept constant as well as in the DC situation, i.e., $E_A = 1000$ V/cm.

Figure 6 shows the results of mixing under DC voltage and different frequencies of AC voltage. When $f_{AC} = 10$ KHz, the mixing performance is stronger than that when the frequencies are 1 MHz and 2 MHz. When DC voltage was applied on electrodes, strongest mixing was achieved in a very short time. However, bubbles were also generated within 1 s since voltage was applied. The channel was finally blocked by these bubbles. C distributions in the transverse direction are shown in Figure 6e for three different AC frequency mixing results. According to this quantitative C distribution, when $f_{AC} = 10$ KHz, concentration distribution reaches relative uniformity at $x = 2.3w$ from the entrance. At the same streamwise position, however, when $f_{AC} = 1$ MHz and 2 MHz, the profiles of the concentration distribution are still far from uniform. From Figure 6f, it can be seen that κ is always the largest at $f_{AC} = 10$ KHz, compared with that under other AC frequencies.

Moreover, mixing results under high frequency were also investigated. It was found that, rapid mixing result can be also achieved at high frequency besides low frequency as long as the $\vec{E}$ is sufficiently strong. Results are shown in Figure 7. Applied frequencies vary from 30 MHz to 40 MHz, and the E_A was increased to 1667 V/cm.

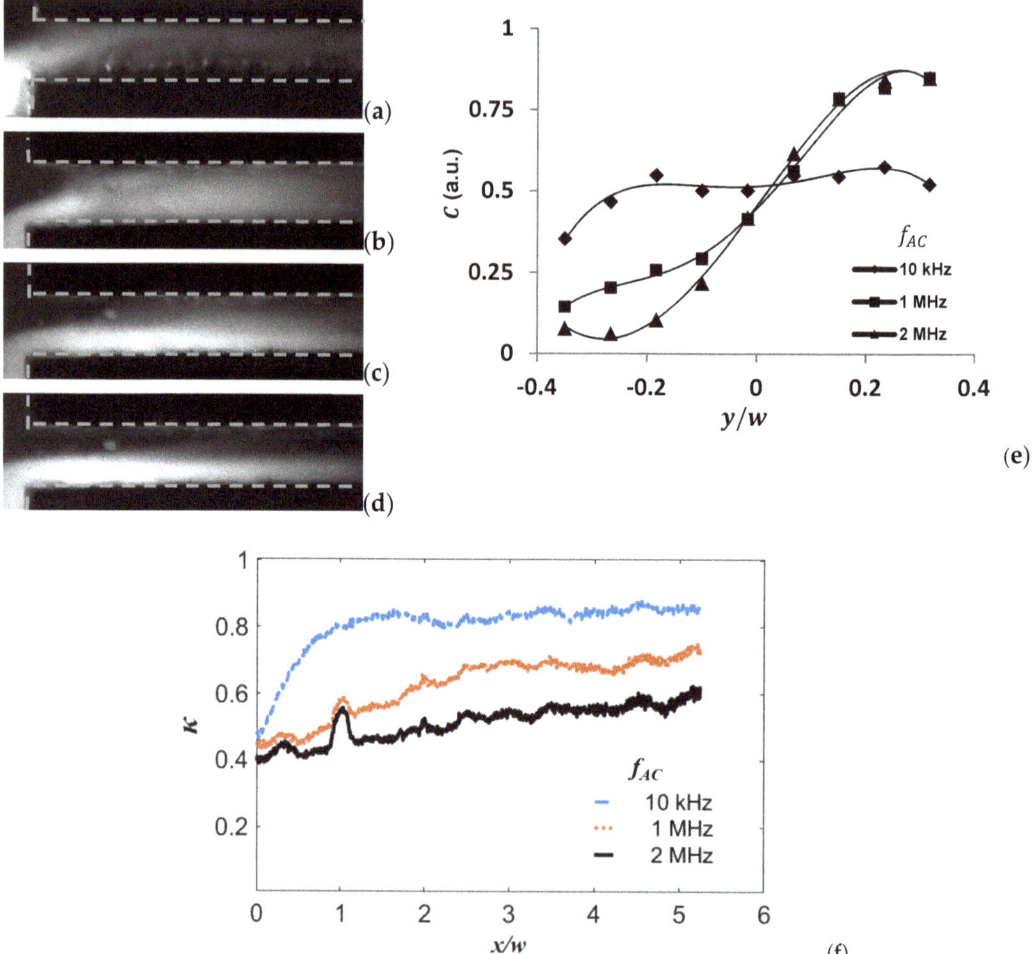

Figure 6. Visualization and comparison of mixing results in microchannels with different frequencies of AC and DC signals. Here, $\gamma = 10$ and E_A = 1000 V/cm. (**a**) DC, (**b**) $f_{AC} = 10$ KHz, (**c**) $f_{AC} = 1$ MHz, (**d**) $f_{AC} = 2$ MHz, and (**e**) comparison of concentration profile in transverse direction at $x = 2.3w$ from the channel entrance with AC signal. (**f**) Mixing index varying along streamwise direction at different AC frequencies.

Figure 7 shows mixing results under each f_{AC}. It obviously shows that mixing is stronger when $f_{AC} = 30$ MHz, than that when $f_{AC} = 35$ MHz and 40 MHz. However, at E_A = 1667 V/cm (limitation of the function generator), 40 MHz was the highest frequency at which we can achieve mixing augmentation in this mixer. It is also an important advantage that mixing can be acquired under high AC frequency electric field, since in many cases, low frequency AC signal could generate bubbles due to electrolysis in microchannels, especially when highly conductive buffer is used. The present new design of the micromixer could significantly reduce the risk of generation of bubbles in the microfluidic device, when the operation AC frequency is increased to higher than 10 kHz. Especially, even in fluids with relatively high conductivity (1000 μS/cm), no bubble was generated.

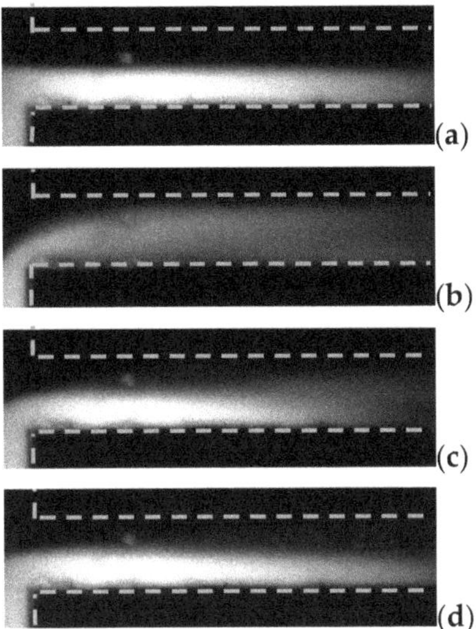

Figure 7. Visualization and comparison of mixing result in microchannels with different f_{AC}. Here, $\gamma = 10$ and E_A = 1667 V/cm. (**a**) No voltage, (**b**) 30 MHz, (**c**) 35 MHz, and (**d**) 40 MHz.

It is known that the EK flow can become unstable or perturbated when applied E_A exceeds a threshold value under a certain frequency. This particular E_A value is called critical E_A, beyond which the interface becomes fluctuating. The relation between critical E_A and frequency was investigated in the micromixer with conductive sidewalls, as plotted in Figure 8.

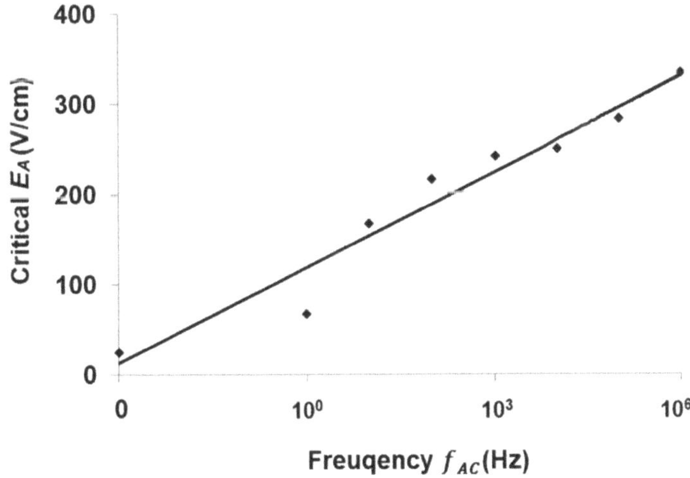

Figure 8. Critical electric field vs. the f_{AC} for mixing enhancement. Flow rate was kept at 5 μL/min with $\gamma = 10$.

Figure 8 suggests that, along with the increasing of frequency, the critical voltage required for the mixing enhancement is also increased. For $f_{AC} = 1$ Hz, E_A = 67 V/cm is sufficiently large to result in mixing augmentation inside the microchannel. When the frequency is increased to 1 MHz, E_A = 333 V/cm is required to enhance the mixing. Since the applied AC frequency covers 6 orders, several different EK mechanisms could exist in the mixing process. Although the general form of electric body force is known, a comprehensive theory for predicting the critical values of local E_A has not been established for the broad frequency range. Nevertheless, in the high frequency regime, i.e., $f_{vc} \ll f_{AC} \ll \langle\sigma\rangle/2\pi\varepsilon$ (where f_{vc} is the cut-off frequency of velocity fluctuation in frequency domain), according to the theory of Zhao and Wang [13], we have approximately:

$$\left|\vec{f}_e\right| \sim -\frac{\varepsilon E_A^2}{w}\left(\frac{\gamma-1}{\gamma+1}\right)\left(1-\beta^2\right) \tag{5}$$

where $\beta = 2\pi f_{AC}\varepsilon/\langle\sigma\rangle$ is a dimensionless AC frequency. As f_{AC} is increased, β increases accordingly, and thus, $\vec{f}_e$ is decreased. To generate sufficiently large $\vec{f}_e$ to disturb the flow, E_A must accordingly increase simultaneously.

According to the theoretical research of Zhao and Wang [12,13], the electric volume force in DC electric field is larger than that in AC electric field, under equivalent electric field magnitudes. The present experimental investigation on frequency effect supports the theoretical conclusion, and the fastest mixing could be achieved under DC electric field in a very short time. However, for practical applications, to avoid bubbles generated by electrolysis, the AC electric field is applied.

3.4. Electric Field Effect

As the EBF plays a key role in the currently designed mixing process, mixing should be directly related to $\vec{E}$. Therefore, voltage effect on mixing result was investigated. In this experiment, frequencies of the applied signals are kept constant, i.e., f_{AC} = 10 kHz. E_A was varied from 0 $V_{p\text{-}p}$ to 1167 V/cm (14 $V_{p\text{-}p}$).

Figure 9 shows mixing performance under different applied E_A. As visualized in Figure 9, despite the molecular diffusion, there is no obvious mixing on the interface of the two streams when no E_A is supplied. However, the mixing can be significantly enhanced when the applied E_A is increased to 500 V/cm. With further increasing E_A to 1167 V/cm, the mixing becomes the strongest.

Figure 9d shows the quantitative concentration C distribution in the transverse direction, at streamwise position $x/w = 3$ away from the entrance. It shows that, C distribution under 1167 V/cm reaches relative uniformity at $x/w = 3$ from the entrance, while at the same streamwise position, the profile of the concentration distribution under E_A of 0 and 500 V/cm are still far from flat. The same consequence can also be concluded from Figure 9e, where higher electric field results in higher mixing index. Especially at $E_A = 1167$ V/cm, κ reaches 0.83 at $x/w = 0.71$, which only costs time in the amount of 14 ms. All the results indicate that mixing is enhanced rapidly with increased electric field in the conductive sidewalls micromixer.

It should be noted that, besides the EK flow generated directly on the interface of electric conductivity gradient, there are also two additional EK flows generated in the electrokinetic micromixer system. One is the induced charge electrokinetic flow adjacent to the electrodes [33], the other is electro-osmotic flow on the top and bottom walls.

When the electric voltage applied is sufficiently large, nonlinear induced charge with vortical structures can be induced adjacent to the electrodes because of concentration polarization. The flow can be chaotic and apparently enhance fluid mixing in the diffusion layer, which is several hundred times of Debye length from electrodes [34]. Therefore, the mixing of fluids can be enhanced by nonlinear EK flow induced near electrodes. Besides, due to the unbalanced electric field on the low and high electric conductivity streams, a large scale vortical flow can be generated by the electro-osmotic flow (EOF) adjacent to the top and bottom walls, as have been investigated by Nan et al. [25]. The vortical flow could significantly enhance the 3D mixing of fluids on large scales. Thus, the fast mixing is achieved as a result of all these EK mechanisms.

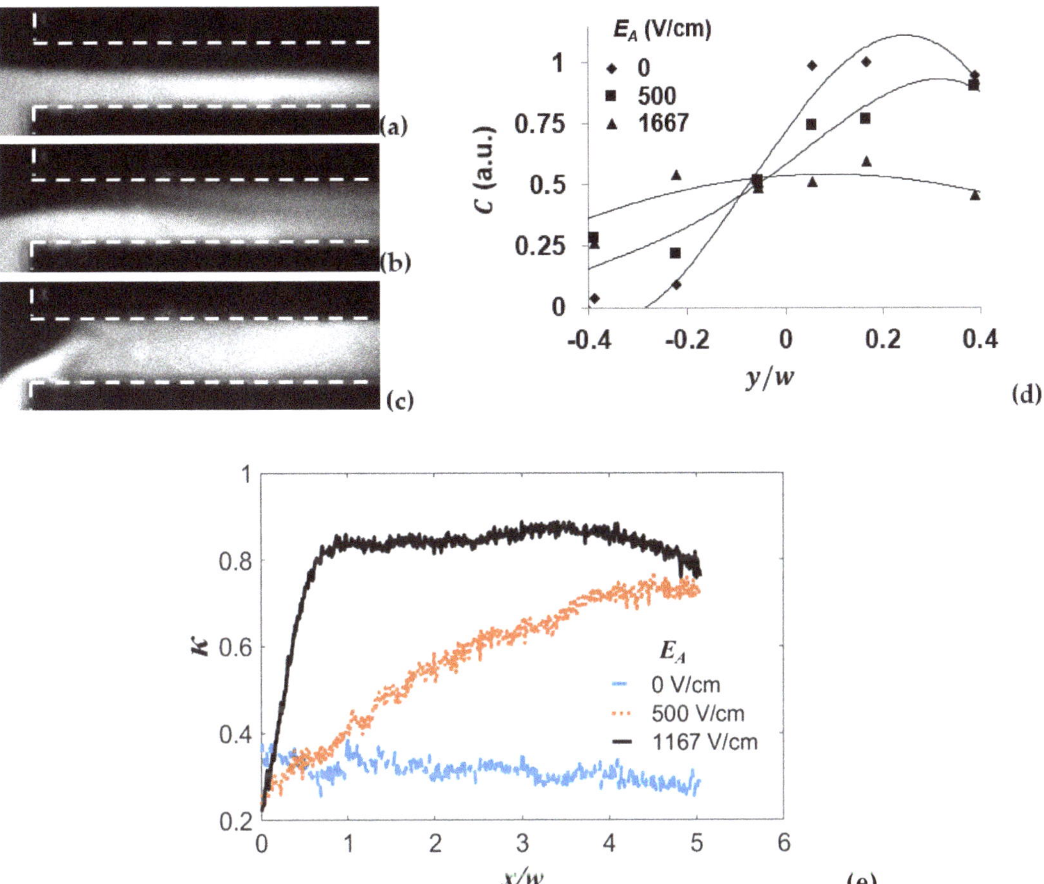

Figure 9. Visualization and comparison of mixing result in microchannels with various E_A. Here, $f_{AC} = 10$ kHz and $\gamma = 10$. (**a**) 0; (**b**) 500 V/cm; (**c**) 1167 V/cm; and (**d**) comparison of concentration profile in transverse direction at $x/w = 3$ from the channel entrance with different AC voltages. (**e**) Mixing index varying along streamwise direction at different E_A.

3.5. Re Number Effect

Different Re number effects on mixing results were investigated as well. In this experiment, frequency of the applied signal is kept constant, i.e., $f_{AC} = 10$ kHz. Applied E_A was kept at 500 V/cm. Flow rate was changed in the range of 1 µL/min to 5 µL/min to increase Re. Three different Re numbers, i.e., 0.1, 0.3, and 0.5 were compared. Results are shown in Figure 10.

As visualized in Figure 10, the mixing is strongest at a given downstream position when the Re number was 0.1, compared with situations where the Re numbers were 0.3 and 0.5. The mixing length (the downstream distance from the inlet of the channel required for the mixing to be achieved in the transverse direction, not the Prandtl mixing length in turbulent flows) is much shorter when low Re number was applied than when high Re number was applied. Note that although the mixing length is shorter at lower Re, the mixing time (required for the mixing to be achieved in transverse direction) is not necessarily shorter because the bulk flow velocity is larger in the higher Re.

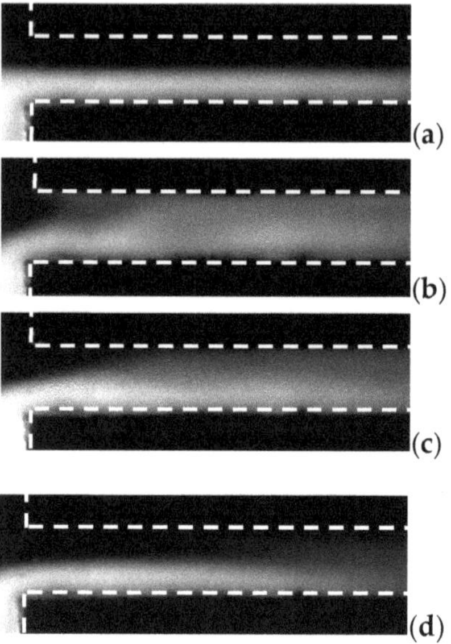

Figure 10. Visualization and comparison of mixing result in the micromixer with different Re numbers. (**a**) No voltage, (**b**) Re = 0.1, (**c**) Re = 0.3, and (**d**) Re = 0.5. Applied E_A was kept at 500 V/cm with $f_{AC} = 10$ kHz.

Here we use mass transport equation to explain the observed effect of Re on mixing. In EK micromixer, the mixing is dominated by the scalar transport due to velocity fluctuations. This can be explained by a convection–diffusion equation, as:

$$\frac{\partial C}{\partial t} + \vec{u} \cdot \nabla C = D\nabla^2 C \quad (6)$$

where D is the diffusion coefficient. Considering a quasi-steady process, i.e., $\vec{u} = \vec{u'} + U$ (where $\vec{u\prime}$ is the velocity fluctuation primarily attributed to EBF), $C = \overline{C} + c'$ and $\partial\overline{C}/\partial t = 0$. $\overline{C}$ and c' are the mean value and fluctuations of concentration, respectively. Subsequently, we have:

$$\frac{\partial c\prime}{\partial t} + \left(\vec{u'} + U\right) \cdot \nabla\left(\overline{C} + c'\right) = D\nabla^2\left(\overline{C} + c'\right) \quad (7)$$

Taking temporal averaging on Equation (6), we have:

$$\overline{\vec{u'} \cdot \nabla c'} + U \cdot \nabla\overline{C} = D\nabla^2\overline{C} \quad (8)$$

By combining Equations (6) and (7), and considering U is only in streamwise direction, we further have the transport equation of $c\prime$, which is:

$$\frac{\partial c\prime}{\partial t} + \vec{u'} \cdot \nabla\overline{C} + \vec{u'} \cdot \nabla c' + U\frac{\partial c'}{\partial x} - \overline{\nabla \cdot \left(c'\vec{u'}\right)} = D\nabla^2 c' \quad (9)$$

Since EBF is perpendicular to the flow direction, $\vec{u'}$ is in transverse direction initially at the interface between the two streams. Normally $c'/\overline{C} \ll 1$, we dimensionally have $\vec{u'} \cdot \nabla\overline{C} \gg \vec{u'} \cdot \nabla c'$. If we only focus on large-scale concentration fluctuations, the influence of the diffusion term can also be ignored. Thus, the initial spreading of the mixing can be approximately determined by:

$$\frac{\partial c\prime}{\partial t} + v'\frac{\partial \overline{C}}{\partial y} + U\frac{\partial c'}{\partial x} - \overline{\nabla \cdot (c'u')} = 0 \tag{10}$$

where v' is the velocity fluctuation component in y direction. The mixing of fluids is primarily determined by two convection terms, which are $v'\partial\overline{C}/\partial y$ and $U\partial c'/\partial x$. Dimensionally, in the EK flow, $v'^2 \sim \left|\vec{f}_e\right|$, and $U \sim Re$. When the electric field intensity and solutions are given, $\left|\vec{f}_e\right|$ could be approximately fixed in the initial stage in this investigation, and thus v'^2 remains approximately unchanged. In addition, in turbulent flows, commonly, $\frac{v'}{U} < 1$. Consequently, as Re is increased, $U\partial c'/\partial x$ convects and transports more mass downstream before they are spreading along transverse direction by the relatively smaller $v'\partial\overline{C}/\partial y$. Hence, mixing in our mixer could have a much shorter mixing length under lower Re number than that under high Re number.

4. Conclusions

In this paper, a novel quasi T-channel micromixer with conductive sidewalls is introduced. Compared with the conventional micromixers, where electrodes are located at the ends of the channel and the electric field and conductivity gradient are orthogonal, the micromixer with conductive sidewalls, where the electric field and conductivity gradient are parallel, can generate faster mixing under the same electric field. In the present device, no amplifier or high voltage supply is required, and a function generator is sufficient to create fast mixing. Furthermore, effects of Re numbers, electric field strength, AC frequency, and conductivity ratio on mixing results have been studied in the conductive sidewall micromixer. The results reveal that the mixing length is shorter with lower Re number and AC frequency, and stronger electric field and higher conductivity ratio. This mixing strategy provides a new and convenient method for enhancing the mixing of two fluids at low Re in microchannels, which is a common key step in sample pretreatment in biomedical and biochemical analysis applications.

Author Contributions: Conceptualization, G.W., C.K. and F.Y.; methodology, F.Y.; software, W.Z.; validation, F.Y. and W.Z.; formal analysis, W.Z.; investigation, F.Y.; resources, G.W.; data curation, W.Z. and F.Y.; writing—original draft preparation, F.Y.; writing—review and editing, G.W.; visualization, F.Y.; supervision, G.W.; project administration, G.W.; funding acquisition, G.W. and F.Y. All authors have read and agreed to the published version of the manuscript.

Funding: This research was partially funded by North American Mixing Forum (NAMF), NSF CAREER CBET-0954977, S&T Development Planning Program of Jilin Province 20190201178JC, 20200301029RQ, respectively.

Conflicts of Interest: The authors declare no conflict of interest.

References

1. Wen, J.; Guillo, C.; Ferrance, J.P.; Landers, J.P. Microfluidic-Based DNA Purification in a Two-Stage, Dual-Phase Microchip Containing a Reversed-Phase and a Photopolymerized Monolith. *Anal. Chem.* **2007**, *2*, 6135–6142. [CrossRef]
2. Bachman, H.; Chen, C.; Rufo, J.; Zhao, S.; Yang, S.; Tian, Z.; Nama, N.; Huang, P.-H.; Huang, T.J. An acoustofluidic device for efficient mixing over a wide range of flow rates. *Lab Chip* **2020**, *2*, 1238–1248. [CrossRef] [PubMed]
3. Shi, H.; Nie, K.; Dong, B.; Long, M.; Xu, H.; Liu, Z. Recent progress of microfluidic reactors for biomedical applications. *Chem. Eng. J.* **2019**, *361*, 635–650. [CrossRef]
4. Ward, K.; Fan, Z.H. Mixing in microfluidic devices and enhancement methods. *J. Micromech. Microeng.* **2015**, *25*, 094001. [CrossRef] [PubMed]
5. Lee, C.-Y.; Wang, W.-T.; Liu, C.-C.; Fu, L.-M. Passive mixers in microfluidic systems: A review. *Chem. Eng. J.* **2016**, *288*, 146–160. [CrossRef]

6. Yesiloz, G.; Boybay, M.S.; Ren, C.L. Effective thermo-capillary mixing in droplet microfluidics integrated with a microwave heater. *Anal. Chem.* **2017**, *89*, 1978–1984. [CrossRef] [PubMed]
7. Xia, Q.; Zhong, S. Liquid mixing enhanced by pulse width modulation in a Y-shaped jet configuration. *Fluid Dyn. Res.* **2013**, *45*, 025504. [CrossRef]
8. El Moctar, A.O.; Aubry, N.; Batton, J. Electro–hydrodynamic micro-fluidic mixer. *Lab Chip* **2003**, *3*, 273–280. [CrossRef]
9. Mavrogiannis, N.; Desmond, M.; Ling, K.; Fu, X.; Gagnon, Z. Microfluidic mixing and analog on-chip concentration control using fluidic dielectrophoresis. *Micromachines* **2016**, *7*, 214. [CrossRef]
10. Yeo, L.Y.; Friend, J.R. Surface Acoustic Wave Microfluidics. *Annu. Rev. Fluid Mech.* **2014**, *2*, 379–406. [CrossRef]
11. Owen, D.; Ballard, M.; Alexeev, A.; Hesketh, P.J. Rapid microfluidic mixing via rotating magnetic microbeads. *Sensors Actuators A Phys.* **2016**, *251*, 84–91. [CrossRef]
12. Zhao, W.; Wang, G. Cascade of turbulent energy and scalar variance in DC electrokinetic turbulence. *Phys. D* **2019**, *399*, 42–50. [CrossRef]
13. Zhao, W.; Wang, G. Scaling of velocity and scalar structure functions in ac electrokinetic turbulence. *Phys. Rev. E* **2017**, *2*, 023111. [CrossRef] [PubMed]
14. Oddy, M.H.; Santiago, J.G.; Mikkelsen, J.C. Electrokinetic Instability Micromixing. *Anal. Chem.* **2001**, *73*, 5822–5832. [CrossRef]
15. Shin, S.M.; Kang, I.S.; Cho, Y.-K. Mixing enhancement by using electrokinetic instability under time-periodic electric field. *J. Micromech. Microeng.* **2005**, *15*, 455–462. [CrossRef]
16. Dubey, K.; Gupta, A.; Bahga, S.S. Coherent structures in electrokinetic instability with orthogonal conductivity gradient and electric field. *Phys. Fluids* **2017**, *29*, 092007. [CrossRef]
17. Jagdale, P.; Song, L.; Liu, Z.; Li, D.; Zhang, C. Electrokinetic instability in microchannel viscoelastic fluid flows with conductivity gradients. *Phys. Fluids* **2019**, *31*, 082001. [CrossRef]
18. Chen, C.-H.; Lin, H.; Lele, S.K.; Santiago, J.G. Convective and absolute electrokinetic instability with conductivity gradients. *J. Fluid Mech.* **2005**, *524*, 263–303. [CrossRef]
19. Posner, J.D.; Santiago, J.G. Convective instability of electrokinetic flows in a cross-shaped microchannel. *J. Fluid Mech.* **2006**, *555*, 1–42. [CrossRef]
20. Posner, J.D.; Perez, C.L.; Santiago, J.G. Electric fields yield chaos in microflows. *Proc. Natl. Acad. Sci. USA* **2012**, *2*, 14353–14356. [CrossRef]
21. Wang, S.-C.; Wei, H.-H.; Chen, H.-P.; Tsai, M.-H.; Yu, C.-C.; Chang, H.-C. Dynamic superconcentration at critical-point double-layer gates of conducting nanoporous granules due to asymmetric tangential fluxes. *Biomicrofluidics* **2008**, *2*, 014102–014109. [CrossRef]
22. Yossifon, G.; Chang, H.-C. Selection of Nonequilibrium Overlimiting Currents: Universal Depletion Layer Formation Dynamics and Vortex Instability. *Phys. Rev. Lett.* **2008**, *2*, 254501. [CrossRef]
23. Coleman, J.T.; McKechnie, J.; Sinton, D. High-efficiency electrokinetic micromixing through symmetric sequential injection and expansion. *Lab Chip* **2006**, *2*, 1033–1039. [CrossRef]
24. Yang, F.; Kuang, C.; Zhao, W.; Wang, G. AC Electrokinetic Fast Mixing in Non-Parallel Microchannels. *Chem. Eng. Commun.* **2017**, *2*, 190–197. [CrossRef]
25. Nan, K.; Hu, Z.; Zhao, W.; Wang, K.; Bai, J.; Wang, G. Large-Scale Flow in Micro Electrokinetic Turbulent Mixer. *Micromachines* **2020**, *2*, 813. [CrossRef] [PubMed]
26. Wang, G.; Yang, F.; Zhao, W. There can be turbulence in microfluidics at low Reynolds number. *Lab Chip* **2014**, *2*, 1452–1458. [CrossRef] [PubMed]
27. Wang, G.; Yang, F.; Zhao, W.; Chen, C.-P. On micro-electrokinetic scalar turbulence in microfluidics at a low Reynolds number. *Lab Chip* **2016**, *2*, 1030–1038. [CrossRef]
28. Lin, H.; Storey, B.D.; Oddy, M.H.; Chen, C.-H.; Santiago, J.G. Instability of electrokinetic microchannel flows with conductivity gradients. *Phys. Fluids* **2004**, *2*, 1922–1935. [CrossRef]
29. Xie, G.; Luo, J.; Liu, S.; Zhang, C.; Lu, X. Micro-Bubble Phenomenon in Nanoscale Water-based Lubricating Film Induced by External Electric Field. *Tribol. Lett.* **2008**, *2*, 169–176. [CrossRef]
30. Park, J.; Shin, S.M.; Huh, K.Y.; Kang, I.S. Application of electrokinetic instability for enhanced mixing in various micro-channel geometries. *Phys. Fluids* **2005**, *17*, 118101. [CrossRef]
31. Hoburg, J.F.; Melcher, J.R. Internal electrohydrodynamic instability and mixing of fluids with orthogonal field and conductivity gradients. *J. Fluid Mech.* **1976**, *2*, 333–351. [CrossRef]
32. Arockiam, S.; Cheng, Y.H.; Armenante, P.M.; Basuray, S. Experimental determination and computational prediction of the mixing efficiency of a simple, continuous, serpentine-channel microdevice. *Chem. Eng. Res. Des.* **2021**, *167*, 303–317. [CrossRef]
33. Olesen, L.H.; Bazant, M.Z.; Bruus, H. Strongly nonlinear dynamics of electrolytes in large ac voltages. *Phys. Rev. E* **2010**, *82*, 011501. [CrossRef] [PubMed]
34. Karatay, E.; Druzgalski, C.L.; Mani, A. Simulation of chaotic electrokinetic transport: Performance of commercial software versus custom-built direct numerical simulation codes. *J. Colloid Interface Sci.* **2015**, *446*, 67–76. [CrossRef]

 *micromachines*

Article

Generation of Concentration Gradients by a Outer-Circumference-Driven On-Chip Mixer

Fumiya Koike *,† and **Toshio Takayama** †

Department of Mechanical Engineering, Tokyo Institute of Technology, 2-12-1, Ookayama, Tokyo 152-8552, Japan; takayama.t.aa@m.titech.ac.jp
* Correspondence: koike.f.ab@m.titech.ac.jp; Tel.: +81-3-5734-2820
† These authors contributed equally to this work.

Abstract: The concentration control of reagents is an important factor in microfluidic devices for cell cultivation and chemical mixing, but it is difficult to realize owing to the characteristics of microfluidic devices. We developed a microfluidic device that can generate concentration gradients among multiple main chambers. Multiple main chambers are connected in parallel to the body channel via the neck channel. The main chamber is subjected to a volume change through a driving chamber that surrounds the main chamber, and agitation is performed on the basis of the inequality of flow caused by expansion or contraction. The neck channel is connected tangentially to the main chamber. When the main chamber expands or contracts, the flow in the main chamber is unequal, and a net vortex is generated. The liquid moving back and forth in the neck channel gradually absorbs the liquid in the body channel into the main chamber. As the concentration in the main chamber changes depending on the pressure applied to the driving chamber, we generated a concentration gradient by arranging chambers along the pressure gradient. This allowed for us to create an environment with different concentrations on a single microchip, which is expected to improve observation efficiency and save space.

Keywords: micromixer; microfluidics; density control; lab on a chip; pneumatically driven

Citation: Koike, F.; Takayama, T. Generation of Concentration Gradients by a Outer-Circumference-Driven On-Chip Mixer. *Micromachines* **2022**, *13*, 68. https://doi.org/10.3390/mi13010068

Academic Editor: Kwang-Yong Kim

Received: 13 December 2021
Accepted: 29 December 2021
Published: 31 December 2021

1. Introduction

Traditionally, cell-culture and chemical-reaction experiments are conducted in containers whose size is similar to that of a Petri dish. However, a Petri dish is used for a very small observation target, and for the aforementioned experiments, a large space is required, as much waste fluid is generated. To solve these problems, research on microfluidic devices and microreactors was conducted to realize a cell culture and chemical mixing on microchips by scaling down experiments that were conducted at a laboratory scale [1–7].

As most microfluidic devices process a low Reynolds number, liquids are mixed using molecular diffusion, which is very time-consuming. To solve this problem, various methods have been proposed, and they can be classified into passive and active mixers. Passive mixers are designed to increase the contact area of two liquids by applying ingenuity to the shape of the flow path, thereby promoting diffusion [8–11]. For instance, Chien-Chong Hong et al. studied mixing using a Tesla valve [12], and Sung-Jin Park studied mixing using a three-dimensional flow channel [13]. Active mixers use electrophoresis, ultrasonic or other waves to forcibly generate a vortex for mixing [11,14–16]. Ahmed et al. proposed a mixing process based on the ultrasonic vibration of microbubbles [17]. Glasgow et al. proposed a mixing method based on pulsed flow [18].

To observe the behavior of cells in various environments on a microfluidic device, it is necessary to realize various concentrations on a microchip. There are two ways to achieve this: one is to use diffusion to generate a concentration gradient and keep supplying it [19–22], and the other is to prepare multiple chambers with different concentrations [23–28]. With regard to the former method, Fukuda et al. succeeded in supplying a stable concentration gradient by using a branching channel and a meandering channel [29]. For the latter, Hung et al. succeeded in creating multiple chambers with different concentrations on a microchip by orthogonally flowing the concentration gradient-generated liquid and perfusion liquid on the microarray [30]. Bo Dai et al. realized a continuously varying concentration gradient in the side chamber of the body channel just through the flow of solution [31].

The authors developed an active mixer (outer-circumference-driven mixer) that uses the inequality of flow during the expansion or contraction of chambers [32–36]. The driving principle of the outer-circumference-driven mixer is shown in Figure 1. The outer-circumference-driven mixer comprises a main chamber that mixes the two liquids, a driving chamber that surrounds the main chamber, a body channel through which the liquid flows, and a neck channel that connects the main chamber to the body channel. When pressure is applied to the driving chamber, the main chamber expands and contracts through the elastic wall. When this process is repeated, the liquid in the body channel is gradually absorbed into the main chamber with a slight back and forth movement through the neck channel. This repeated expansion or contraction mixes with the liquid in the main chamber and reaches the concentration in the body channel. When the main chamber reaches the desired concentration, air can be injected into the body channel to maintain the concentration in the main chamber.

The mixing speed of this mixer is proportional to the amplitude of the liquid moving back and forth in the neck channel, and the amplitude of the neck channel is proportional to the expansion or contraction of the main chamber. In this study, we generated a concentration gradient by applying a pressure gradient to each driving chamber to differentiate the volume of the main chamber. If we can create chambers with different concentrations on a single microchip, we can save space, reduce the amount of waste fluid, and improve the observation efficiency.

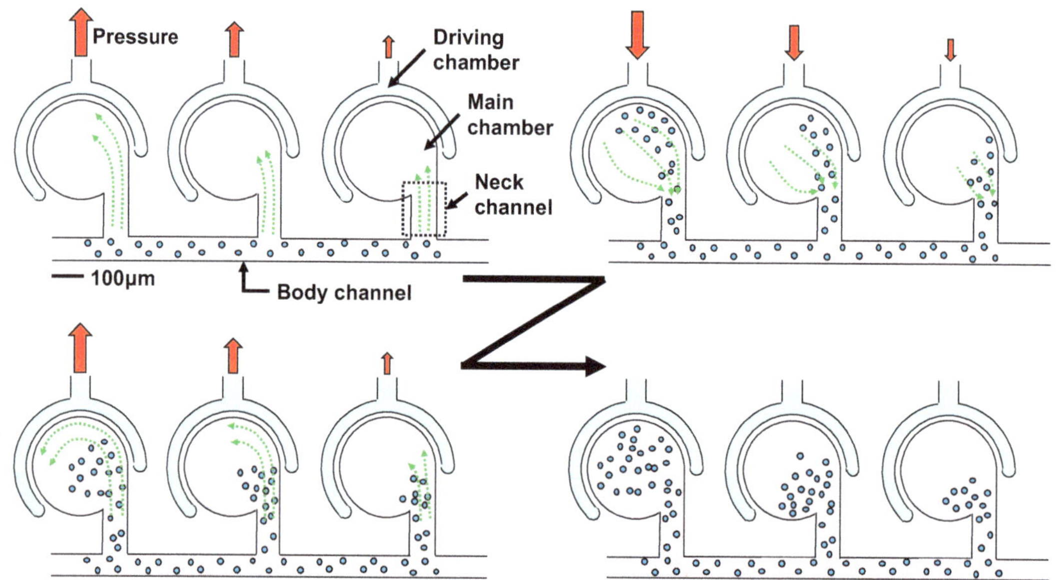

Figure 1. Driving principle of outer-circumference-driven mixer. The liquid in the main channel is gradually siphoned into the main chamber. Mixing speed can be changed by driving pressure.

2. Materials and Methods

2.1. Experimental Method and Channel Design

In a previous study [32–36], a piezoelectric actuator was used to drive the driving chamber through water; however, in this study, air pressure was used. The transmission channel to the driving chamber was filled with water, and the water was vibrated using a piezoelectric actuator. The piezoelectric actuator can powerfully drive the driving chamber because it can force the volume change; however, the volume change is limited. Therefore, as the number of main chambers increases, the volume change supplied to the driving chamber decreases, and there is a concern that the same results cannot be obtained when the number of chambers is small. Since a drive source such as a piezoelectric actuator is required to vibrate water at high speed, it is difficult to achieve a large scale with water in the transmission channel. Therefore, we decided to use air pressure, because it is possible to increase the pressure even when there is a large number of main chambers.

As the absorption rate of the main chamber can be adjusted by the pressure applied to the driving chamber, the concentration gradient can be generated by arranging chambers along the pressure gradient. The designed channel is shown in Figure 2. The process of why this channel design was chosen is described in Appendix A. A tube was connected to the air inlet, as shown in Figure 2a, a solenoid valve was connected to a tube, and a compressor was connected to a solenoid valve. Moreover, the syringe was connected to the liquid inlet to supply the liquid to body channel. The farther the distance from the air pressure source is, the narrower the flow path at the top of the driving chamber becomes. Therefore, the narrower the channel is, the more difficult it is to transmit pressure, and concentration gradients can thus be generated.

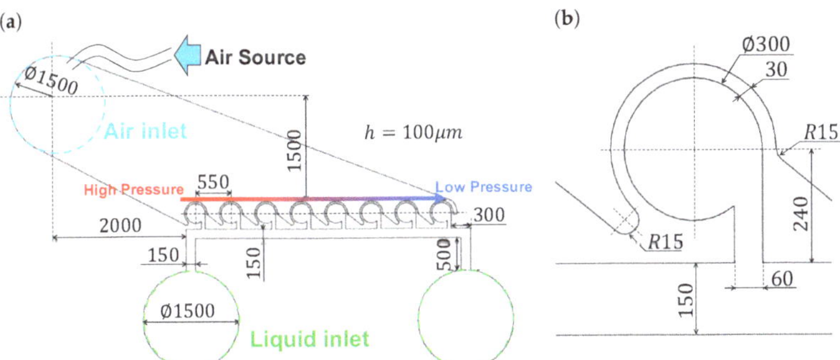

Figure 2. Design of the microchip. (**a**) Overall dimensions of flow channel. Air pressure is supplied from the air source. Arrow above the driving chamber indicates the expected pressure supplied to the driving chamber. (**b**) Detailed dimensions of the main chamber, which were determined on the basis of previous studies [33].

In Figure 2b, the main chamber dimensions were determined on the basis of previous studies [33]. In this study, we used the same dimensions, but there is room for debate as to whether the dimensions are suitable for using air pressure. Determining the optimal dimensions of the main chamber when air pressure is used is a future task.

To confirm the principle of this mixer, that is, whether flows are different between expansion and contraction, we conducted simulations using Autodesk CFD2020. Since this mixer is driven by the wall deformation at high frequency, it is difficult to simulate the flow caused by it. Therefore, we assumed that pressure is applied to the wall of the main chamber and verified whether the streamlines are different during expansion and contraction.

The result of simulation is shown in Figure 3. Figure 3a denotes the model, Figure 3b denotes how to apply the pressure, and Figure 3c,d denote the results that are the streamlines of expansion and contraction. Since the surface to which pressure is applied must be flat, the flow channel outline was polygonal. The streamlines were different between expansion and contraction, confirming the inequality of flows, which is the principle of this mixer.

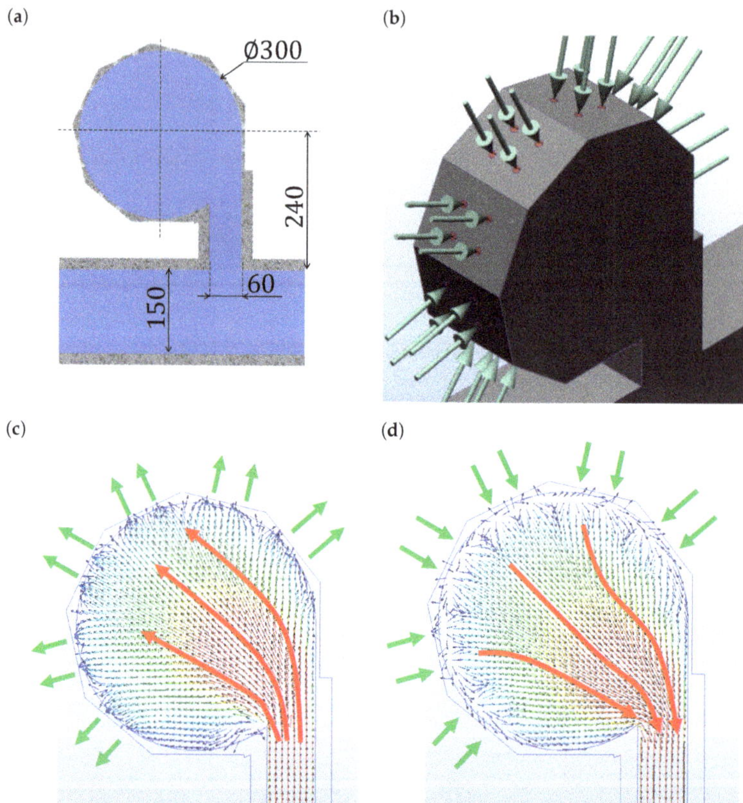

Figure 3. Simulation results. (**a**) Main chamber model. Because the surface to which pressure is applied must be flat, the outer shape was polygonal. (**b**) How to apply pressure to the main chamber. Since it was not possible to simulate the change in flow owing to the deformation of the wall, it was assumed that pressure is applied. (**c**,**d**) Streamlines during expansion and contraction. Streamlines were different between expansion and contraction; therefore, the net vortex was created by repeating this process.

2.2. Microchip Fabrication

A silicon wafer was spin-coated with SU8-3050 (MicroChem Inc., Japan), and prebaked (95 °C, 45 min). A mask with a channel pattern was placed on SU8 and irradiated with UV light (exposure energy, 250 mJ/cm^2, 10 s). After that, a mold with the channel pattern was developed with a thinner. Polydimethylsiloxane (PDMS, SILPOT184, Dow Inc., Japan, base:curing agent = 9:1 (mass ratio)) was poured into the mold. After deaeration and curing, PDMS that was transferred into the channel pattern was bonded to the slide glass using plasma treatment to create a flow channel.

2.3. Experimental Setup

Figure 4 shows the experimental apparatus and microchip drive unit. The air was supplied from the compressor (AK-T20R, Max Co., Ltd., Japan.) to the solenoid valve, which was controlled by the microcomputer (AIO-160802AY-USB, CONTEC Co., Ltd., Japan) through the regulator. This air was supplied to the driving chamber, and the air pressure causes a volume change in the driving chamber. The main chamber was filled with pure water, and a mixture of 3 μm microbeads (Polybead Polystyrene 3.0 Microspheres, Polysciences Inc., Philadelphia, PA, USA) and pure water flowed in the body channel. We used 3 μm microbeads to improve the streamlines more easily to see and increase the contrast for easier analysis.

In this study, air pressure was 0.30 MPa, the solenoid valve operated at 50 Hz, and duty ratio was 50%. This condition was experimentally determined to maximize absorption speed. Although there may be more optimal conditions by adjusting frequency, pulse width, etc., we used this condition in the experiment. The exploration of the ideal parameters is a future task.

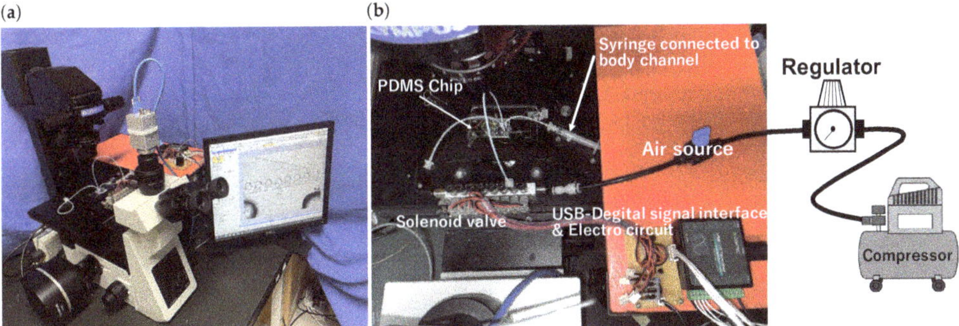

Figure 4. (**a**) Experimental setup. (**b**) Experimental apparatus.

2.4. Evaluation

The concentration was quantitatively evaluated using the luminance value. We recorded the experiment with a microscope (OLYMPUS, IX73P1F, Japan), measured the change in luminance value of each main chamber on the basis of the program, and evaluated the concentration using the following formula:

$$L_n(t) = \sum^{A_n} \frac{Pxl(t,x,y)}{S_n} \quad (1)$$

$$MI = \frac{L_n(t)}{L_n(0)} \quad (2)$$

where MI denotes the mixing index, n denotes the number of main chambers, A_n denotes the area of the n-th chamber, S_n denotes the number of pixels in A_n, and $Pxl(t,x,y)$ denotes the pixel value. The concentration of each main chamber was measured in the area shown in Figure 5. Moreover, Python and OpenCV were used to calculate the luminance values and measure the concentrations at each frame.

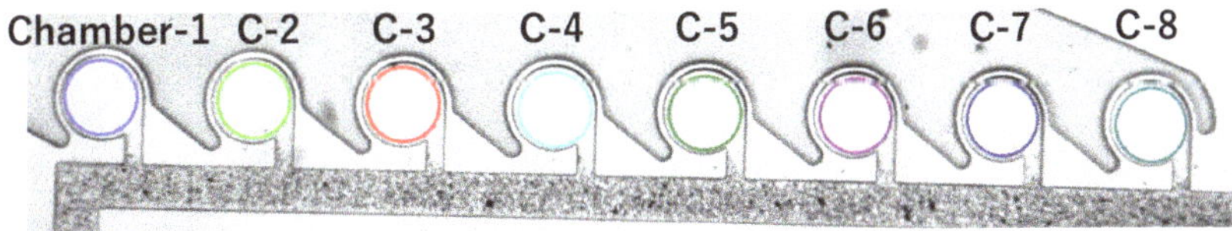

Figure 5. Measured concentration. Numbering from left to right is 1, 2, ..., 8. Calculations were performed using Python and OpenCV.

We produced three chips and performed three experiments using each chip. We evaluated whether the concentration gradient was generated on the basis of MI in the nine experiments. The chamber that was supplied with high pressure absorbed and agitated the liquid. The more the chamber absorbed the liquid, the lower the luminance value and consequently the lower the *MI* were. Therefore, the lower the *MI* was, the more absorption and agitation occurred in the chamber.

3. Result

The experimental results are shown in Figure 6, where Figure 6a denotes the time variation of an experiment, and Figure 6b denotes the results obtained 9 s after the start of the experiment. The time variation of the concentration until 9 s after the application of air pressure is shown in Figure 7, where C-n indicates the n-th main chamber from the left. The raw data are very difficult to see, and the graph was drawn using 3 points of a simple moving average (raw data and why a simple moving average was applied are in Appendix B). The *MI* indicates the degree of mixing; the lower the *MI* is, the more the beads are absorbed and mixed. Similar results were obtained in all experiments, with a gradual decrease in concentration starting from the left chamber. Figure 8 is a graph showing the mean and standard error of *MI* after 9 s for the nine experimental results. The standard error of C-1 was large (standard errors of C-1 and C-2 were 0.0199 and 0.169).

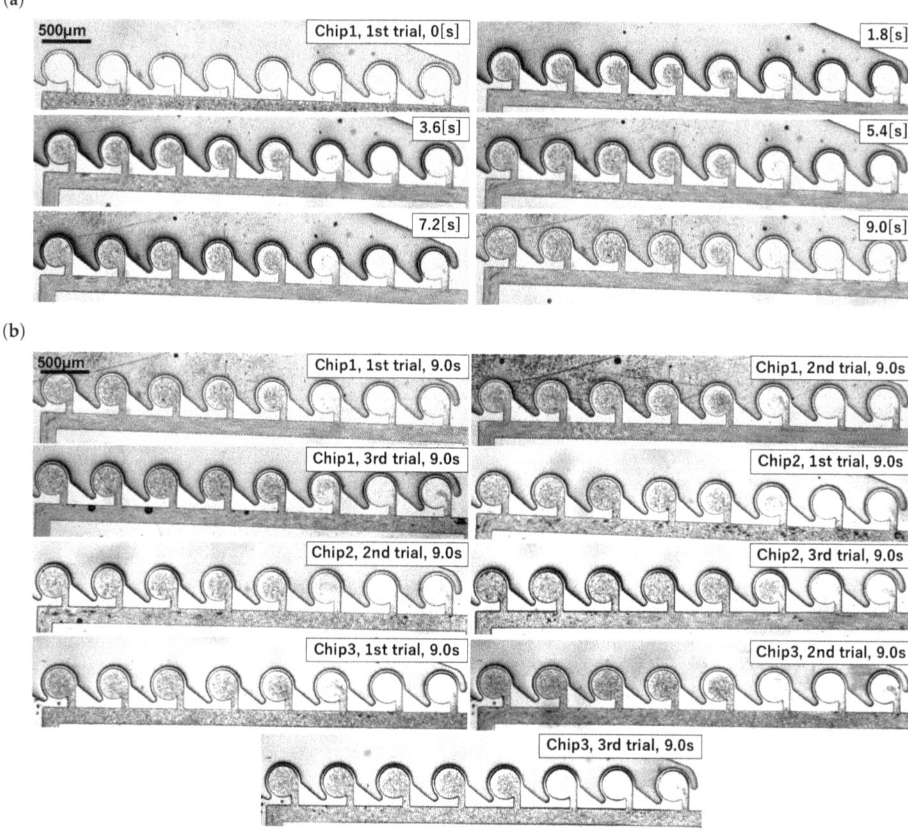

Figure 6. Generation of concentration gradient using an outer-circumference-driven mixer. (**a**) Time variation of an experiment (Chip 1, first trial). (**b**) Experimental results. Similar results were obtained from nine experiments.

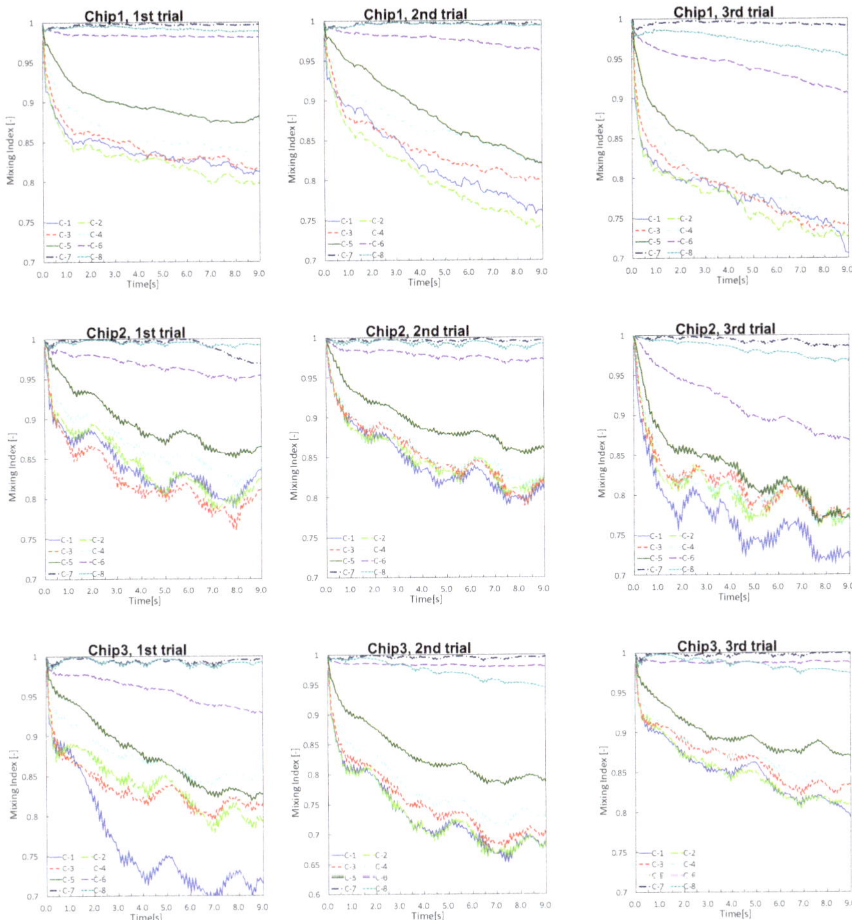

Figure 7. Time variation of mixing index until 9 s after application of air pressure. Graph was drawn using 3 points of a simple moving average. Mixing index indicates the degree of mixing; the lower the mixing index was, the more the beads were absorbed and mixed.

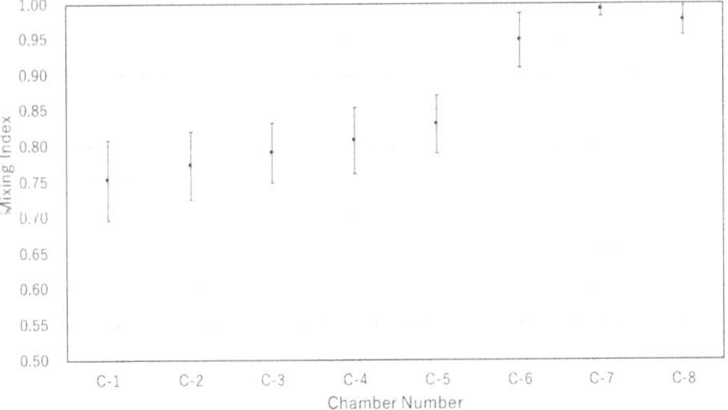

Figure 8. Graph of average of concentrations after 9 s for 9 experiments. Error bars represent standard error. The concentration gradient iwass generated roughly along the pressure gradient.

4. Discussion

4.1. Experiment Evaluation

In order to comprehensively check whether the concentration gradient was generated, the average value of the nine experimental results was calculated, and a simple moving average was applied, as shown in Figure 9.

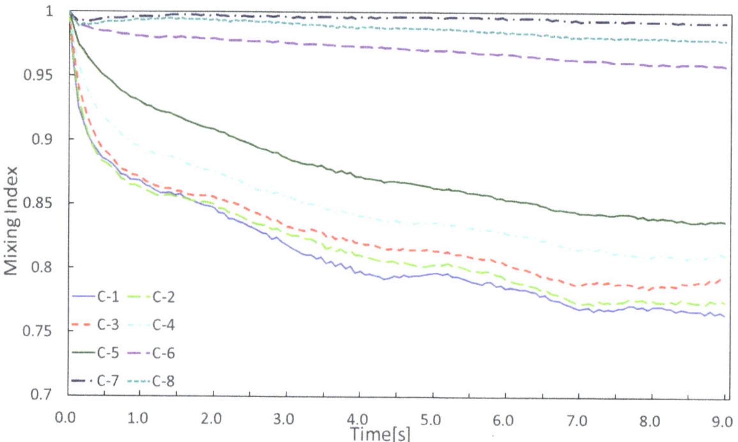

Figure 9. Average of nine experiments was calculated, and graph shows the simple moving average. Final concentrations were C-1, -2, -3, -4, -5, -6, -8, and -7.

Concentration gradients were generated along the pressure gradient, but chambers at both ends did not follow the pressure gradient; C-1 was almost the same absorption speed and mixing index as those of C-2, and C-8 was larger than the mixing index of C-7. This may have been caused by wall friction loss. Except for the chambers at both ends (C-1 and C-8), concentrations became thinner in the order of C-2, -3, -4, ..., -7, and the concentration gradient was generated along the pressure gradient. However, there was a large difference in the mixing index between C-5 and C-6. A detailed design method for generating a uniform concentration gradient has not been established, and it is necessary to explore the design of the driving chamber for generating a uniform concentration gradient in the future.

On average, we were successful in generating concentration gradients, but the order of the concentration gradients varied in each. In the future, it is necessary to pursue reproducibility so that the same concentration gradient could be obtained in all experiments.

4.2. Experiments Using Colored Water

In a previous experiment, we used 3 μm microbeads to evaluate the concentration. To confirm that the same results could be obtained with liquid–liquid mixing, we conducted an experiment using colored water.

The experimental results using colored water are shown in Figures 10 and 11, where Figure 10 denotes the time variation in an experiment, and Figure 11 denotes the time variation in the mixing index. We obtained a similar result to that in the previous experiments.

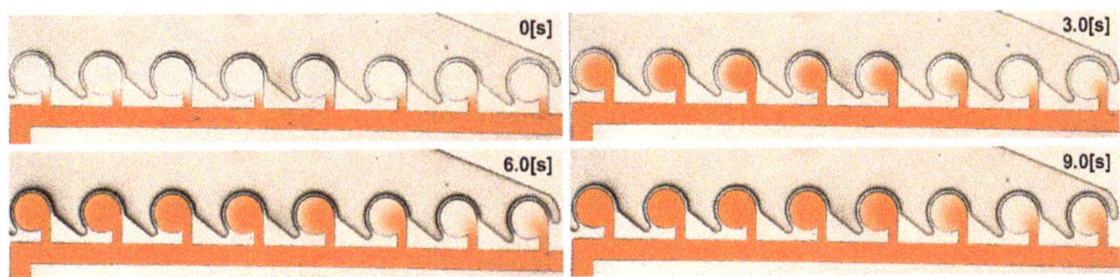

Figure 10. Time variation of experiment using colored water.

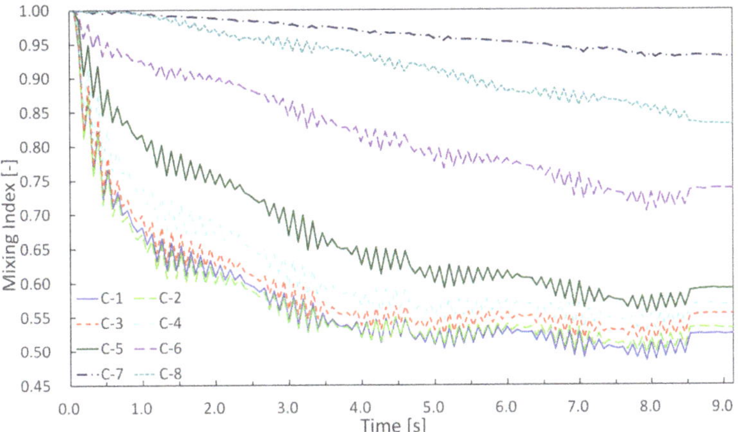

Figure 11. Time variation of mixing index in experiments using colored water. We obtained s similar result as that in experiments using 3 μm microbeads. Final mixing indices were C-1, -2, -3, -4, -5, -6, -8, -7.

4.3. Evaluation of Pressure Gradient

The mixer generated a concentration gradient owing to the pressure gradient. To confirm that the mixer was functioning properly, we measured the deformation of the wall. The measurement method and results are shown in Figure 12, where Figure 12a,b denote the measuring area, and Figure 12c denotes the graph of the amount of wall deformation. When air pressure was applied, the wall between the driving chamber and the main chamber was significantly deformed (Figure 12a). Since the amount of deformation depends on the supplied pressure, we could evaluate the pressure gradient by comparing the amount of deformation. To measure the wall deformation, the main chamber was divided into three regions as shown in Figure 12b. Region A is the entire main chamber and the wall area, Region B is the main chamber area, and Region C is the wall area. Since the pixel value in Region C represents the amount of deformation of the wall, the pixel value in Region C was calculated using the pixel values in Regions A and B. The wall deformation was evaluated with the following formula;

$$P_X(t) = \sum^{X} Pxl(t, x, y) \tag{3}$$

$$ND = 1 - \frac{P_C(t_{expansion})}{P_C(0)S_C} = 1 - \frac{P_A(t_{expansion}) - P_B(t_{expansion})}{(P_A(0) - P_B(0))(S_A - S_B)} \tag{4}$$

where ND denotes normalization deformation, X denotes each region, P_X denotes the total pixel values of the region, $Pxl(x, y)$ denotes the pixel value at x, y, and S_X denotes the number of pixels of region X. Figure 12c shows the normalized deformation of the wall. It was confirmed that the pressure gradient was generally generated in the order of the concentration gradient. The standard error of C-1 was also larger than that of other regions (standard errors of C-1 and C-2 were 0.0431 and 0.348, respectively); therefore, as shown in Figure 7, the concentration of C-1 was unstable, so it was difficult to adjust the concentration according to the pressure gradient.

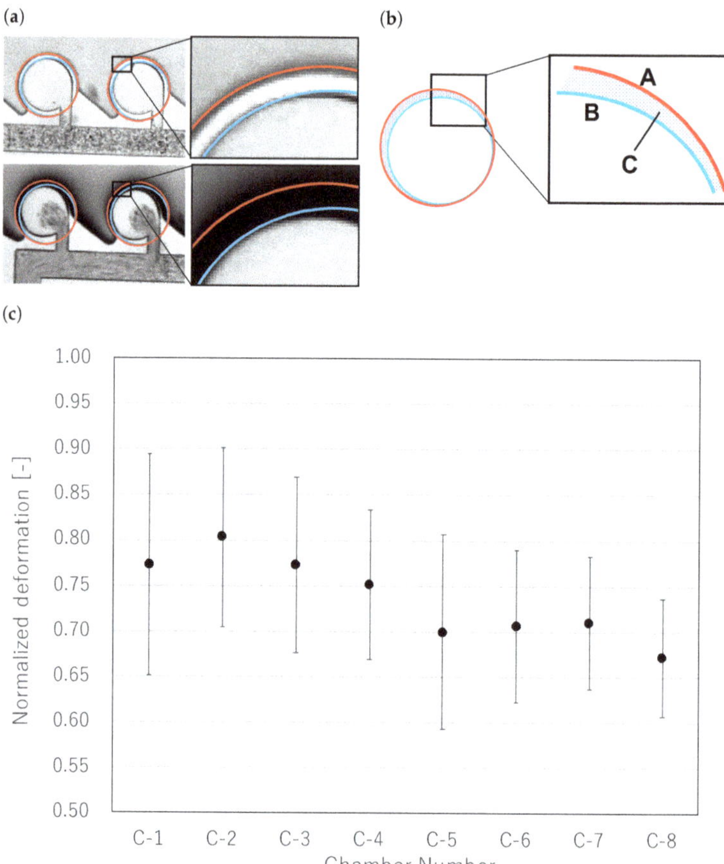

Figure 12. Evaluation of pressure gradient. (**a**) Wall deformation during expansion. Since the amount of deformation varied with the magnitude of pressure, it was used to evaluate the pressure gradient. (**b**) Details of evaluation region. Pixel values in Regions A and B were used to evaluate wall deformation in Region C. Equation (4) was used for evaluation. (**c**) Wall deformation in each chamber. Pressure gradients were generated approximately in the order of the concentration gradient.

5. Conclusions

The flow path shown in Figure 6 was successful in generating a concentration gradient, but the chambers at both ends did not adhere to the gradient due to wall friction. If the goal is to generate the concentration gradient, the liquid in the chambers at both ends can be discarded, and the concentration gradient can be realized in the other main chambers. The volume of the main chamber of the channel designed in this study was 7.06 nL; therefore, even if the liquid in the chambers at both ends was discarded, the advantage of reducing the amount of liquid waste was not lost.

We attempted to generate a concentration gradient in an outer-circumference-driven mixer. We succeeded in generating a concentration gradient using the difference in the pressure applied to the driving chambers, which decayed as the distance from the air pressure source increased. As the micromixer was driven by air pressure, the channel could be expanded by increasing the applied pressure. However, although we succeeded in generating a concentration gradient, it was difficult to adjust the concentration in the main chambers according to the expected value (i.e., there was a large difference between C-5 and C-6). In the future, we aim to determine parameters that could fine-tune the concentration, and to further increase the scale of the system.

Author Contributions: F.K., experiments, data analysis, and editing; T.T., conceptualization and data analysis. All authors have read and agreed to the published version of the manuscript.

Funding: This work was supported by JSPS KAKENHI grant no. 21H03837.

Data Availability Statement: The datasets used and/or analyzed during the current study are available from the corresponding author on reasonable request.

Conflicts of Interest: The authors declare no conflict of interest.

Appendix A. Determination of Channel Design

Appendix A.1. Concentration Gradient Generation Method

In this study, air was used. Therefore, the driving frequency cannot be increased owing to compressibility, but large deformation occurred in the driving chamber by increasing the driving pressure; therefore, the main chamber could be the same stirring speed as that in a previous study [32–36].

When an outer-circumference-driven mixer is driven using piezoelectric actuators, it was confirmed in previous studies [32–36] that the concentration in the main chamber could be adjusted depending on how much the driving chamber surrounds the main chamber (enclosure angle) and the pressure applied to the driving chamber. Therefore, there are two processes to generate the concentration gradient:

1. Equalize pressure in each driving chamber and adjust the concentration on the basis of enclosure angle.
2. Make the enclosure angle of each driving chamber uniform, and adjust the concentration on the basis of pressure.

The concept of each is shown in Figure A1. To investigate whether the concentration gradient using air could be realized in these channels, we designed a verification channel. As the first process requires uniform pressure supply, a linearly symmetrical branch channel was used to verify whether the concentration can be made uniform. In the second process, a serial-type channel was used to verify if a concentration gradient could be generated. If it was not possible to generate a concentration gradient, it was necessary to devise a new channel shape to drive the device on the basis of air pressure.

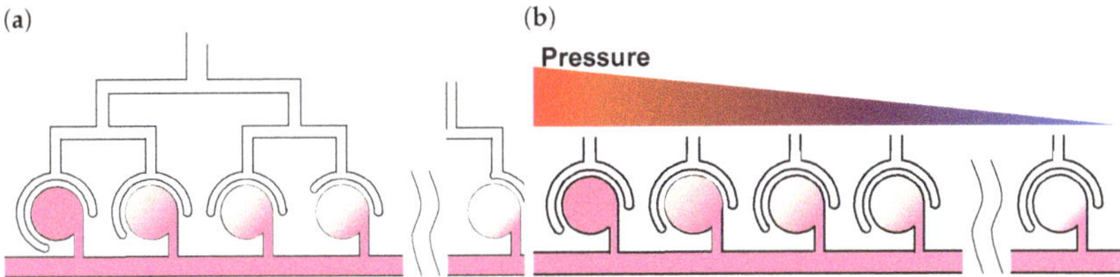

Figure A1. How to generate concentration gradient by outer-circumference-driven on-chip mixer. (**a**) Generation of concentration gradient by enclosure angle. Flow path was designed so that all driving chambers were equally pressurized, and the concentration was adjusted by enclosure angle. (**b**) Generation of concentration gradient by pressure gradient. All driving chambers had the same enclosure angle, and the concentration gradient was generated by the pressure supplied to the driving chambers.

Appendix A.2. Verification of Ability to Generate Chambers with Uniform Concentration Using Branch Channels

To generate a concentration gradient based on the enclosure angle, the pressure must be uniformly supplied to each driving chamber. To verify this, an experiment was conducted using the flow path shown in Figure A2a. The driving chambers were arranged independently, so that they were not affected by the neighboring chambers, and the driving chambers were designed in a treelike pattern, so that there was no difference in flow length among the driving chambers. If concentrations in all the main chambers were equal (Figure A2a), concentration could be adjusted by the enclosure angle. The experimental results are shown in Figure A2b. Concentrations in the main chambers are not uniform, and it is difficult to adjust them on the basis of enclosure angle. When pneumatic pressure is used, the pressure may not be evenly transmitted if the flow path is narrow. Therefore, the application of air pressure to such a branch channel was not appropriate to generate a concentration gradient.

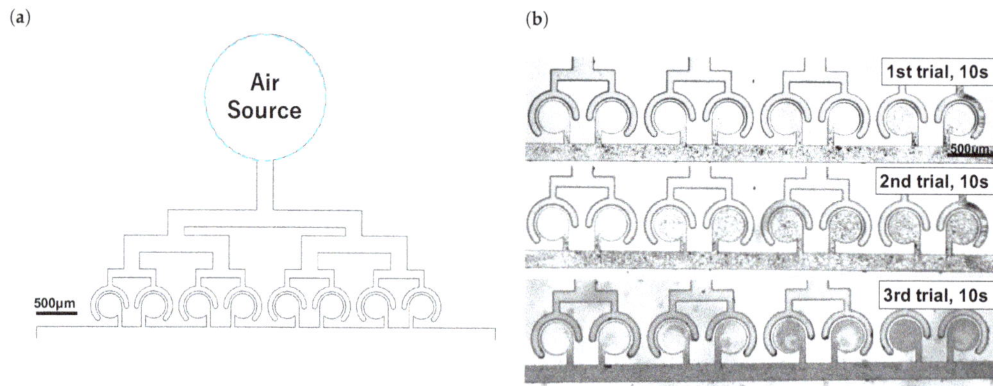

Figure A2. Verification of pressure uniformity. (**a**) To generate a concentration gradient using the enclosing angle, pressure must be uniform to generate a concentration gradient using the enclosing angle. (**b**) Experimental results. Concentrations were not uniform.

Appendix A.3. Generation of Concentration Gradients Based on Serial-Type Flow Paths

As mentioned in the previous section, it is difficult to evenly supply pressure to the driving chamber; therefore, we attempted to generate a concentration gradient based on pressure gradient. The flow path is shown in Figure A3a. The farther the distance from the air source, the more pressure loss occurs owing wall friction, and the more pressure gradient is generated in the driving chamber. The experimental results are shown in Figure A3b. No stirring occurred at all owing to the effect of the channel width, as described in the previous section.

Assuming that the channel width affects agitation, we designed the channel to generate the concentration gradient based on the pressure gradient by removing the effect of the channel width and conducted the experiment. The flow path shown in Figure A4a was designed in a way that the width of the flow path between the pneumatic source and each driving chamber was maximal. Experimental results are shown in Figure A4b. Although a concentration gradient was generated, the chambers located at both ends of the flow path exhibited a low concentration because of the pressure loss caused by wall friction.

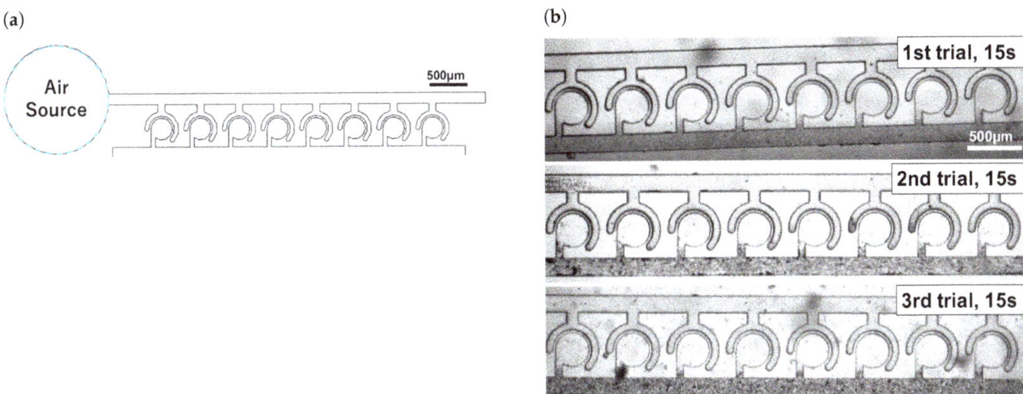

Figure A3. Use of pressure gradient. (**a**) Air was supplied from the left side of the figure, and the driving chamber was deformed by air pressure. The pressure gradient was generated by the pressure drop through the flow path, which generated the concentration gradient. (**b**) Experimental results of (**a**). Each chamber was driven for 15 s, but no stirring occurred.

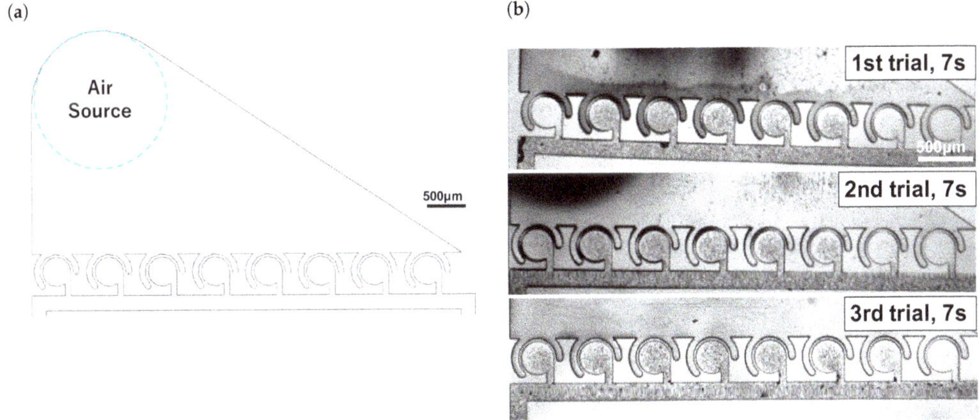

Figure A4. Flow path without the effect of channel width. (**a**) Concentration gradient was generated by the pressure gradient after removing the effect of the channel width. (**b**) Experimental results. Concentration gradient was generated, but concentrations in the chambers at both ends were low.

The aforementioned results imply that it is difficult to supply pressure to the driving chamber as expected when the channel width of the driving chamber is narrow. Therefore, when air is used for pressure transmission and the frequency of expansion/contraction is high, it is better to increase the channel width of the driving chamber as much as possible.

Appendix A.4. Influence of Deformation of Wall

Another possible reason for the diluted concentration in the chambers at both ends is the difference in the deformation of the driving chamber. Figure A5 shows how the wall is deformed when the driving chamber is driven using a high-speed camera. There is a difference in the deformation of the driving chamber depending on the positional relationship between the air pressure source and driving chamber. It is not clear whether this affects the concentration or not, but the parameters related to the concentration can be reduced by removing this effect.

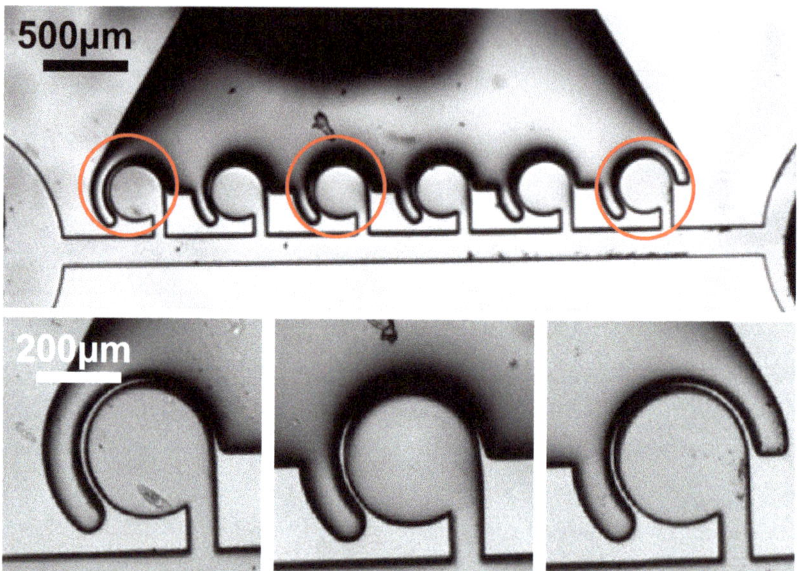

Figure A5. Wall deformation in relation to the position of the pneumatic source. Deformation differeed depending on the position of the chamber.

Appendix B. Raw Data of the Experiments

Figure A7 shows raw data of Figure 7. In Figure A7, the graph exhibits rattling behavior owning to the recording frame rate. Figure A6 shows the n-th, N + 1st, and N + 2nd frames in the experiments. The main chamber repeatedly expanded and contracted, which caused the liquid to be gradually absorbed as it moved back and forth between the neck chambers, resulting in rattling luminance values during the driving chamber operation.

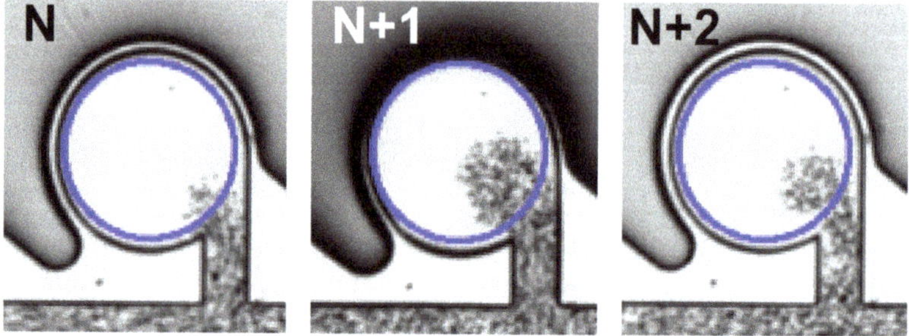

Figure A6. N-th to N + 2nd frame of an experiments. Main chamber repeatedly expanded and contracted, which caused rattling luminance values.

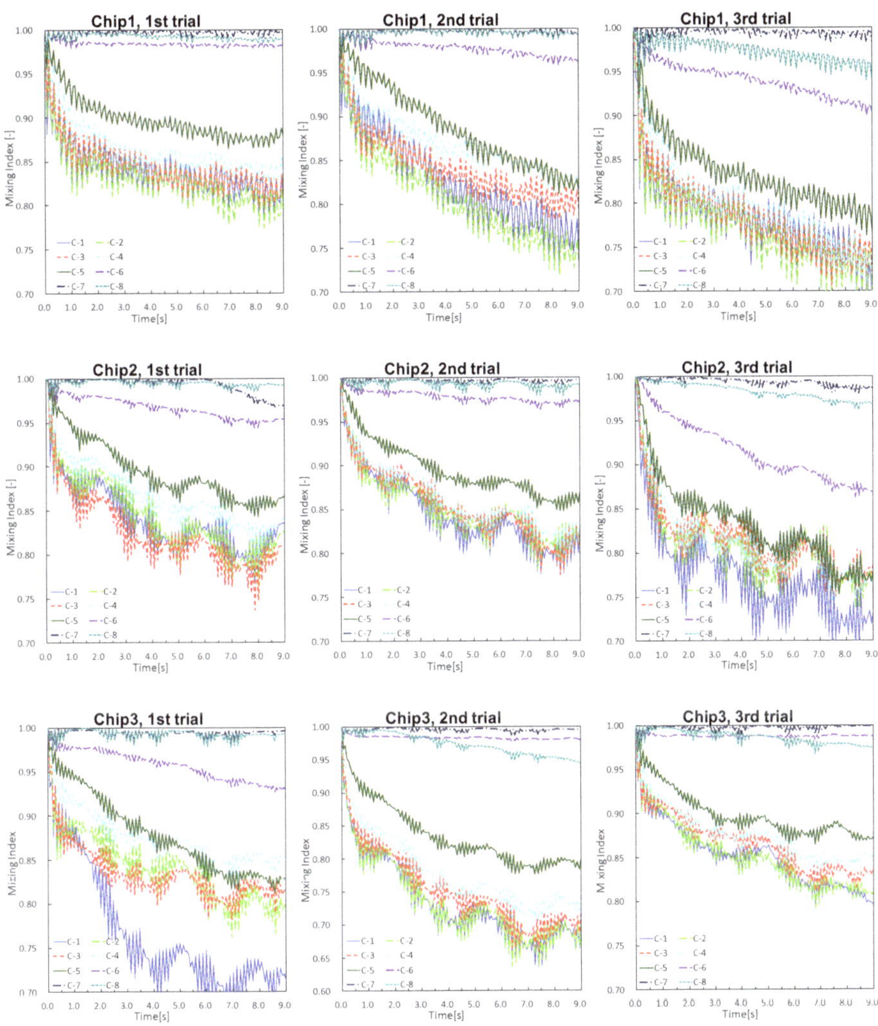

Figure A7. Time variation in mixing index until 9 s after the application of air pressure.

References

1. Lee, C.Y.; Fu, L.M. Recent advances and applications of micromixers. *Sens. Actuators B Chem.* **2018**, *259*, 677–702. [CrossRef]
2. Kawai, K.; Arima, K.; Morita, M.; Shoji, S. Microfluidic valve array control system integrating a fluid demultiplexer circuit. *J. Micromech. Microeng.* **2015**, *25*, 065016. [CrossRef]
3. Hardt, S.; Drese, K.; Hessel, V.; Schonfeld, F. Passive micro mixers for applications in the micro reactor and tas. In Proceedings of the Field the 2nd International Conference on Microchannels and Minichannels, New York, NY, USA, 17–19 June 2004.
4. Chen, G.; Zhu, X.; Liao, Q.; Chen, R.; Ye, D.; Liu, M.; Wang, K. A novel structured foam microreactor with controllable gas and liquid flow path: Hydrodynamics and nitrobenzene conversion. *Chem. Eng. Sci.* **2021**, *229*, 116004. [CrossRef]
5. Wu, Z.; Willing, B.; Bjerketorp, J.; Jansson, J.K.; Hjort, K. Soft inertial microfluidics for high throughput separation of bacteria from human blood cells. *Lab Chip* **2009**, *9*, 1193–1199. [CrossRef] [PubMed]
6. Sassa, F.; Fukuda, J.; Suzuki, H. Microprocessing of Liquid Plugs for Bio/chemical Analyses. *Anal. Chem.* **2008**, *80*, 6206–6213. [CrossRef] [PubMed]
7. Lee, W.S.; Jambovane, S.; Kim, D.; Hong, J.W. Predictive model on micro droplet generation through mechanical cutting. *Microfluid Nanofluid* **2009**, *7*, 431–438. [CrossRef]

8. Alam, A.; Afzal, A.; Kim, K.-Y. Mixing performance of a planar micromixer with circular obstruct ions in a curved microchannel. *Chem. Eng. Res. Des.* **2014**, *92*, 423–434. [CrossRef]
9. Huang, P.; Xie, Y.; Ahmed, D.; Rufo, J.; Nama, N.; Chen, Y.; Chan, C.Y.; Huang, T.J. An acoustofluidic micromixer based on oscillating sidewall sharp-edges. *Lab Chip* **2013**, *13*, 3847–3852. [CrossRef]
10. Raza, W.; Hossain, S.; Kim, K.W. Review of Passive Micromixers with a Comparative Analysis. *Micromachines* **2020**, *11*, 455. [CrossRef]
11. Bayareh, M.; Ashani, M.N.; Usefian, A. Active and passive micromixers: A comprehensive review. *Chem. Eng. Process.* **2020**, *147*, 107771. [CrossRef]
12. Hong, H.; Choi, I.; Ahn, C.H. A novel in-plane passive microfluidic mixer with modified Tesla structures. *Lab Chip* **2004**, *4*, 109–113. [CrossRef]
13. Park, S.; Kim, J.K.; Park, J.; Chung, S.; Chung, C.; Chang, J.K. Rapid three-dimensional passive rotation micromixer using the breakup process. *J. Micromech. Microeng.* **2003**, *14*, 6–14. [CrossRef]
14. Zhou, T.; Wang, H.; Shi, L.; Liu, Z.; Joo, S.W. An enhanced electroosmotic micromixer with an efficient asymmetric lateral structure. *Micromachines* **2016**, *7*, 218. [CrossRef] [PubMed]
15. Owen, D.; Ballard, M.; Alexeev, A.; Hesketh, P.J. Rapid microfluidic mixing via rotating magnetic microbeads. *Sens. Actuators A Phys.* **2016**, *251*, 84–91. [CrossRef]
16. Abbas, Y.; Miwa, J.; Zengerle, R.; von Stetten, F. Active continuous-flow micromixer using an external braille pin actuator array. *Micromachines* **2013**, *4*, 80–89. [CrossRef]
17. Ahmed, D.; Mao, X.; Juluri, B.K.; Huang, T.J. A fast microfluidic mixer based on acoustically driven sidewall-trapped microbubbles. *Microfluid Nanofluid* **2009**, *7*, 727–731. [CrossRef]
18. Glasgow, I.; Lieber, S.; Aubry, N. Parameters Influencing Pulsed Flow Mixing in Microchannels. *Anal. Chem.* **2004**, *76*, 4825–4832. [CrossRef] [PubMed]
19. Hong, B.; Xue, P.; Wu, Y.; Bao, J.; Chuah, Y.J.; Kang, Y. A concentration gradient generator on a paper-based microfluidic chip coupled with cell culture microarray for high-throughput drug screening. *Biomed. Microdevices* **2016**, *18*, 21. [CrossRef]
20. Irimia, D. Dan A Geba and Mehmet Toner, Universal Microfluidic Gradient Generator. *Anal. Chem.* **2006**, *78*, 3472–3477. [CrossRef] [PubMed]
21. Shourabi, A.Y.; Kashaninejad, N.; Saidy, M.S. An integrated microfluidic concentration gradient generator for mechanical stimulation and drug delivery. *J. Sci. Adv. Mater. Devices* **2021**, *6*, 280–290. [CrossRef]
22. Zhou, B.; Gao, Y.; Tian, J.; Tong, R.; Wu, J.; Wen, W. Preparation of orthogonal physicochemical gradients on PDMS surface using microfluidic concentration gradient generator. *Appl. Surf. Sci.* **2019**, *471*, 213–221. [CrossRef]
23. Alam, A.; Kim, K.W. Mixing performance of a planar micromixer with circular chambers and crossing constriction channels. *Sens. Actuators B Chem.* **2013**, *176*, 639–652. [CrossRef]
24. Sundaram, N.; Tafti, D.K. Evaluation of Microchamber Geometries and Surface Conditions for Electrokinetic Driven Mixing. *Anal. Chem.* **2004**, *76*, 3785–3793. [CrossRef]
25. Yang, C.; Xu, Z.; Lee, A.P.; Wang, J. A microfluidic concentration-gradient droplet array generator for the production of multi-color nanoparticles. *Lab Chip* **2013**, *13*, 2815–2820. [CrossRef] [PubMed]
26. Guo, Y.; Gao, Z.; Liu, Y.; Li, S.; Zhu, J.; Chen, P.; Liu, B. Multichannel synchronous hydrodynamic gating coupling with concentration gradient generator for high-throughput probing dynamic signaling of single cells. *Anal. Chem.* **2020**, *92*, 12062–12070. [CrossRef] [PubMed]
27. Tang, M.; Huang, X.; Chu, Q.; Ning, X.; Wang, Y.; Kong, S.; Zhang, X.; Wang, G.; Ho, H. A linear concentration gradient generator based on multilayered centrifugal microfluidics and its application in antimicrobial susceptibility testing. *Lab Chip* **2018**, *18*, 1452–1460. [CrossRef]
28. Lim, W.; Park, S. A Microfluidic Spheroid Culture Device with a Concentration Gradient Generator for High-Throughput Screening of Drug Efficacy. *Molecules* **2018**, *23*, 3355. [CrossRef]
29. Okuyama, T.; Yamazoe, H.; Seto, Y.; Suzuki, H.; Fukuda, J. Cell Micropatterning inside a microchannel and assays under a stable concentration gradient. *J. Biosci. Bioeng.* **2010**, *110*, 230–237. [CrossRef]
30. Hung, P.J.; Lee, P.J.; Sabounchi, P.; Lin, R.; Lee, L.P. Continuous Perfusion Microfluidic Cell Culture Array for High-Throughput Cell-Based Assays. *Biotechnol. Bioeng.* **2005**, *89*, 1–8. [CrossRef]
31. Dai, B.; Long, Y.; Wu, J.; Huang, S.; Zhao, Y.; Zheng, L.; Tao, C.; Guo, S.; Lin, F.; Fu, Y.; et al. Generation of flow and droplets with an ultra-long-range linear concentration gradient. *Lab Chip* **2021**, *21*, 4390–4400. [CrossRef]
32. Takayama, T.; Horade, M.; Tsai, C.D.; Kaneko, M. Push/Pull Inequality Based On-Chip Density Mixer with Active Enhancer. In Proceedings of the 22nd International Conference on Miniaturized Systems for Chemistry and Life Sciences, Kaohsiung, Taiwan, 11–15 November 2018.
33. Takayama, T.; Kaneko, M.; Tsai, C.D. On-Chip Micro Mixer Driven by Elastic Wall with Virtual Actuator. *Micromachines* **2021**, *12*, 217. [CrossRef] [PubMed]
34. Takayama, T.; Hosokawa, N.; Tsai, C.; Kaneko, M. Push/Pull Inequality Based High-Speed On-Chip Mixer Enhanced by Wettability. *Micromachines* **2020**, *11*, 950. [CrossRef]

35. Takayama, T.; Miyashiro, H.; Tsai, C.D.; Ito, H.; Kaneko, M. On-chip density mixer enhanced by air chamber. *Biomicrofluidics* **2018**, *12*, 044108. [CrossRef] [PubMed]
36. Tsai, C.D.; Takayama, T.; Shimozyo, Y.; Akai, T.; Kaneko, M. Virtual vortex gear: Unique flow patterns driven by microfluidic inertia leading to pinpoint injection. *Biomicrofluidics* **2018**, *12*, 034114. [CrossRef] [PubMed]

 micromachines

Article

A Numerical Investigation of the Mixing Performance in a Y-Junction Microchannel Induced by Acoustic Streaming

Sintayehu Assefa Endaylalu and Wei-Hsin Tien *

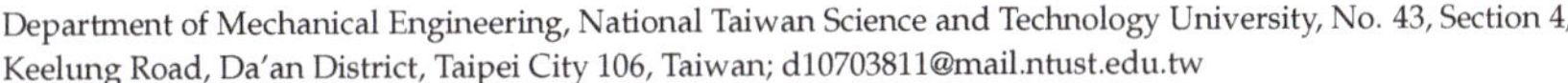

Department of Mechanical Engineering, National Taiwan Science and Technology University, No. 43, Section 4, Keelung Road, Da'an District, Taipei City 106, Taiwan; d10703811@mail.ntust.edu.tw
* Correspondence: whtien@mail.ntust.edu.tw; Tel.: +886-2-27376443

Abstract: In this study, the mixing performance in a Y-junction microchannel with acoustic streaming was investigated through numerical simulation. The acoustic streaming is created by inducing triangular structures at the junction and sidewalls regions. The numerical model utilizes Navier–Stokes equations in conjunction with the convection–diffusion equations. The parameters investigated were inlet velocities ranging from 4.46 to 55.6 μm/s, triangular structure's vertex angles ranging from 22° to 90° oscillation amplitude ranging from 3 to 6 μm, and an oscillation frequency set to 13 kHz. The results show that at the junction region, a pair of counter-rotating streaming vortices were formed, and unsymmetrical or one-sided vortices were formed when additional triangles were added along the sidewalls. These streaming flows significantly increase the vorticity compared with the case without the acoustic stream. Mixing performances were found to have improved with the generation of the acoustic stream. The mixing performance was evaluated at various inlet velocities, the vertex angles of the triangular structure, and oscillation amplitudes. The numerical results show that adding the triangular structure at the junction region considerably improved the mixing efficiency due to the generation of acoustic streaming, and further improvements can be achieved at lower inlet velocity, sharper vertex angle, and higher oscillation amplitude. Integrating with more triangular structures at the sidewall regions also improves the mixing performance within the laminar flow regime in the Y-microchannel. At Y = 2.30 mm, oscillation amplitude of 6 μm, and flow inlet velocity of 55.6 μm/s, with all three triangles integrated and the triangles' vertex angles fixed to 30°, the mixing index can achieve the best results of 0.9981, which is better than 0.8355 in the case of using only the triangle at the junction, and 0.6642 in the case without acoustic streaming. This is equal to an improvement of 50.27% in the case of using both the junction and the two sidewall triangles, and 25.79% in the case of simply using a junction triangle.

Keywords: acoustic streaming; micromixer; acoustofluidics; microfluidics; computational fluid dynamics

Citation: Endaylalu, S.A.; Tien, W.-H. A Numerical Investigation of the Mixing Performance in a Y-Junction Microchannel Induced by Acoustic Streaming. *Micromachines* **2022**, *13*, 338. https://doi.org/10.3390/mi13020338

Academic Editor: Kwang-Yong Kim

Received: 27 January 2022
Accepted: 17 February 2022
Published: 21 February 2022

1. Introduction

Fluid mixing is a phenomenon process in microfluidics that involves the combination of various materials. Homogeneous and rapid mixing in microscales are critical for various applications such as drug delivery, chemical process industries, biomedical diagnostics, etc. The main mixing mechanism is governed by molecular diffusion. Micromixers are thus integrated into microfluidic devices to achieve quick mixing by facilitating mass transfer between the flow streams. They can be categorized as either active or passive depending on the amount of energy consumed [1]. Passive micromixers are easy to set up and integrate with other microfluidic components. However, this type of micromixer has less performance compared with active micromixers. The different geometry of the microchannel section is modified in passive micromixers to induce flow disturbance as well as pressure difference, which then promotes mixing quality. The most common geometrical features used as passive micromixers are swirl inducing [2], obstructions [3], the placement of obstacles [4], etc. Active micromixers perform mixing to enhance quality and speed by utilizing external power sources such as acoustic field [5], pressure-driven [6], electric field [7], and magnetic-field [8]. Active micromixers can be activated based on the user's needs and provide controllable mixing with electrical voltage, pressure gradient, and integrated elements. In general, active micromixers improve mixing by magnetically, mechanically, electrically, and acoustically swirling the fluid streams.

Acoustofluidics is a combination of acoustics and microfluidics. It has many application areas such as microfluidic pump [9], bimodal signal amplification [10], mixing and migration of micro-size particles [11], etc. Acoustic streaming occurs as a result of the time-averaged nonlinear dynamics of the solid–fluid interaction in a viscous flow [12–14]. It is induced by the second order time-averaged streaming fluctuating components in any fluid flow. Ovchinnikov et al., studied numerically the flow pattern and scaling around a single sharp edge and found that it generated a high Reynold body force when compared with its non-sharp counterparts [15]. Similarly, Doinikov et al., recently reported that sharp edge structures were used to generate acoustic streaming in a circular microfluidic device of the fluid domain [16]. Both studies indicated that the sharp edge of the solid body is used as the origin of acoustic streaming to oscillate the flow. The acoustic streaming production is strong enough near the apex with a small vertex angle and is also directly related to the applied amplitude [17]. Its generation is also affected by the fluid's Reynold number flow characteristics. Generally, the induced streaming flow pattern is stronger at low Reynold number flows than at high Reynold number flows [18]. The acoustic stream is produced more efficiently when the radius of the curvature of the apex is small in comparison with the thickness of the viscous boundary layer [19]. Acoustic streaming has been recognized as an important and non-invasive solution for different applications such as mixing [20,21], heat transfer [22], synthesis of organic nanoparticles [23], and particle patterning [24]. Nama et al., numerically investigated the performance of micromixing by inducing acoustic oscillated sharp edges in a sidewalls rectangular computational domain [25]. In general, active and passive mixing in Y-microchannels has previously been performed such as acoustically induced bubbles [26,27], rotating magnetic fields [28], placement of obstacles in the microchannels [29,30], electronically driven flow [31], acoustic streaming around the sharp triangular structure in the sidewall directions [20], adding split and recombination features [32], side lateral obstructions [33], and symmetrical cylindrical grooves [34].

The Y-microchannel is a micromixer configuration used in microfluidics devices. Different geometries such as a circle, square, rectangle, various inlet and outlet section regions, and channel angles are developed, but its configuration is mainly defined by two inlets and one outlet, with the single outlet acting as a micromixer at various angles [35]. The mixing of two or more fluids is a critical process inside the Y-microchannel. Since the fluid flow inside the small dimension of the Y-microchannel is mostly laminar, it is dependent on its outlet effective length and diffusion disturbance to achieve the maximum optimum species mixture concentration at the common microchannel outlet region. Thus, fluid flow pattern disturbance is required before passing through the junction region to minimize the effective outlet length and also reduce the microchannel manufacturing costs.

The main objective of this study is to investigate numerically the mixing performance of acoustic streaming in a Y-microchannel by inducing a triangular structure starting from the microchannel junction region. It can achieve quick mixing and improve homogeneity within the microchannel mixing section by disrupting the flow pattern starting from the junction region and generating a vortex streaming flow. The paper is organized as follows: methods are presented in Section 2, results and discussion are presented in Section 3, and Section 4 concludes with the important findings and future work.

2. Numerical Methods

2.1. Microchannel Geometry and Numerical Scheme

Figure 1 demonstrates the CAD design of the microchannel used in the current study. The Y-junction microchannel segments are separated by a 120° angle. Figure 1a illustrates the dimensions and configuration of the common standard Y-junction microchannel. The new microchannel configuration includes fillet dimensions to reduce the microchannel edge effect and enlarge the junction region. The triangle's apex inside the microchannel is rounded with 0.30 μm, which is below the viscous boundary layer thickness. Similarly, the triangle has a characteristic height, h, around the junctions of 0.35 mm, 0.25 mm, and 0.17 mm at the microchannel outlet sidewalls, which is equal to the distance between the triangle's tip and the beginning of its base. Figure 1b,c show the new microchannel designs for numerical modeling. It was designed with an induced single triangular structure only at the junction region and the addition of two more triangles to the microchannel sidewalls at the outlet direction. As shown in Figure 1c, the first sidewall triangle induced 0.40 mm away from the apex of the junction triangle, while the second triangle induced in the other side at a distance of S away from the first sidewall triangle's apex. The triangle's apex is located between the two edges of the outlet microchannel, as shown in Figure 1b, but below 0.10 mm in the case of the 0.25 mm triangle height, h, as shown in Figure 1c. The base edges of the triangular structure inside the microchannel are rounded with 0.10 mm and 0.02 mm, as shown in Figure 1b,c.

The Y-junction microchannel numerical model is in the coordinate rectangular region ranging in x-axis (0.4398, 0.2598) to (2.4255, 0.2598) and in y-axis (0.5898 and 2.2755, 0) to (1.1326 and 1.7326, 2.3484) as shown in Figure 2. The origin of the coordinate system is set to be at the lower-left corner of Figure 2. Throughout this study, the origin is fixed at this point and the coordinate system is unified in order to express all the data and results. The Y-shaped microchannel has two inlets and one outlet microchannel for mixing purpose. The lower and inlet side microchannel edges of the 2D modeling are 30° and 60° from the x-axis, respectively. Similarly, the upper and inlet side microchannel edges are 60° and 30° from the y-axis, respectively.

The numerical model scheme is used to solve the two-dimensional water domain of the Y-microchannel cross section, which has two inlets and one outlet segment. The material properties used in the numerical study are listed in Table 1. Figure 2 shows the geometry of the numerical model discretization of the new Y-junction microchannel. The numerical model was implemented and the governing equation solved using the finite element software COMSOL Multiphysics [36]. Three sets of governing equations are used to obtain the results in this study. Based on the temporal and spatial scales, the acoustic velocity field is first calculated by using the thermoviscous acoustics module in frequency domain. The streaming flow velocity field is then calculated by applying the laminar flow physics module. Lastly, the transport of diluted species module is used to solve the convective transport of a solute species from one inlet side to the outlet's common mixing region in the Y-junction microchannel. Table 2 shows the numerical study input parameters.

Table 1. Material properties.

Water Properties at T = 25 °C [37]	Value	Units
Viscous dynamics viscosity, μ	890	μPas
Specific heat capacity, C_p	4180	J/kg·K
Density, ρ_o	997	kg/m^3
Speed of sound, c_o	1497	m/s
Compressibility, $k_o = 1/(\rho_o c_o{}^2)$	4.47×10^{-10}	1/Pa
Specific heat capacity ratio, γ	1.012	
Thermal conductivity, k_{th}	0.61	W/m·K
Thermal expansion coefficient, $\alpha = \sqrt{C_P(\gamma - 1)/(Tc_o{}^2)}$	2.74×10^{-4}	1/K
Thermal diffusivity, $D_{th} = k_{th}/(\rho_o c_p)$	1.464×10^{-7}	m^2/s
Bulk dynamic viscosity, μ_b	2.47	mPas

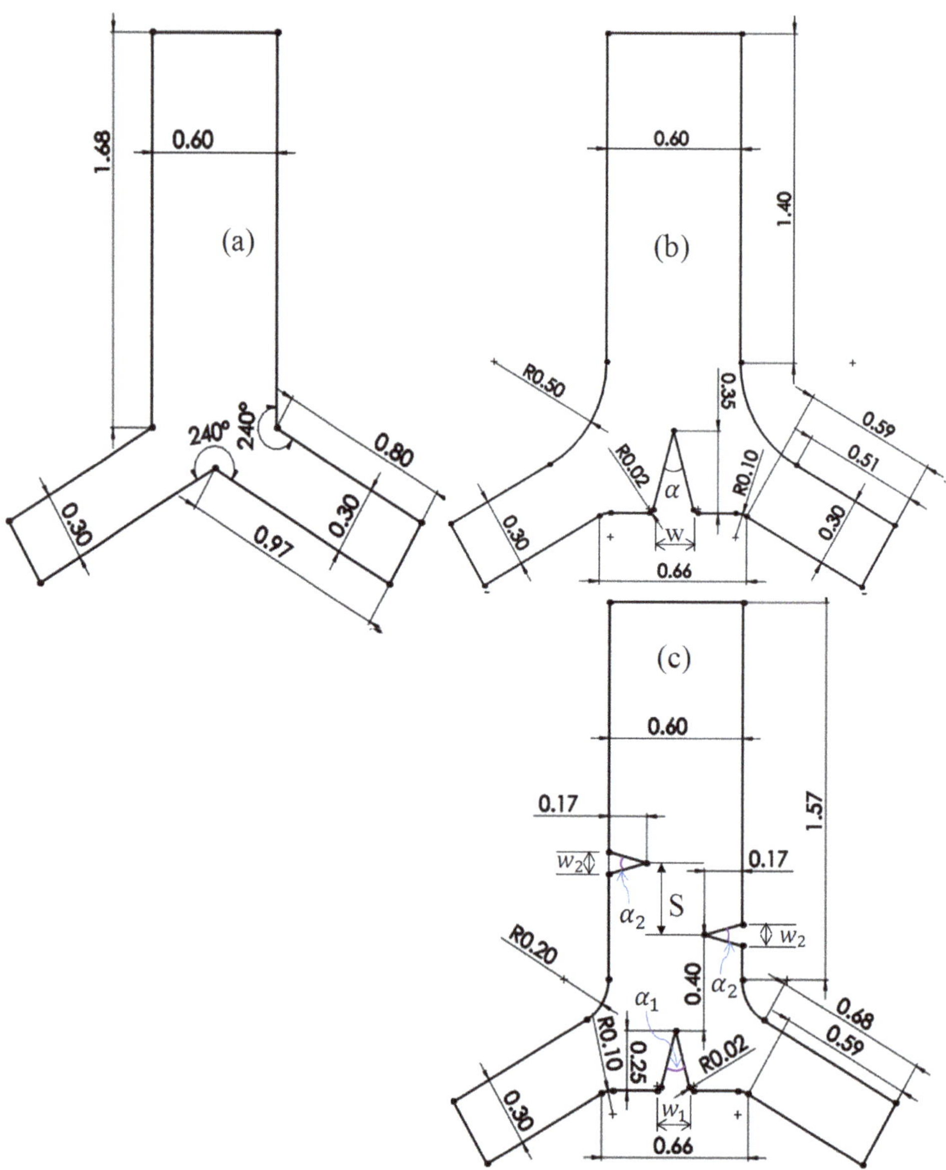

Figure 1. Designs of Y-microchannel configurations of the present study: (**a**) normal Y-channel without triangle structures, (**b**) triangular structure for acoustic streaming at the junction only, and (**c**) triangular structures in the junction and the channel sidewalls. All dimensions are marked in mm. In (**b**), where α is the triangle's vertex angle, w is the width of the triangle; in (**c**), where S is the sidewall triangle's spacing, α_1 and α_2 are the vertex angles of the triangles at the junction and the sidewalls, respectively. w_1 and w_2 are the widths of the junction and sidewall triangles, respectively.

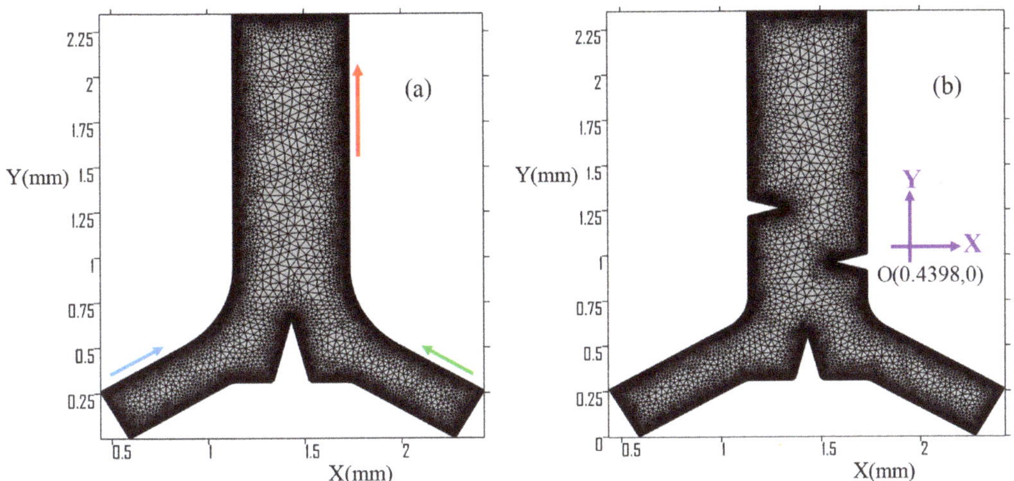

Figure 2. Geometrical mesh; (**a**) with only one junction triangle; (**b**) with three triangles.

Table 2. Study input parameters.

Parameters	Values	Units
Inlets velocities	4.46, 8.89, 30, 55.6	μm/s
Oscillation frequency	13	kHz
Oscillation amplitude	3–6	μm
Vertex angles	$\alpha = 22–60°$, $\alpha_1 = 30°$, $\alpha_2 = 22–90°$	
Spacing gap S of the side walls triangles	0.15, 0.30, 0.40, 0.60, 0.90	mm
Diffusion coefficient (fluorescein sodium salt)	10^{-9}	m^2/s

2.2. Governing Equations and Boundary Conditions

Acoustic streaming fundamental governing equations have been presented in various studies [38,39]. The perturbation theory is used to solve microchannel flow problems in the context of "weak disturbance". It is an effective tool for reducing the Navier–Stokes equation, which includes the nonlinear terms that couple the acoustic and streaming velocity fields. The general governing equations are continuity, momentum, and energy equations. The numerical model was performed in three steps: (i) solving the wave equation to compute the acoustic velocity fields, (ii) computing the streaming flow, and (iii) solving the flux of the concentration profile in the microchannel.

In the thermoviscous acoustics module, governing equations are obtained from the linearized Navier–Stokes equations, which are used to solve the continuity and momentum in the microchannel system. All governing equations fields and sources are assumed to be harmonic with $e^{i\omega t}$. Therefore, the acoustic velocity field was determined in the following ways:

$$i\omega\rho_a + \nabla.(\rho_o \mathrm{v_a}) = 0 \tag{1}$$

$$i\omega\rho_o \mathrm{v_a} = \nabla.\left[-PI + \mu\left(\nabla \mathrm{v_a} + (\nabla \mathrm{v_a})^T\right) - \left(\frac{2}{3}\mu - \mu_B\right)(\nabla.\mathrm{v_a})\mathrm{I}\right] \tag{2}$$

where ω is the frequency of actuation, v_a is the acoustic velocity field, and ρ_a is the density at temperature T_a:

$$\rho_a = \rho_o(\kappa_T p_a - \alpha T_a) \tag{3}$$

$$\kappa_T = \frac{1}{\rho}\left(\frac{\partial\rho}{\partial p}\right)_T = \frac{1}{\rho_o {c_o}^2} \tag{4}$$

c_o is the speed of sound in fluid, κ_T is the isothermal compressibility coefficient, and α is the coefficient of thermal expansion.

$$\alpha = \frac{1}{\rho}\left(\frac{\partial \rho}{\partial T}\right)_P = \frac{1}{c}\sqrt{\frac{C_p(\gamma-1)}{T}} \tag{5}$$

μ and μ_b are the viscous and bulk dynamic viscosities, respectively. I is the identity matrix. The walls of the microchannel are solid surfaces with no-slip and isothermal boundary conditions. The thermoviscous acoustics length scale is defined by viscous and thermal boundary layer thicknesses, but due to the low oscillation frequency in kHz level, it is sufficient to describe it with only the viscous boundary layer thickness. Therefore, the viscous penetration depth (viscous boundary layer thickness) is given by:

$$\delta = \delta_v = \sqrt{\frac{2\mu}{\omega\rho}} = \sqrt{\frac{\mu}{\pi f \rho}} = \sqrt{\frac{\nu}{\pi f}} \tag{6}$$

According to Equation (6), the thickness of the viscous boundary layer decreases as the frequency, f, of the oscillation increases, but it can also increase as the frequency, f, decreases.

The ratio of inertia to viscous forces, i.e., the Reynold number $(R_e = (\rho v D_h)/\mu)$, is small, indicating that viscous forces dominate and damp out all disturbance in the microchannel. Similarly, acoustic streaming at the sharp edge has a dominant body force over other driving forces in the microchannel. As a result, the fluid flow inside the microchannel is in a single phase and operates in a laminar flow regime. Likewise, the acoustic wavelength ($\lambda = c/f$) is much larger than the width dimension of the microchannel, where c is the speed of sound in water. The dimensionless number of a fluid flow Mach number ($\mathrm{Ma} = v_a/c \ll 1$) and the temperature variation in the microchannel are very small, and the density is also nearly constant. As a result, the laminar fluid motion is assumed to be incompressible, and the governing equations of the continuity and Navier–Stokes momentum equations are as follows in Equations (7) and (8), respectively:

$$\rho \nabla . v = 0 \tag{7}$$

$$\frac{\partial v}{\partial t} + (v.\nabla)v = \frac{1}{\rho}\left(\nabla.(-pI + \mu\left(\nabla v + \left(\nabla v\right)^T\right)\right) + F\right) \tag{8}$$

where $F = \langle \rho_a \frac{\partial v}{\partial t} \rangle + \rho_o \langle (v.\nabla)v \rangle$ is the volume force ($\mathrm{N/m^3}$) and v is the streaming velocity field.

As stated previously, the microchannel has a solid wall and is stationary. The wall boundary condition is therefore no-slip and zero velocity is assumed. Similarly, the fluid flow at both inlets is considered as fully developed and uniform, so the inflow boundary condition is assigned by the average velocity. The pressure boundary condition is equal to zero at inlets and absolute pressure p_o at the outlet.

The transport of fluorescein sodium salt species in the microchannel via convection-diffusion is solved using Equation (9):

$$\frac{\partial c_i}{\partial t} + \nabla.J_i + v.\nabla c_i = R_i \tag{9}$$

where c_i is the concentration of the species ($\mathrm{mol/m^3}$), R_i is a reaction rate expression for the species ($\mathrm{mol/(m^3 \cdot s)}$), and v is the streaming fluid averaged velocity vector (m/s). J_i denotes the mass flux factors that define the diffusive flux vector and are solved using Equation (10):

$$J_i = -D_i \nabla c_i \tag{10}$$

where D_i is the diffusion coefficient of fluorescein sodium salt species ($\mathrm{m^2/s}$).

The initial boundary condition concentration c_o of the material species was set to 1 mol/m^3 in one inlet direction and 0 in the other inlet side.

The mixing index or degree of mixedness is investigated based on the statistical method. It is calculated at any cross section of the microchannel width from the standard deviation of the fluorescein sodium salt species concentration using Equation (11) [40]:

$$M = 1 - \sqrt{\frac{\sigma^2}{\sigma^2{}_{max}}} \tag{11}$$

$$\sigma = \sqrt{\frac{1}{n}\sum\nolimits_{i=1}^{n}(c_i - \bar{c})} \tag{12}$$

$$\sigma_{max} = \sqrt{\bar{c}(1-\bar{c})} \tag{13}$$

where σ is the standard deviation of the concentration species c_i in each point on the width of the cross section, $\bar{c}$ is the mean of the concentration c_i on the width cross section, and $\sigma_{\rm max}$ is the maximum standard deviation of the species concentration at the specified width cross section of the microchannel.

2.3. Mesh Independence Test

The mesh independence test is required in the simulation to find the optimal mesh. An optimal mesh is one in which the computational time is minimized while the accuracy of the result is maintained. The independence test was evaluated by considering the acoustic velocity fields and streaming velocity magnitude at various mesh element sizes. The computational mesh is created by combining a maximum element size length $\rm d_{mesh}$ at the domain boundaries and bulk of the fluid. The bulk water domain region mesh size $\rm d_{mesh,dk}$ is kept constant with 12δ, where δ is the thickness of viscous boundary layer thickness. Then, the mesh independent test was verified by different boundary element sizes, $\rm d_{mesh,db}$. Moreover, the smallest mesh element size $\rm d_{mesh,db} = 0.25\delta$ is taken as a reference computational mesh element size and evaluated the mesh independent test in terms of relative convergence parameter C(g) using Equation (14) as follows [41]:

$$\mathrm{C(g)} = \sqrt{\frac{\int \left(g - g_{(ref)}\right)^2 dxdy}{\int \left(g_{(ref)}\right)^2 dxdy}} \tag{14}$$

where $g(ref)$ is the reference value at the smallest mesh element size.

The evaluation was performed at the junction tip edge triangles up to 10 µm from the numerical model, as shown in Figure 2a. Figure 3 shows the relative convergence parameter C vs. $\delta/\rm d_{mesh,db}$, demonstrating that the value of the parameter C(g) is very small in all ranges of mesh element size. As a result, $\delta/\rm d_{mesh,db} = 1.33$ is chosen for all succeeding study cases. The velocity fields $\rm v_{1x}$ and $\rm v_{1y}$ are for acoustic velocity components, and $\rm v_2$ is the velocity magnitude for streaming flow.

When the size of the boundary mesh element is reduced to a minimum, the relative convergence parameter approaches zero as shown in Figure 3.

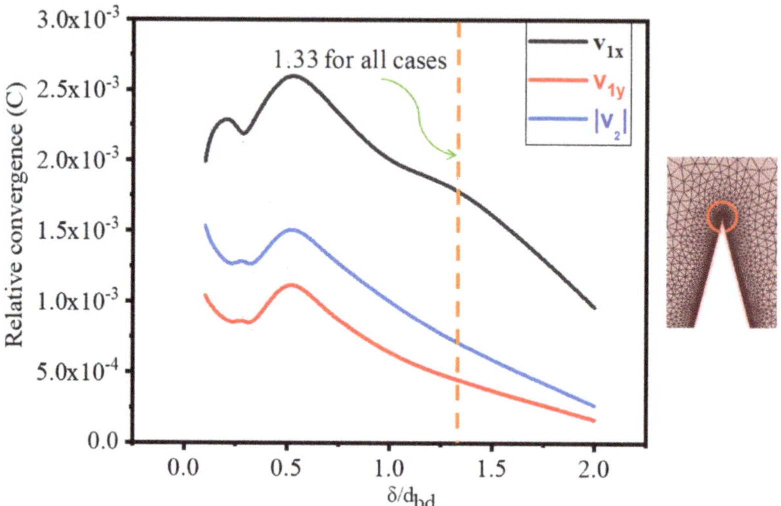

Figure 3. Mesh independence test.

3. Results and Discussion

3.1. Normal Y-Junction Microchannel

Figure 4 illustrates the flow vector plot, velocity magnitude, and species concentration distribution in the original Y-junction microchannel without the triangle structure and acoustic field. The fluid flow from the inlet side of the microchannel followed its path to the microchannel outlet as demonstrated in Figure 4a. Because of the layered stream formed between the two side fluid streams, concentrated species were difficult to diffuse and convect to the other sides of the microchannel outlet region as shown in Figure 4b.

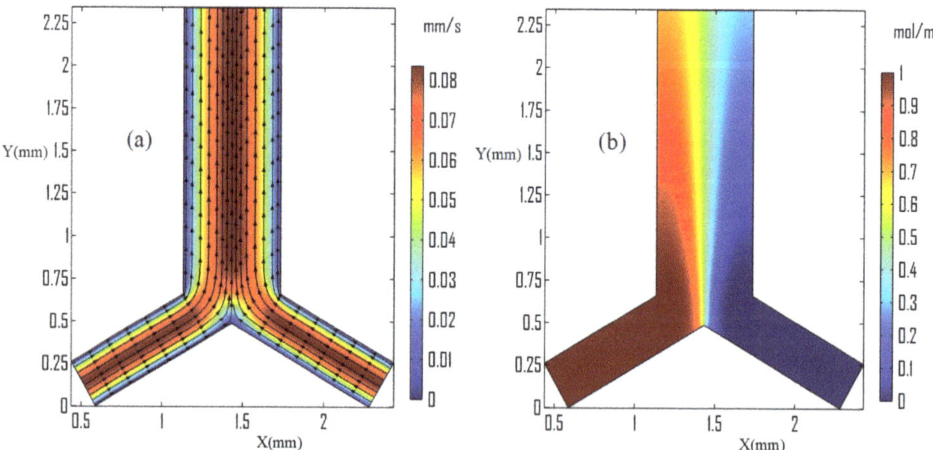

Figure 4. Visualization in standard Y-junction microchannel at 55.6 μm/s; (**a**) velocity vector and magnitude; and (**b**) species concentration distribution.

3.2. Acoustic Streaming Generation

The Y-junction microchannel has an induced triangle edged in the junction and also integrated with the sidewall triangle's edges, which are positioned in an oscillation fluid to create a vortex stream in both clockwise and anti-clockwise rotation. The generated acoustic field and streaming flow problems are addressed using perturbation theory. The viscous boundary thickness was much less than the height dimensions of the microchannel-induced triangle's sharp edge structure. The triangle's sharp edge is rounded to avoid a numerical solution discontinuity purpose. However, its radius curvature of the round is sufficiently sharper in comparison with the viscous boundary layer thickness dimension, i.e., $r \ll \delta$. This implied that the length scale was enough to create acoustic streaming flow.

Figure 5 shows the quiver plot and velocity magnitudes after inducing triangular structures in different positions. The streaming flow originated from the two inlet sides, whereas acoustic streaming flows began from the triangle's sharp edges. Acoustic stream line patterns followed a revolving circular structure and produced strong vorticity on both sides of the induced triangular structure. Figure 5a shows the streaming flow and revolving caused by inducing only the triangular structure around the junction region. Streaming velocity has a high magnitude of streaming velocity around the sharp tip edges, but produces two strong counter-rotating vortices on both sides of the triangle regions. The strength of the vortices expressed interims of vorticity. It was extracted and plotted on both the left and right sides at equal distances from the tip edge, then compared in different junction triangle vertex angles.

The side walls of the isosceles triangle vibrated highly in the study conducted in two-dimensional X- and Y-directions. The results indicated that two sides streaming layers were created starting from the tip sharp edge during the induction of one triangular structure at the junction region, as shown in Figure 5a. When the sidewalls triangles were induced, the layer's curve length size was minimized and rotated in a similar manner into two circular counter vortices, as shown in Figure 5b–f. First and foremost, the effect of the placement of the gap, S, between the two sidewall triangles was considered and investigated. When the spacing between the sidewall triangles was reduced to 0.15 mm, it produced a short curved revolving streaming from the first right sidewall triangle's apex to the second left sidewall triangle edge, as shown in Figure 5b, but when the space, S, was increased to 0.90 mm, the stream pattern became less curved and the second left sidewall triangle created two unsymmetrical streaming vortices, as shown in Figure 5f.

Figure 6b shows symmetrical vortices and high vorticity on both sides of the junction-induced triangle, as plotted by cutting plane lines X = 1.283 mm, X = 1.583 mm, and Y = 0.485 to 2.30 mm. There are clockwise and anti-clockwise vortices on both sides of the triangular structure. Therefore, there is a higher tip velocity maximum as well as vorticity at the very sharp-edged tip of 22° than 60°. It has a maximum velocity of around 5 mm/s on a 30° triangle tip sharp edge angle, as shown in Figure 5. Figure 5c also shows the vorticity generation at a specified geometry coordinate region of both sides of the junction triangle in different oscillation frequencies from 3 to 13 kHz. The 13 kHz oscillation frequency has better vorticity generation performance. Therefore, this frequency is selected for the input parameter to evaluate all numerical study work.

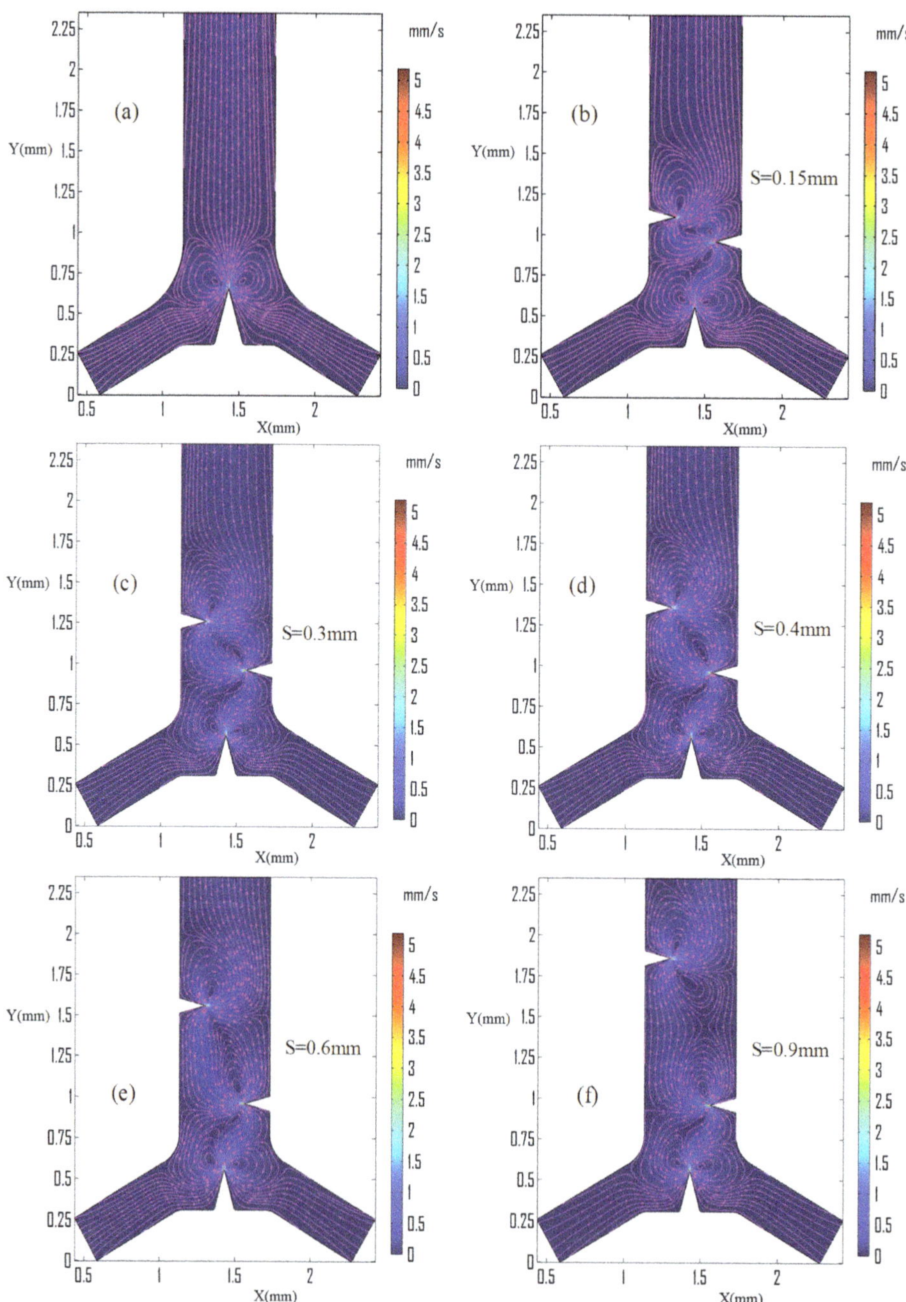

Figure 5. Acoustic streaming generation at 13 kHz, 6 µm amplitude, 4.46 µm/s inlet velocities on $\alpha = 30^{\circ}$, $\alpha_1 = 30^{\circ}$, $\alpha_2 = 30^{\circ}$; (**a**) only inducing the triangular structure on junction region; and (**b**–**f**) at different sidewalls triangles spacing.

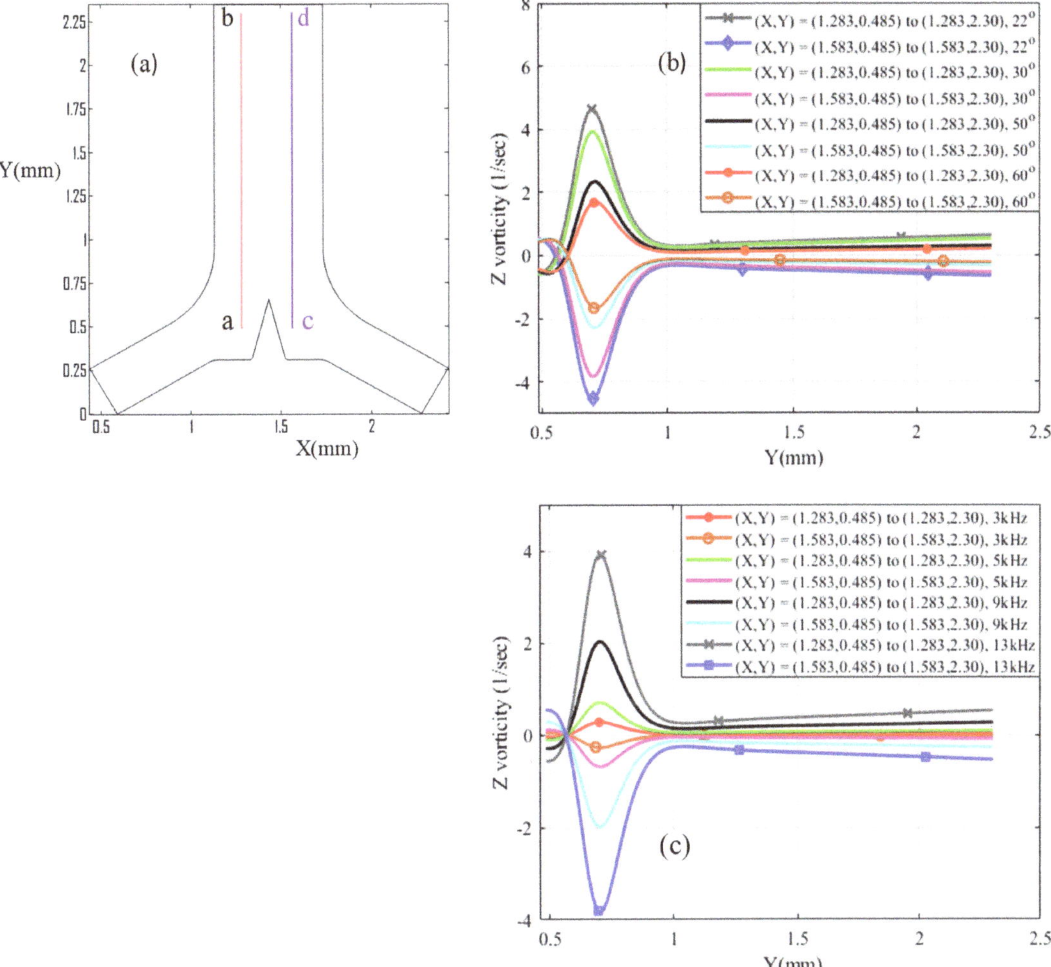

Figure 6. Vorticity comparison: (**a**) cutting plane line $\overline{ab}$ and $\overline{cd}$ position in (X,Y) coordinate system; (**b**) Z vorticity comparison using 13 kHz, 6 µm, and 4.46 µm/s in different α of junction triangle's vertex angles; and (**c**) Z vorticity using 6 µm and 4.46 µm/s on $\alpha = 30°$ junction triangle vertex angle in different oscillation frequency.

3.3. Mixing Performance

The acoustic streaming produced increased the disturbance of the concentrated species beginning at the junction of the Y-microchannel region as shown in Figure 7.

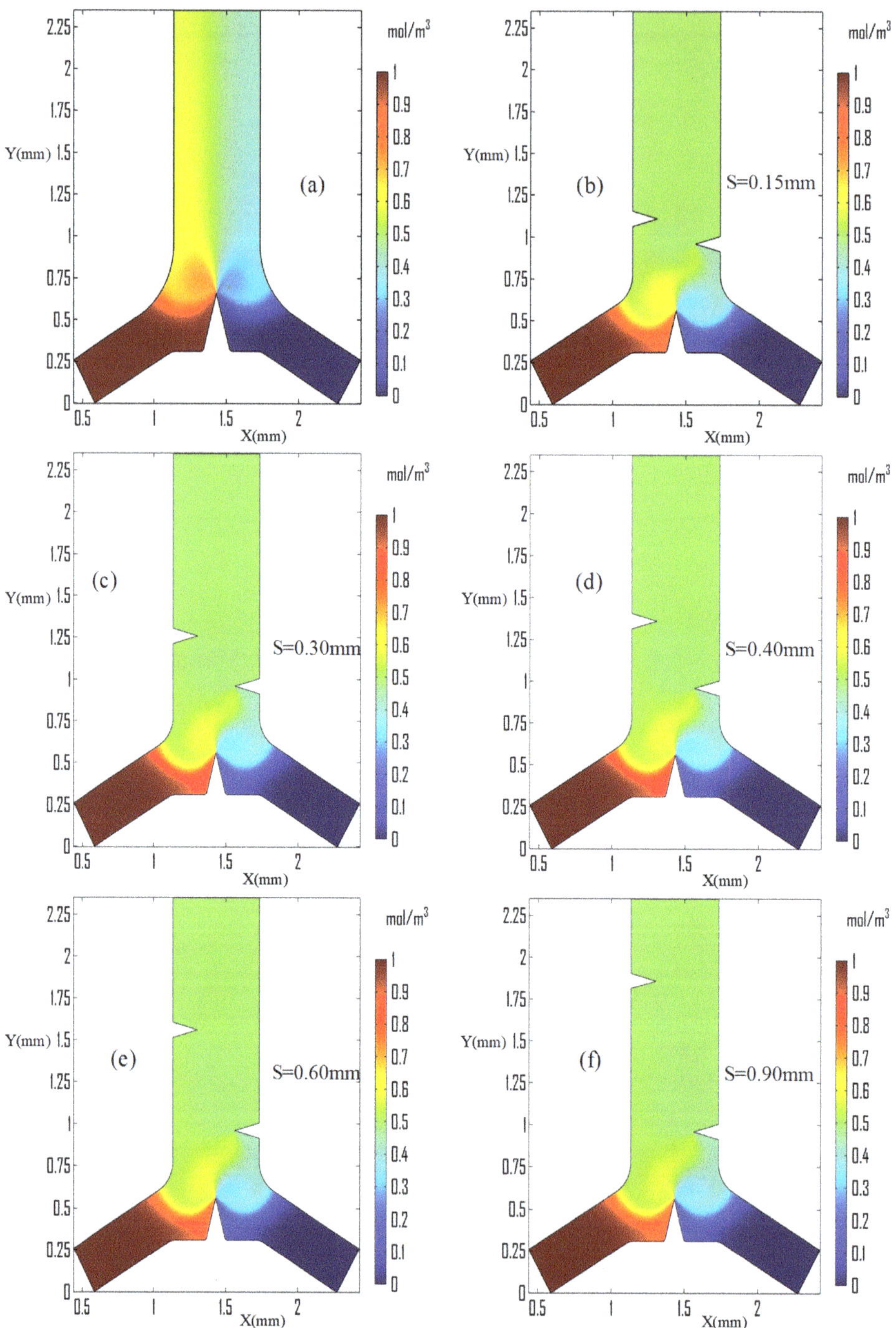

Figure 7. Mixing performance visualization at 13 kHz, 6 μm amplitude, 4.46 μm/s inlet velocity on $\alpha = 30°$, $\alpha_1 = 30°$, $\alpha_2 = 30°$; (**a**) only inducing the triangular structure on junction region; and (**b–f**) at different sidewalls triangles spacing.

The diffusional movement of concentrated species followed the streaming pattern and more disturbances originated from the junction triangle's apex. Due to the high streaming velocity around the sharp edge compared with other regions, it created a high-streaming curved layer. Therefore, with only a sharp edge at the Y-junction, mixing was incomplete and did not to fully achieve the desired objective. Another microchannel design and the number of integrated triangles was used to improve the mixing performance of the microchannel. Figure 7b–f shows the visualization of successive mixing performance after integrating two sidewall triangles and shortening the two strong layers created in the junction triangle's apex.

The effect of streaming velocity and vortices strength on the concentration species distribution in the microchannel outlet region was evaluated using only the junction region induced triangular structure, as shown in Figure 7a. The mixing performance was evaluated using the concentration flux in comparison with the expected optimum concentration of the species across the width of the outlet microchannel and also the dimensionless parameter mixing index along the flow direction in the measurement region as shown in Figure 8. The width of the microchannel outlet ranges of from X = 1.1326 to 1.7326 mm along the horizontal axis.

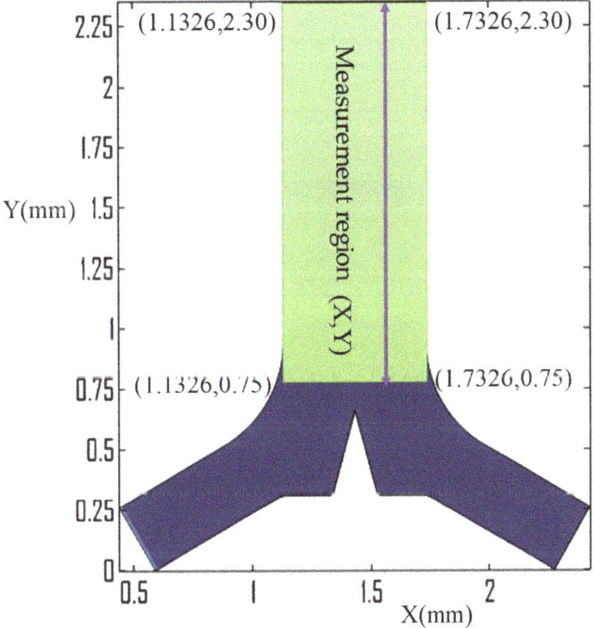

Figure 8. Mixing evaluation region in the Y-microchannel.

The performance of acoustic streaming on mixing is affected by variables such as inlet velocity, oscillation amplitude, and the vertex angles. Considering the evaluation mechanism of mixing effectiveness is a concentration profile in the microchannel perpendicular to the flow direction in comparison with the expected optimum species concentration. At the end of the mixing process, the expected optimum concentration species is 0.50 mol/m^3 inside the outlet microchannel. Figure 9 illustrates the species concentration profile at Y = 2.30 mm perpendicular to the flow direction with various parameters. As shown in Figure 9b, the concentration profile resulted in less than the expected optimum concentration in all cases due to the low inlet velocity as well as the junction and sidewalls with the same vertex angle. However, it was improved at high inlet velocity, as shown in Figure 9c.

Similarly, it improved sufficiently at low inlet velocity when the junction triangle's vertex angle was more than the sidewall triangle tip angles, as displayed in Figure 9d.

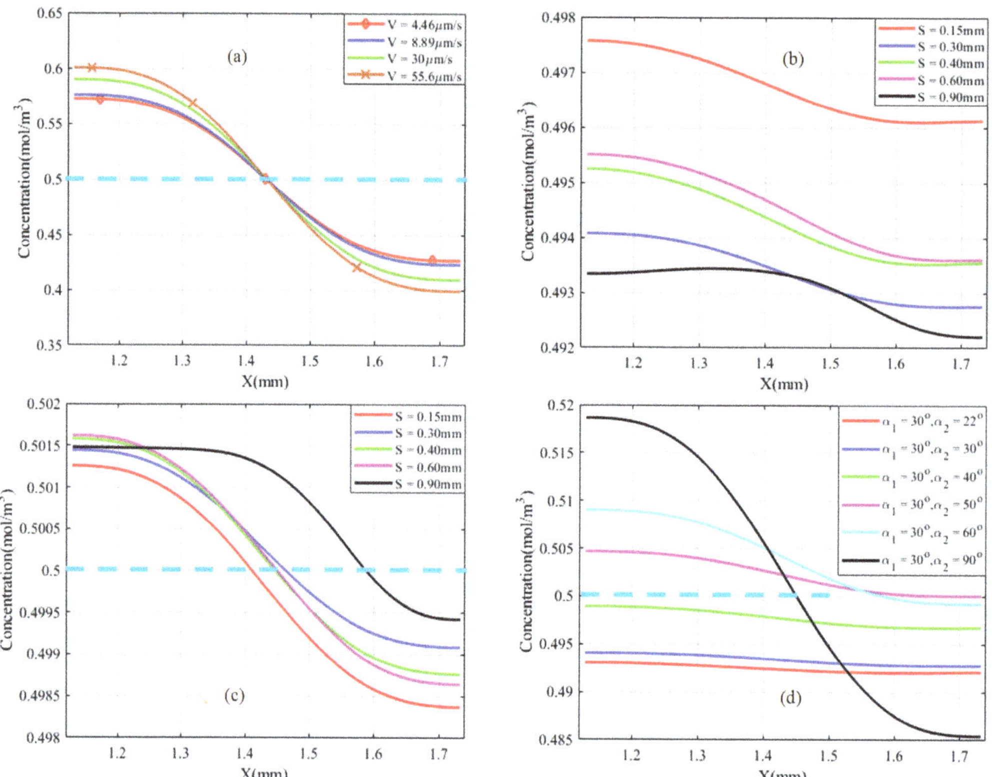

Figure 9. Species concentration profile comparison with 13 kHz, 6 μm oscillation amplitude at Y = 2.30 mm; (**a**) only junction $\alpha = 30°$; (**b**) at $\alpha_1 = 30°$, $\alpha_2 = 30°$, at 4.46 μm/s with different S; (**c**) at $\alpha_1 = 30°$, $\alpha_2 = 30°$, at 55.6 μm/s inlet velocity with different S; and (**d**) at 4.46 μm/s inlet velocity and S = 0.30 mm.

Acoustic streaming produces more body force around sharp edges than other non-microchannel parts, which is expected to result in a 3D flow phenomenon. However, the mixing process takes place in small microchannel dimensions and follows the bulk movement of fluids. The Y-microchannel must always have an inlet and an outlet with a continuous flow process. As a result, the mixing performance is always evaluated far from the high disturbance and around the outlet cross section. Therefore, it is one of the primary reasons to focus on mixing performance evaluation in 2D. Figure 10 illustrates the concentration profile when the sidewall triangle's vertex angle is reduced from $\alpha_2 = 30°$ to 50°. The results show that the sidewall triangles at a distance of S = 0.15 mm and 0.30 mm delivered close to the expected optimum concentration flux profile compared with other dimensions, as shown in Figure 10a. Similarly, low inlet velocity has better output performance than high inlet velocity, as displayed in Figure 10b, but has worse output performance at lower oscillation amplitude, as seen in Figure 10c.

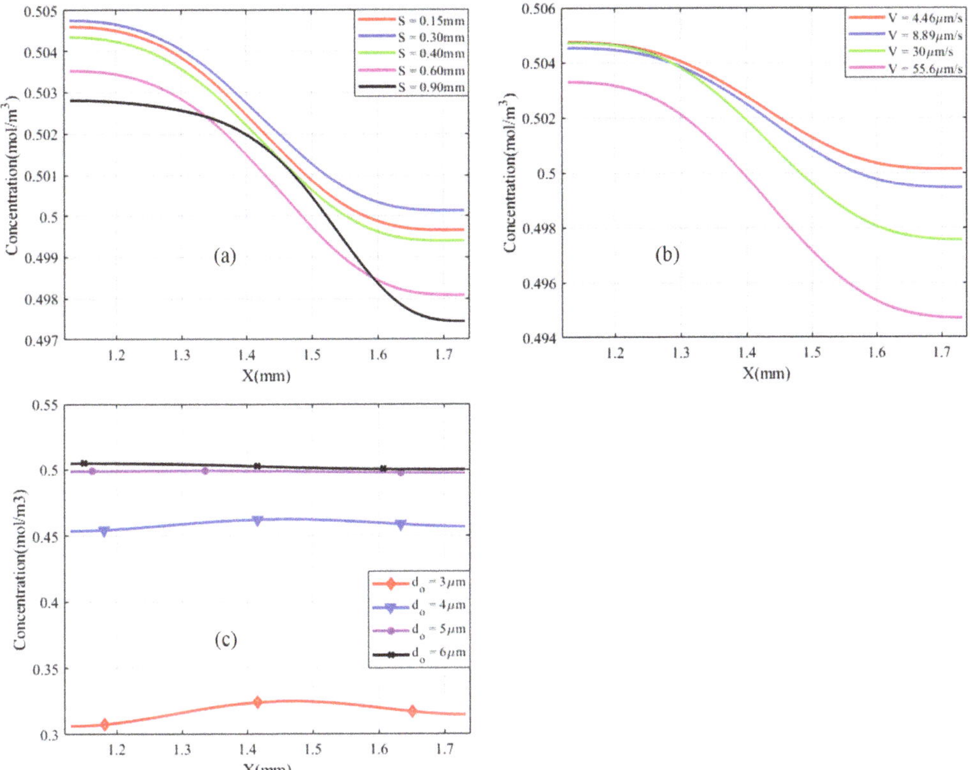

Figure 10. Concentration profile comparison with 13 kHz at Y = 2.30 mm, and $\alpha_1 = 30°$, $\alpha_2 = 50°$; (**a**) at 4.46 µm/s with different S; (**b**) $\alpha_1 = 30°$, $\alpha_2 = 50°$, S = 0.30 mm, and at different inlet velocity; and (**c**) $\alpha_1 = 30°$, $\alpha_2 = 50°$, S = 0.30 mm, 4.46 µm/s inlet velocity, and at different oscillation amplitudes.

The mixing index M is determined using Equation (11) by a concentration of species taken in the measurement region between y = 0.75 to 2.30 mm in all cases, as shown in Figure 8. It is increased along the flow direction and its acoustic streaming performance is also more effective at low inlet velocity, as indicated in Figure 11a,d,e. Figure 11a illustrates a mixing index profile ranging from 0.631 to 0.88 after inducing a single triangle at the junction region with an inlet velocity of 4.46 m/s, however, when the inlet velocity increased to 55.6 m/s, the performance, M, profile changed from 0.45 to 0.84.

Mixing performance is improved with the same inlet after sidewall triangles are induced and integrated. The distance, S, between the sidewall triangles was investigated with different values, as shown in Figures 5, 7, 9b,c and 10a. The distance, S, was better at 0.30 mm based on the streaming vortices profile shown in Figure 5 and the expected optimum concentration profile shown in Figure 9b. Moreover, the dimensionless mixing index, M, was investigated for all values of S, as shown in Figure 11b,c. However, the mixing index M value did not result in an exaggerated result at the microchannel outlet in all S value cases except around the sidewall triangles regions. In general, the sharp edge design of three integrated triangles performed better than the sharp edge design of a single junction triangle. As shown in Figure 11f, the sharper edge of the sidewall provided a better mixing index than the junction triangle's apex, but its concentration profile was lower than the expected optimum concentration value at Y = 2.30 mm plane cutting line. As a result, to satisfy those conditions at low inlet velocity, the junction triangle's vertex angle should be sharper than the sidewall triangles'.

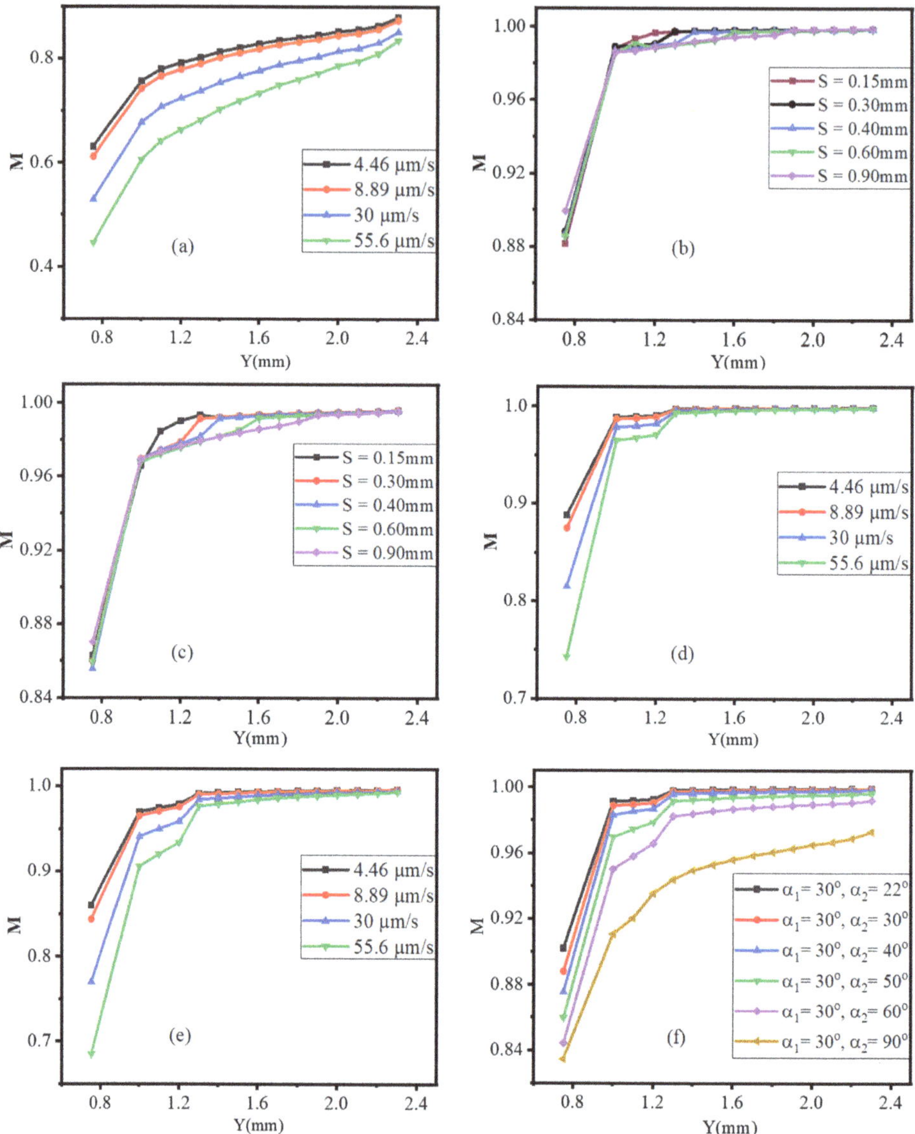

Figure 11. Mixing index with 13 kHz, 6 μm, (**a**) only induced triangular structure at junction with $\alpha = 30°$; (**b**) at $\alpha_1 = 30°, \alpha_2 = 30°$, at 4.46 μm/s with different sidewalls triangles spacing; (**c**) at $\alpha_1 = 30°, \alpha_2 = 50°$, at 4.46 μm/s with different sidewalls triangles spacing; (**d**) at $\alpha_1 = 30°$, $\alpha_2 = 30°$, S = 0.30 mm and different inlet velocities; (**e**) at $\alpha_1 = 30°$, $\alpha_2 = 50°$, S = 0.30 mm, and different inlet velocities; and (**f**) at 4.46 μm/s, S = 0.30 mm, and different α_2.

In general, when compared with a standard Y-junction microchannel, the generated acoustic streaming disturbance has a significant effect on fluid disturbance and improved mixing performance. It performed well in both the concentration profile with a reference to the expected optimum concentration of 0.50 mol/m^3, as indicated in Figure 12, and the mixing index M, as shown in Figure 13 below.

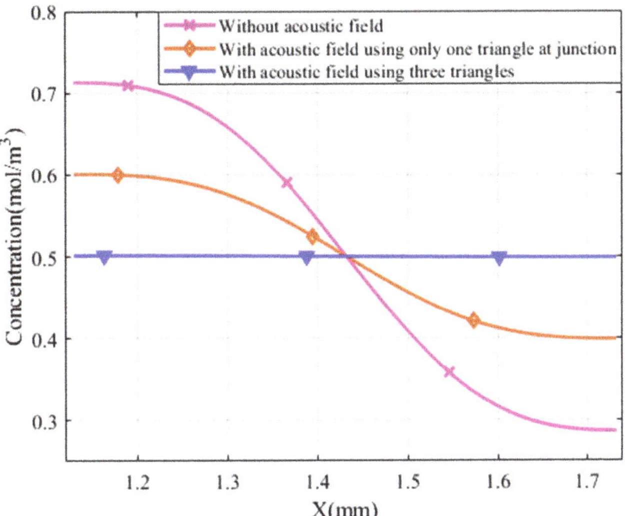

Figure 12. Species concentration profile of inlet velocity 55.6 μm/s at Y = 2.30 mm, $\alpha = 30°$, $\alpha_1 = 30°$, $\alpha_2 = 30°$, S = 0.30 mm, and 6 μm oscillation amplitude.

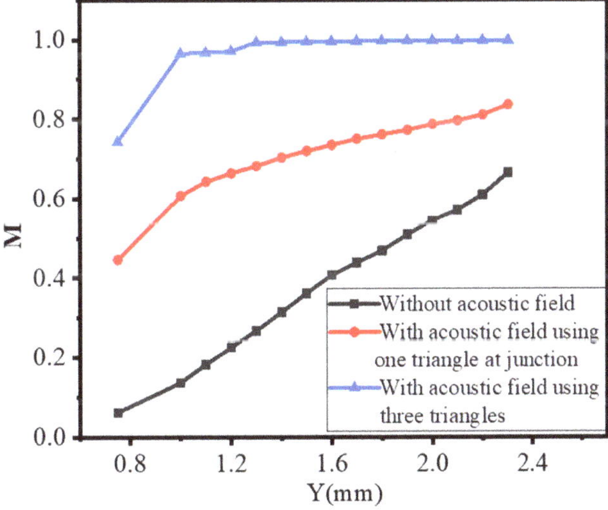

Figure 13. Mixing index comparison with and without acoustic field with inlet velocity 55.6 μm/s $\alpha = 30°$, $\alpha_1 = 30°$, $\alpha_2 = 30°$, S = 0.30 mm, and 6 μm oscillation amplitude.

The concentration of species profile varied between 0.7129 and 0.2871 mol/m^3 in the normal Y- junction microchannel cases, whereas with one triangle at the junction, the species concentration profile ranged between 0.6008 and 0.3989 mol/m^3, but using three triangles improved its range between 0.5015 and 0.4991 mol/m^3.

Similarly, at Y = 2.30 mm, the mixing index M performance of a normal Y-junction microchannel, with only one triangle at the junction, and an integrated three triangles is 0.6642, 0.8355, and 0.9981, respectively.

4. Conclusions

In this study, a new configuration of acoustofluidics was proposed to improve the mixing performances of a Y-junction micromixer. By introducing acoustic streaming with triangular structures at the junction and sidewall regions, the acoustic streaming vortices created by the structure at the junction can improve the mixing efficiency, and the streaming vortices generated by this triangular structure at the junction region can be further integrated with two additional ones at the channel sidewalls to produce successive vortices in the mixing channel to elongate the mixing enhancement. Through numerical simulations, the strength of the vortices is evaluated in terms of the Z-vorticity magnitude, whereas the mixing effectiveness is measured by the concentration profile across the width of the outlet microchannel section and the dimensionless mixing index, M, in the measurement region X = 1.1326 to 1.7326 mm and Y = 0.75 to 2.30 mm along the flow direction. Conclusions can be drawn from the obtained results:

First, introducing the acoustic streaming considerably increased the vorticity in the flow field, and therefore increased the mixing index. In the present study, the Y-junction microchannel without acoustic streaming had a mixing index of 0.6642. By introducing the acoustic streaming with a single triangular structure at the junction region, the mixing performance improved to 0.8355. Three triangular structures at the junction and sidewall regions further improved the result to 0.9981 at Y = 2.30 mm with an inlet velocity of 55.6 μm/s. This corresponds to an increase of 25.79% and 50.27% in the case without acoustic streaming, respectively.

Second, the strength of the streaming vortices and mixing performance are influenced by the parameters such as inlet velocity and the vertex angles of the triangular structure. Higher inlet velocities result in slightly worse mixing performance. In the present study, inlet velocities were tested from 4.46 μm/s to 55.6 μm/s. The mixing index, M, gradually decreased from 0.880 at 4.46 μm/s inlet to 0.8355 at inlet velocity 55.6 μm/s. The same trends can be observed for the cases with three triangular structures, decreasing from 0.9989 to 0.9981 at low and high inlet velocities, respectively. Sharper vertex angles produced higher vorticities and more uniform species distributions in the microchannel. To achieve even better performances in terms of both concentration profile and mixing index, the junction triangle's apex should be sharper than the sidewall triangles'. The sharper case, $\alpha_2 = 30^\circ$, has a mixing index of 0.9989, while the case of $\alpha_2 = 90^\circ$ only has a mixing index of 0.9729. Further optimization of the geometrical details might be achieved by adopting methods such as topology optimization, if certain design restrictions or requirements can be specified to form the objective function. However, the results might be sensitive around the apexes of the triangular structures due to the higher curvature.

Finally, the Z-vorticity calculated from the velocity fields correlated with the mixing performance well. The improvement of performances, namely, the species concentration profile with respect to optimum concentration and mixing index, M, due to acoustic streaming, can be explained by the increase in the vorticity magnitude in the vorticity map. This correlation can be applied to experimental studies to evaluate the mixing performance using velocimetry techniques such as micro-particle image velocimetry ($\mu - PIV$). More ongoing works are currently conducted to investigate the flow patterns experimentally and take the three-dimensional boundary effects into account, such as the channel height and vertical triangular structure.

Author Contributions: Conceptualization, W.-H.T. and S.A.E.; methodology, S.A.E.; formal analysis, S.A.E.; writing—original draft preparation, S.A.E.; writing—review and editing, W.-H.T.; supervision, W.-H.T.; funding acquisition, W.-H.T. All authors have read and agreed to the published version of the manuscript.

Funding: This work was supported by the Ministry of Science and Technology (MOST), Taiwan, under Grant No. MOST-110-2221-E-011-072-.

Data Availability Statement: The data that support the findings of this study are available from the corresponding author upon reasonable request.

Conflicts of Interest: The authors declare no conflict of interest.

References

1. Bayareh, M.; Ashani, M.N.; Usefian, A. Active and passive micromixers: A comprehensive review. *Chem. Eng. Process. Process Intensif.* **2020**, *147*, 107771. [CrossRef]
2. Matsunaga, T.; Nishino, K. Swirl-inducing inlet for passive micromixers. *RSC Adv.* **2014**, *4*, 824–829. [CrossRef]
3. Alam, A.; Afzal, A.; Kim, K.-Y. Mixing performance of a planar micromixer with circular obstructions in a curved microchannel. *Chem. Eng. Res. Des.* **2014**, *92*, 423–434. [CrossRef]
4. Antognoli, M.; Stoecklein, D.; Galletti, C.; Brunazzi, E.; Di Carlo, D. Optimized design of obstacle sequences for microfluidic mixing in an inertial regime. *Lab Chip* **2021**, *21*, 3910–3923. [CrossRef]
5. Luong, T.-D.; Phan, V.-N.; Nguyen, N.-T. High-throughput micromixers based on acoustic streaming induced by surface acoustic wave. *Microfluid. Nanofluid.* **2011**, *10*, 619–625. [CrossRef]
6. Xia, Q.; Zhong, S. Liquid mixing enhanced by pulse width modulation in a Y-shaped jet configuration. *Fluid Dyn. Res.* **2013**, *45*, 025504. [CrossRef]
7. Usefian, A.; Bayareh, M. Numerical and experimental study on mixing performance of a novel electro-osmotic micro-mixer. *Meccanica* **2019**, *54*, 1149–1162. [CrossRef]
8. Hejazian, M.; Nguyen, N.-T. A Rapid Magnetofluidic Micromixer Using Diluted Ferrofluid. *Micromachines* **2017**, *8*, 37. [CrossRef]
9. Ozcelik, A.; Aslan, Z. A practical microfluidic pump enabled by acoustofluidics and 3D printing. *Microfluid. Nanofluidics* **2021**, *25*, 10. [CrossRef]
10. Hao, N.; Pei, Z.; Liu, P.; Bachman, H.; Naquin, T.D.; Zhang, P.; Zhang, J.; Shen, L.; Yang, S.; Yang, K.; et al. Acoustofluidics-Assisted Fluorescence-SERS Bimodal Biosensors. *Small* **2020**, *16*, 11. [CrossRef]
11. Gelin, P.; Sukas, O.S.; Hellemans, K.; Maes, D.; De Malsche, W. Study on the mixing and migration behavior of micron-size particles in acoustofluidics. *Chem. Eng. J.* **2019**, *369*, 370–375. [CrossRef]
12. Bach, J.S.; Bruus, H.J.P.R.L. Suppression of acoustic streaming in shape-optimized channels. *Phys. Rev. Lett.* **2020**, *124*, 214501. [CrossRef] [PubMed]
13. Marmottant, P.; Raven, J.P.; Gardeniers, H.; Bomer, J.G.; Hilgenfeldt, S. Microfluidics with ultrasound-driven bubbles. *J. Fluid Mech.* **2006**, *568*, 109–118. [CrossRef]
14. Lutz, B.R.; Chen, J.; Schwartz, D.T. Microscopic steady streaming eddies created around short cylinders in a channel: Flow visualization and Stokes layer scaling. *Phys. Fluids* **2005**, *17*, 023601. [CrossRef]
15. Ovchinnikov, M.; Zhou, J.; Yalamanchili, S. Acoustic streaming of a sharp edge. *J. Acoust. Soc. Am.* **2014**, *136*, 22–29. [CrossRef] [PubMed]
16. Doinikov, A.A.; Gerlt, M.S.; Pavlic, A.; Dual, J. Acoustic streaming produced by sharp-edge structures in microfluidic devices. *Microfluid. Nanofluid.* **2020**, *24*, 13. [CrossRef]
17. Liou, Y.-S.; Kang, X.-J.; Tien, W.-H. Particle aggregation and flow patterns induced by ultrasonic standing wave and acoustic streaming: An experimental study by PIV and PTV. *Exp. Therm. Fluid Sci.* **2019**, *106*, 78–86. [CrossRef]
18. Nama, N.; Barnkob, R.; Mao, Z.; Kähler, C.J.; Costanzo, F.; Huang, T.J. Numerical study of acoustophoretic motion of particles in a PDMS microchannel driven by surface acoustic waves. *Lab Chip* **2015**, *15*, 2700–2709. [CrossRef]
19. Zhang, C.; Guo, X.; Royon, L.; Brunet, P. Unveiling of the mechanisms of acoustic streaming induced by sharp edges. *Phys. Rev. E* **2020**, *102*, 043110. [CrossRef]
20. Zhang, C.; Guo, X.; Brunet, P.; Costalonga, M.; Royon, L. Acoustic streaming near a sharp structure and its mixing performance characterization. *Microfluid. Nanofluidics* **2019**, *23*, 104. [CrossRef]
21. Huang, P.H.; Xie, Y.; Ahmed, D.; Rufo, J.; Nama, N.; Chen, Y.; Chan, C.Y.; Huang, T.J. An acoustofluidic micromixer based on oscillating sidewall sharp-edges. *Lab Chip* **2013**, *13*, 3847–3852. [CrossRef] [PubMed]
22. Legay, M.; Simony, B.; Boldo, P.; Gondrexon, N.; Le Person, S.; Bontemps, A. Improvement of heat transfer by means of ultrasound: Application to a double-tube heat exchanger. *Ultrason. Sonochem.* **2012**, *19*, 1194–1200. [CrossRef] [PubMed]
23. Rasouli, M.R.; Tabrizian, M. An ultra-rapid acoustic micromixer for synthesis of organic nanoparticles. *Lab Chip* **2019**, *19*, 3316–3325. [CrossRef] [PubMed]
24. Vuillermet, G.; Gires, P.-Y.; Casset, F.; Poulain, C. Chladni Patterns in a Liquid at Microscale. *Phys. Rev. Lett.* **2016**, *116*, 184501. [CrossRef] [PubMed]
25. Nama, N.; Huang, P.-H.; Huang, T.J.; Costanzo, F. Investigation of micromixing by acoustically oscillated sharp-edges. *Biomicrofluidics* **2016**, *10*, 024124. [CrossRef] [PubMed]
26. Ozcelik, A.; Ahmed, D.; Xie, Y.; Nama, N.; Qu, Z.; Nawaz, A.A.; Huang, T.J. An Acoustofluidic Micromixer via Bubble Inception and Cavitation from Microchannel Sidewalls. *Anal. Chem.* **2014**, *86*, 5083–5088. [CrossRef]
27. Wang, S.S.; Jiao, Z.J.; Huang, X.Y.; Yang, C.; Nguyen, N.T. Acoustically induced bubbles in a microfluidic channel for mixing enhancement. *Microfluid. Nanofluid.* **2009**, *6*, 847–852. [CrossRef]
28. Lee, S.H.; van Noort, D.; Lee, J.Y.; Zhang, B.-T.; Park, T.H. Effective mixing in a microfluidic chip using magnetic particles. *Lab Chip* **2009**, *9*, 479–482. [CrossRef]
29. Lee, C. Optimum design of a y-channel micromixer for enhanced mixing according to the configuration of obstacles. *Adv. Appl. Fluid Mech.* **2016**, *19*, 1–22. [CrossRef]

30. Wang, C.-T.; Hu, Y.-C. Mixing of Liquids Using Obstacles in Y-Type Microchannels. *Tamkang J. Sci. Eng.* **2010**, *13*, 385–394.
31. Wu, Z.; Nguyen, N.-T.; Huang, X. Nonlinear diffusive mixing in microchannels: Theory and experiments. *J. Micromech. Microeng.* **2004**, *14*, 604–611. [CrossRef]
32. Shah, I.; Kim, S.W.; Kim, K.; Doh, Y.H.; Choi, K.H. Experimental and numerical analysis of Y-shaped split and recombination micro-mixer with different mixing units. *Chem. Eng. J.* **2019**, *358*, 691–706. [CrossRef]
33. Sahu, P.K.; Golia, A.; Sen, A.K. Investigations into mixing of fluids in microchannels with lateral obstructions. *Microsyst. Technol.* **2013**, *19*, 493–501. [CrossRef]
34. Wang, L.; Ma, S.; Han, X. Micromixing enhancement in a novel passive mixer with symmetrical cylindrical grooves. *Asia-Pac. J. Chem. Eng.* **2015**, *10*, 201–209. [CrossRef]
35. Hsieh, S.-S.; Lin, J.-W.; Chen, J.-H. Mixing efficiency of Y-type micromixers with different angles. *Int. J. Heat Fluid Flow* **2013**, *44*, 130–139. [CrossRef]
36. COMSOL Multiphysics 5.6. Available online: www.comsol.com (accessed on 23 September 2021).
37. CRCnetBASE Product. *Handbook of Chemistry and Physics*, 97th ed.; CRC Press: Boca Raton, FL, USA, 2016–2017.
38. Bruus, H. Acoustofluidics 2: Perturbation theory and ultrasound resonance modes. *Lab Chip* **2012**, *12*, 20–28. [CrossRef]
39. Friend, J.; Yeo, L.Y. Microscale acoustofluidics: Microfluidics driven via acoustics and ultrasonics. *Rev. Mod. Phys.* **2011**, *83*, 647–704. [CrossRef]
40. Rafeie, M.; Welleweerd, M.; Hassanzadeh-Barforoushi, A.; Asadnia, M.; Olthuis, W.; Warkiani, M.E. An easily fabricated three-dimensional threaded lemniscate-shaped micromixer for a wide range of flow rates. *Biomicrofluidics* **2017**, *11*, 014108. [CrossRef]
41. Muller, P.B.; Barnkob, R.; Jensen, M.J.H.; Bruus, H. A numerical study of microparticle acoustophoresis driven by acoustic radiation forces and streaming-induced drag forces. *Lab Chip* **2012**, *12*, 4617–4627. [CrossRef]

Article

Characterization of Mixing Performance Induced by Double Curved Passive Mixing Structures in Microfluidic Channels

Ingrid H. Oevreeide [1], Andreas Zoellner [2] and Bjørn T. Stokke [1,*]

[1] Division of Biophysics and Medical Technology, Department of Physics, NTNU The Norwegian University of Science and Technology, NO-7491 Trondheim, Norway; ingrid.h.ovreeide@ntnu.no

[2] Independent Researcher, Palo Alto, CA 94301, USA; amz@alumni.stanford.edu

* Correspondence: bjorn.stokke@ntnu.no

Abstract: Functionalized sensor surfaces combined with microfluidic channels are becoming increasingly important in realizing efficient biosensing devices applicable to small sample volumes. Relaxing the limitations imposed by laminar flow of the microfluidic channels by passive mixing structures to enhance analyte mass transfer to the sensing area will further improve the performance of these devices. In this paper, we characterize the flow performance in a group of microfluidic flow channels with novel double curved passive mixing structures (DCMS) fabricated in the ceiling. The experimental strategy includes confocal imaging to monitor the stationary flow patterns downstream from the inlet where a fluorophore is included in one of the inlets in a Y-channel microfluidic device. Analyses of the fluorescence pattern projected both along the channel and transverse to the flow direction monitored details in the developing homogenization. The mixing index (MI) as a function of the channel length was found to be well accounted for by a double-exponential equilibration process, where the different parameters of the DCMS were found to affect the extent and length of the initial mixing component. The range of MI for a 1 cm channel length for the DCMS was 0.75–0.98, which is a range of MI comparable to micromixers with herringbone structures. Overall, this indicates that the DCMS is a high performing passive micromixer, but the sensitivity to geometric parameter values calls for the selection of certain values for the most efficient mixing.

Keywords: passive mixing; curved mixing structure; confocal microscopy; mixing efficiency

Citation: Oevreeide, I.H.; Zoellner, A.; Stokke, B.T. Characterization of Mixing Performance Induced by Double Curved Passive Mixing Structures in Microfluidic Channels. *Micromachines* **2021**, *12*, 556. https://doi.org/10.3390/mi12050556

Academic Editor: Kwang-Yong Kim

Received: 22 April 2021
Accepted: 11 May 2021
Published: 13 May 2021

1. Introduction

Integration of microfluidics and microfluidic channels within several scientific domains have increased over the past few decades. This merger has led to the development of devices such as Lab-on-Chip, Organ-on-Chip, and sensors for the determination of various biological, chemical, and medical analytes, as well as applications as reactors [1–4]. Substantial advantages associated with working in the miniaturized regime, including microfluidics, motivates such developments. An example of favorable features includes the use of a reduced sample volume and read out time and an improved limit-of-detection (LOD). Although microfluidics come with several advantages, there are also challenges with operating in the micro-regime. One of the major disadvantages is the laminar flow regime encountered for fluid flow at the microscale. This reduces the impact of inertial forces over the viscous forces and yields low Reynold's numbers. Typically, with a laminar flow, the only mixing that occur is by diffusion, which is a relatively inefficient mixing process. This limitation of laminar flows has led to the development of different mixing techniques to enhance the rate of mixing. The various fluid mixing strategies are grouped into processes referred to as either active or passive, where the active mixers require external energy input.

Active mixers use peripheral devices to drive the mechanism of the mixing process, e.g., based on acoustics, electrostatics, or other principles [5–8]. The peripherals thus deliver energy to the devices that is dissipated, which may lead to heating. Although the resultant mixing efficiencies are often over 90% for active mixers [9,10], the additional equipment increases the complexity of the system, cost, size and can in cases be harmful to biological samples due to the energy dissipation. These drawbacks have led to an increased focus on the development of passive micromixers and their optimization.

Passive mixers do not require energy supplied by peripherals, while at the same time they show a substantial increased mixing efficiency compared to diffusion driven process. Several different strategies have been implemented ranging from changing the 2D geometry of the channels [11–15], including obstructions [16–20] or adding different structures to one or more of the channel surfaces [21–30], amongst others [31–36]. The slanted groove mixer (SGM) [37] exploiting fabricated structures in the ceiling of the microfluidic channels are among the designs that have attracted the most attention.

The SGM involves the addition of slanted bar structures localized at one of the walls of a microfluidic channel [37]. Stroock et al. (2002) expanded on this design by the addition of an asymmetric arm to the SGM, resulting in the well-known staggered herringbone mixer (SHM) [38]. In their original work, they reported that addition of an asymmetric element resulted in a more efficient passive mixer due to the development of mixing via chaotic advection, compared to channels containing no mixing elements or the SGM.

Following the initial publication of the SHM, several studies have been performed focusing on the variation of different geometrical parameters of the passive mixing structure to enhance their performance. Yang et al. (2005) found that the height ratio and asymmetry index of the grooves (length of long arm to short arm) were the key parameters in influencing the mixing performance of the SHM [35]. An asymmetry index of 2/3 has been shown to be optimal [38–40] and a deeper groove yielded increased mixing efficiencies [35,41–43]. Furthermore, the groove depth was also identified as an important parameter, where a wider groove increased the mixing efficiency induced by the structure [44].

Stroock et al. (2002) also predicted that the use of SHM would increase the rate of a diffusion limited reaction, through the stirring of the boundary layer [38], which was further investigated by other research groups [45,46]. An application of this would be surface-based sensors, such as biosensors. Biosensors often require a certain minimum concentration of analyte receptor binding for the analyte to be sensed by an underlying readout principle, e.g., changes in the mass or refractive index adjacent to or connected to the sensor [47–49]. Passive mixing structures in microfluidic channels have been used in conjunction with surface biosensors to enhance analyte concentration at the sensor [50], thus decreasing the LOD and readout time. Lynn et al. implemented an SHM in the ceiling of a microfluidic channel, which, combined with a surface plasmon resonance (SPR) sensor, showed an increase in sensitivity, where different parameters varied the efficiency of the sensor [51–55].

An increase in the rate of surface coverage of analyte as induced by a microfluidic channel with mixing structures will decrease the channel length needed to increase the probability of an analyte reaching a surface sensor. Comparing this to its efficiency in mixing an optimal channel for mass transport and mixing, e.g., for diluting samples, can be calculated. A shorter channel and increased surface coverage rate would also allow for a reduction in the sample volume required for the minimum concentration necessary to obtain a signal beyond the signal-to-noise ratio of certain types of biosensors within reasonable time frames.

We previously reported on a comparison of the mixing efficiency of a curved passive mixing structure (CMS) with that of herringbone structures (HBS) and found that the CMS represented a more robust group of design geometries in maintaining efficient mixing than the HBS [56]. Specifically, we reported a mixing index (MI) 1 cm downstream from the inlet in the range of 0.85–0.99 for various parameter values of the designs in the CMS group, compared to the HBS group that resulted in an MI range of 0.74–0.98. These finding motivated further studies of passive mixing structure geometries based on our initial CMS design by the addition of an asymmetric element. The resulting structure had two curved structures, thus also resembling the HBS designs with its two branches, and with the location of the mixing elements along the channel, reverting the asymmetry location every second structure. In the following section, we describe the experimental characterization of the mixing performance, the change in MI and mass transfer abilities, of the adapted double curved mixing structure (DCMS) fabricated in the ceiling of microfluidic channels.

2. Materials and Methods

2.1. Fabrication

Microfluidic mixing channels were fabricated using a two-layer photo-lithographic method following the procedure previously outlined [56]. In brief, the master for the soft lithography was fabricated by spin-coating a negative photoresist (MrDWL40, Micro Resist Technologies) on a silicon wafer, baked at 90 °C following a stepwise 5 °C temperature ramping, patterning of the flow channel by exposing to a laser at 405 nm (MLA150 Maskless Aligner, Heidelberg) and post-exposure baked (PEB) at 90 °C. The second resist layer was spun onto layer one after cooling, and the mixing design exposed using the same procedure. The final mold was obtained by developing the exposed structure (MrDev 600, Micro Resist Technology) and was silanized to allow for easier removal of the PDMS. The PDMS (Sylgard 184, Dow Corning) was added at a 1:10 curing agent to elastomer weight, degassed, poured on the mold and then baked at 65 °C for 3 h. The final channels were peeled from the master, cut, and outlet and inlets were punched (Ø1.0 mm, Miltex Biopsy Punch). To complete the device, a microscope slide and the PDMS channel were activated using oxygen plasma and bonded together.

2.2. Passive Mixer Designs

Two different groups of passive mixing designs were studied. These groups were a curved mixing structure (CMS) as introduced previously [56], and a double curved mixing structure (DCMS) extending from the CMS by adding a second curve to the overall geometry (Figure 1). A total of 16 different channels with passive mixing designs were fabricated for each group. The DCMS was divided into cycles of two structures, alternating the direction of the long and short curved structure for each cycle. Due to the varying groove pitch of the mixing structures, the number of structures within the overall channel length (1 cm) differed for each channel. The geometrical parameters are shown in Figure 1.

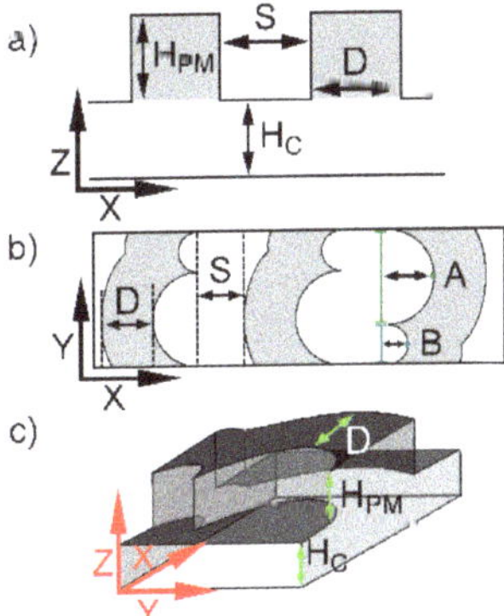

Figure 1. Schematic illustration of microfluidic channel designs showing relevant parameters. The overall transverse widths of each channel were 100 µm. (**a**) Side view of a channel design with passive mixing structures in the ceiling where the design parameters are as follows: H_c depicts the channel height; H_{PM} is the passive mixer height. (**b**) Top view depicting the passive mixer depth, D, and passive mixer spacing, S, for DCMS, where the short arm to perpendicular intersection is 30 µm (teal line) and long arm is 70 µm (green line). A = 35 µm and B = 15 µm. (**c**) 3D view of DCMS also depicting the location of H_c, H_{PM} and D.

The various channels with the DCMS design were realized by selecting different channel heights (H_C), and the passive mixing structure depth (D), spacing (S) and height (H_{PM}) (Table 1).

Table 1. Parameter values of microfluidic channels with DCMS in the ceiling of the microfluidic channel. The table lists all combinations of the parameters implemented by the lithography process. The design parameters are schematically illustrated in Figure 1. The master molds on the four wafers were realized for fabrication of channels with total heights: Wafer 1 (W_1) = 40 μm, Wafer 2 (W_2) = 60 μm, Wafer 3 (W_3) = 80 μm and Wafer 4 (W_4) = 100 μm. Sixteen different channels with design parameter variables were fabricated. All channels had a width (along the y-direction) of 100 μm, and a mixing channel length (x-direction) of 1 cm.

Parameters	**Wafer 1 (W_1)**			**Wafer 2 (W_2)**			**Wafer 3 (W_3)**			**Wafer 4 (W_4)**		
Channel height (H_C)	20			20			40			40		
Passive mixer (PM) height (H_{PM})	20			40			40			60		
Parameter Design Varaible (PDV)	PM Depth (D)	PM Spacing (S)	PM Pitch (D + S)	PM Depth (D)	PM Spacing (S)	PM Pitch (D + S)	PM Depth (D)	PM Spacing (S)	PM Pitch (D + S)	PM Depth (D)	PM Spacing (S)	PM Pitch (D + S)
1	20	20	40	40	40	80	40	40	80	60	60	120
2	20	40	60	40	80	120	40	80	120	60	120	180
3	40	40	80	80	80	160	80	80	160	120	120	240
4	40	80	120	80	160	240	80	160	240	120	240	360

In the following we refer to the parameters in the various mixing structure designs using the wafer they were produced on (W_1–W_4), the passive mixing design used (DCMS) followed by the PDV (1–4), e.g., W_1 DCMS 1 would have the dimensions H_C = 20 μm, H_{PM} = 20 μm, and the Double Curved Mixing Structures (DCMS) would have D = 20 μm and S = 20 μm. Whilst W_4 DCMS 4 would have the dimensions 40 μm, 60 μm, 120 μm and 240 μm, respectively.

2.3. Confocal Imaging

The flow pattern in the microfluidic channels with the various mixing structures were determined using confocal microscopy (Leica TCS SP5), employing fluorescein (Fluorescein sodium salt SigmaAldrich) included in the aqueous solutions injected to one of the Y-inlets as a reporter. The fluorescein solution and deionized water were added to two separate 5 mL syringes and were simultaneously injected (10 μL/min, Harvard apparatus syringe pump) via plastic tubes in the two separate inlets of the Y-channel. This resulted in a mixing flow rate of 20 μL/min and a Reynold's number of ~4. XYZT-stack images were obtained for specific locations downstream from the inlet employing a 10× (NA = 0.4) HC PL Apo CS dry objective. The image acquisition process is further detailed in our previous study [56].

2.4. Data Processing

The acquired confocal XYZT-stacks along the microfluidic channel were analyzed using custom-designed MATLAB scripts. A virtual sensor volume was chosen between two mixing structures for each stack and the lowest and highest Z-level within the microfluidic channel was determined. This was also used as a basis for the analysis within the fluid layer adjacent to a most likely location of a transducing sensor element. After calculating the standard deviation of the fluorescence intensity at the inlet, the pixel intensities were normalized to 1, and each consecutive measurement for the same channel was then normalized to the inlet signal. We calculated the mixing index (MI) from the standard deviation of the normalized fluorescence intensity profile for each image volume along each channel using Equation (1):

$$MI = 1 - \frac{\sqrt{\frac{1}{N-1}\sum_{i=1}^{N}(c_i - \bar{c})^2}}{\sqrt{\frac{1}{N_{in}-1}\sum_{i=1}^{N_{in}}(c_i - \bar{c})^2}_{Inlet}} \tag{1}$$

In Equation (1), N is the number of pixels, c_i is the pixel intensity, c is the mean pixel intensity, and N_{in} is the number of pixels at the inlet. A completely unmixed channel has an MI of 0, whilst a completely mixed channel has a value of 1. The MI was empirically observed to equilibrate towards the completely mixed state with the increasing channel length. To facilitate an easier comparison of the main trends of the mixing performance between the various mixing structures, we fitted the experimental MI(x), x from 0 to 1 cm, with a two-exponential equation to describe the equilibration process. Thus, modeling the MI as a double exponential increase to saturation using four parameters follows Equation (2):

$$MI = a(1 - e^{-bx}) + c(1 - e^{-dx}), \tag{2}$$

The parameters a, b, c and d were obtained by fitting MI(x), Equation (2), to the experimentally observed data using the constraints, a, b, c and d greater than 0 and either a + c < 1 or a + c = 1 (SigmaPlot). The analysis shows that a double exponential function with 4 parameters, where a + c < 1, was adequate to account for the trends in MI(x) while at the same time limiting the number of parameters. Further aspects of the selection of this procedure was reported previously [56]. The image processing, calculation of the MI based on the observed fluorescence distributions and the further representation of MI using Equation (2), were conducted for the layers adjacent to a possible sensor location (termed sensor layer, SL, adjacent to the bottom in Figure 1c) or were based on the mean over the height of the channel (channel layers, CL). Additionally, we refer to all layers, i.e., the channel layers with the extension of the grooves, as all layers (AL).

2.5. Statistics

Differences between micromixer design groups with respect to robustness in mixing performance when varying parameter dimensions were tested statistically based on the variance of MI between the groups. Thus, an F-test was used, rejecting null hypotheses when the ratio of the variances, F, were larger than $F_{\alpha/2, N1\text{-}1, N2\text{-}1}$, where α is the significance level and N1-1 and N2-1 are the degrees of freedom.

3. Results and Discussion

To validate the mixing and surface coverage efficiencies of the designs, a qualitative analysis of the flow pattern (Figure 2) and a quantitative analysis for the MI (Figure 3) were employed. The development of the flow patterns was realized in both the transverse and the axial direction, resulting in the possibility of a pseudo 3D analysis. The addition of the double curve design changes the flow pattern drastically, compared to the CMS previously studied [56], in both dimensions (Figure 2).

In the following, the flow pattern in the DCMS describes the case where the Y-channel inlet initially exposes the fluid flow with the fluorescein solution to the long curve of the DCMS (Figure 2b). The DCMS mixing structure induces change in the observed flow pattern for all Z-layers as follows. The DCMS, W_2 DCMS 1, induces a distance dependent enhancement of the fluorescence on the side initially loaded with water and no enhanced fluorophore within the grooves on that side. This gradual spreading, as observed from the pattern in the XY planes, is further exemplified after 5 and 10 structures where the fluorescently depleted regions remain on the top for the DCMS (Figure 2b). Halfway downstream of the channel (e.g., 50%, Figure 2), there is still an observable region of lower intensity in the projections along the channels, which no longer exists at the outlet, where homogeneous distributions are observed (Figure 2b).

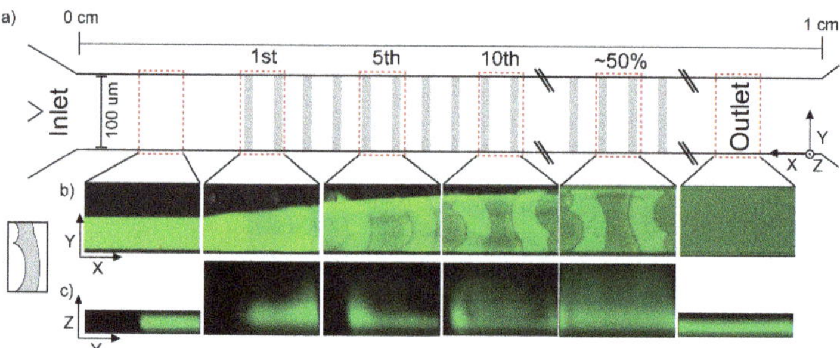

Figure 2. (**a**) Schematic illustration of a mixing channel with arbitrary mixers (grey rectangles) for identification of location of mixing structures from the inlet (left) to the outlet (right) with the location of image acquisition regions (red dashed rectangles). (**b**) Confocal projection of fluorescence intensity along the W_2 DCMS 1 channel within the selected regions (XY view with intensity averaged over all Z layers) and (**c**) projection of fluorophore perpendicular to the channel direction (sum of intensities along X in each region leading to a YZ-projection). The fluorescence intensity projections were deduced by the custom-designed image processing routine of the confocal XYZT-stacks acquired.

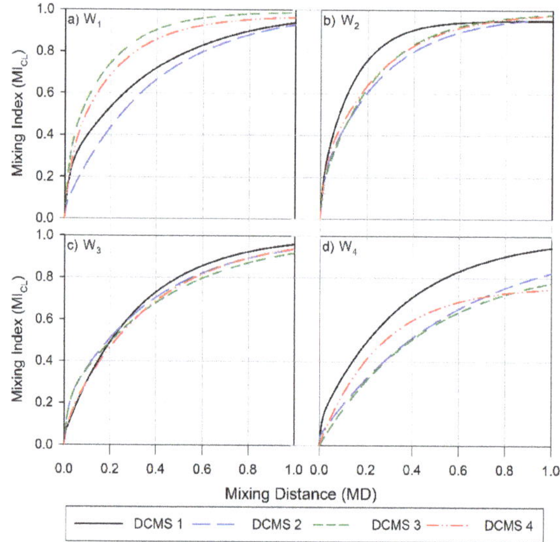

Figure 3. Mixing index for the channel layers (CL) versus the channel length for (**a**) Wafer 1, (**b**) Wafer 2, (**c**) Wafer 3, and (**d**) Wafer 4. Data obtained for channels with parameter design variables (PDV, Table 1) are depicted as follow: PDV 1 is a black solid line, PDV 2 is a blue long-dashed line, PDV 3 green is three short-dashed lines and PDV 4 is a red dashed-dot-dot line.

The developing flow pattern along the channels can also be observed in the YZ projections (Figure 2c), where the DCMS structures induces a combined migration (1st mixing structure) and a region of increased fluorescence in the part of the channel opposite the injection side (5th mixing structure). The latter could indicate that a swirling motion has taken place. Such a rotational flow is more evident after 10 structures, although a more heterogeneous distribution remains after 50% channel length. A similar analysis was performed for the surface layer (SL) (Figure A1) of the DCMS structures as a basis for comparison to the bulk flow. The developing homogenization process, as viewed in the XY projections, are considered to be similar for the SL as compared to the bulk movement. While the above development of the flow pattern is observed when the Y-channel inlet guiding the fluorescently labelled stream to the initial long curve is observed, reverting the labelled fluid to the alternative inlet will generate the "mirror" image of the observation reported in Figure 2b. Thus, the unlabeled part of the developing pattern reflects the initial distance-dependent reduction of the spread across the channel (Figure 2b 1st) for the DCMS structures.

Some differences in the developing flow pattern of DCMS as compared to that of the previously reported single-curved mixing structures (CMS) [56] include an earlier indication of swirling flow already within the 1st mixing structure of CMS as compared to the 5th mixing structure for the DCMS. The initial distance-dependent depletion of fluorescence intensity from the side of the channel where the fluorophore was injected in the case of CMS was compensated by the swirling motion, yielding an overall more efficient homogenization as a function of distance than the DCMS.

The flow visualization provides a qualitative approach to compare efficiencies between different structures. Quantitative determination of the MI was also performed as a further expansion to be used for comparing mixing performance. Employing Equations (1) and (2), we gathered the MI for the various DCMS micromixers with various parameter values (Table 1). These are presented in terms of which H_C:H_{PM} ratio they belong to (W_1, W_2, W_3 or W_4) and whether we have used the average bulk flow, represented as the channel layers (CL), or the fluid layer adjacent to the channel surface, represented as the surface layer (SL), as a basis for the determination of MI (Figures 3 and A2).

The efficiency of the designs to enhance mixing (Figure 3) and to facilitate surface coverage (Figure A2) was found to a depend on the height ratios of the channel. Increasing this ratio from 1:1 (W_1, Figure 3a) to 1:2 (W_2, Figure 3b) increased the overall average efficiencies realized. The MI observed at the outlet, 1 cm downstream from the inlet, was 0.96 (range 0.06) for W_1 (Figure 3a), 0.92 (range 0.03) for W_2 (Figure 3b),0.94 (range 0.04) for W_3 (Figure 3c) and 0.82 (range 0.20) for W_4 (Figure 3d). A similar trend in the average efficiencies was also observed for the surface layers (SL) (Figure A2), where the MI realized for W_4 showed the lowest average MI at the outlet and the largest range. However, this dependence was not as prominent as for the previously reported CMS [56].

The pitch of the structures (S + D, Table 1) was also found to strongly influence the length dependence of the MI. The more strongly distance-dependent increase in mixing and MI at the outlet (1 cm from the inlet) observed for DCMS 3 on W_1 (Figure 3a) as compared to DCMS 1 and 2 on the same wafer correlates with the changes in pitch between these structures. This low MI and high MD observed in this case for DCMS 1 and 2 could be due to the structures having a too-narrow pitch (S + D), resulting in inefficient flow distributions. As the structures becomes wider and further apart, there is a large increase in MI and a decrease in MD, and this trend was also observed for the CMS design [56]. These efficient designs were comparable to those realized in W_2 and W_3 in achieving efficient mixing.

However, a further increase in the pitch, as observed on W_4 (Table 1, Figure 3d) resulted in a large decrease in MI and an increase in the range of observed efficiencies. In this instance, only W_4 DCMS 1 resulted in an MI comparable to the most efficient design (W_1 DCMS 3). This decrease in efficiency, as the structures become wider and deeper, could be due to the formation of dead volumes within the structures, thereby lowering their ability to perform as efficient mixers. However, this is not in line with that reported for the CMS group, where the least efficient channels were those realized on wafers with geometry as W_1, and here especially, CMS 1 and CMS 2 [56].

The channels showed a similar MI at the outlet for all DCMS 1 designs (mean 0.95, range 0.02), where the ratio between H_{PM}: D: S was 1:1:1. A similar trend in MI was also observed for all designs where the pitch was 120 μm (Table 1) (mean 0.95, range 0.03), whereas an increase in pitch to 240 μm resulted in an average MI of 0.90 (range 0.20). This indicates that there is not a particular design parameter that can be universally applied to achieve the best mixing performance. From the designs evaluated, the three main parameters to consider are the height (H_{PM}), depth (D), and spacing (S) of the DCMS.

A similar relationship between H_{PM}:D:S was also observed for the surface layer (SL) (Figure A2), thus indicating an apparent similar developing fluorescence distribution at the surface and throughout the channel. However, a notable difference was observed for DCMS 2 from W_1 and W_3. In the case of W_1 DCMS 2, the rate of homogenization of the fluid layer at the surface (Figure A2a) was lower than for the rest of the bulk fluid (Figure 3a), thereby resulting in lower MI values at the same distance from the inlet (i.e., $MD_{0.6}$ = 0.49 cm (SL) and 0.33 cm (CL), whilst at 1 cm, the MI = 0.84 (SL) and 0.93 (CL)). The SL for W_3 DCMS 2, on the other hand, showed a slightly faster initial rate of fluid homogenization and a larger MI at the outlet compared to the bulk flow in CL. In this instance, the homogenization throughout the channel required an extra 0.06 cm to reach an MI of 0.6 compared to the SL, although the MI at 1 cm was 0.94 for both.

These variations in MI and mixing rate (as determined by the variations in the mixing distance required to reach an MI of 0.6) show the difficulty in predicting flow behavior depending on changes to different geometrical parameters, as well as the relationship between the fluid environment at the surface of the channel compared to an average of the bulk flow throughout the channel volume.

The data obtained showed a strong correlation between the mixing distance (MD) required to achieve an MI of 0.6 ($MD_{0.6}$), and the MI reached at the outlet, where a smaller $MD_{0.6}$ typically leads to a larger MI at the outlet. Since a short $MD_{0.6}$ and a large MI at the outlet both indicate efficient mixing, such a correlation can be expected. However, this was not always the case, as seen for W_4 DCMS 2 and W4 DCMS 4 (Figure 4d). In this instance, W_4 DCMS 2 and W_4 DCMS 4 resulted in an outlet MI of 0.82 and 0.75, respectively, which was unexpected as the corresponding $MD_{0.6}$ values were 0.52 cm and 0.40 cm. Therefore, although W_4 DCMS 4 possessed a more rapid initial mixing rate, the resulting MI at the outlet was lower than for other structures with similar initial rapid mixing.

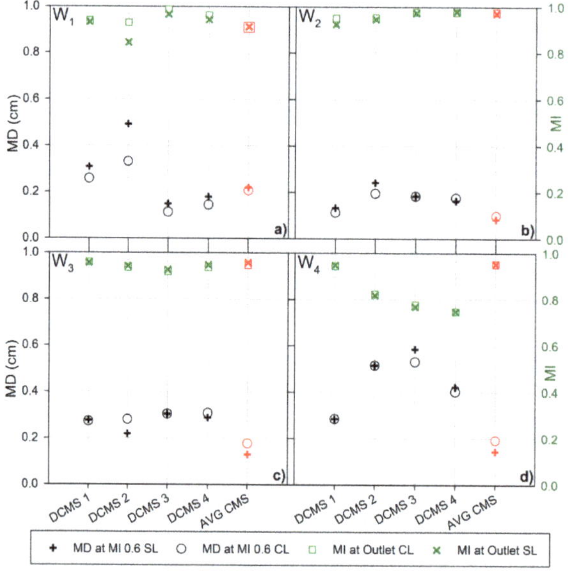

Figure 4. Mixing distance to achieve an MI of 0.6 (60%) (left Y-axis, black) for the channel layers (black ○) and the sensor layer (black +) and the mixing index at the outlet (right Y-axis, green) for the channel layers (green □) and the sensor layer (green ×) for the microfluidic channels with passive mixing structures, as indicated. The parameter values for the microfluidic channels with mixing structures with parameter design variables (Table 1) as prepared from (**a**) Wafer 1 (W_1), (**b**) Wafer 2 (W_2), (**c**) Wafer 3 (W_3) and (**d**) Wafer 4 (W_4), respectively. The average values for the previously reported curved mixing structure (AVG CMS) [56] is represented in red.

A similar trend was also observed for W_2 DCMS 1, where a rapid initial increase in MI resulting in an MI of 0.9 after roughly 0.3 cm, produced the lowest MI at the outlet, for the same wafer designs (Figure 3b). This indicates the presence of different length dependencies of the processes, inducing the homogenization throughout the channel or different optimal sensor locations. For example, comparing the MI for CL and SL illustrates this, where W_1 DCMS 2 (Figure 4a) clearly indicate a difference in the MI and $MD_{0.6}$ analyzed based on CL or SL. Nevertheless, the trend in performance still holds true, as the CL shows a shorter $MD_{0.6}$ and a larger MI compared to the SL.

The average values for the previously reported CMS design provide an easy comparison between the two design groups, as the DCMS is a direct evolution of the CMS design. The passive mixing channels fabricated based on W_2 and W_3 (Table 1, Figure 4b,c) resulted in the lowest range of efficiencies, largely due to the low performance seen for the DCMS group on W_4 (Figure 4d). The low performance for the DCMS group was surprising due to the otherwise low spread in the high efficiencies for the CMS and HBS previously reported [56].

Overall, the passive mixing channels with the CMS group yielded an MI after 1 cm (outlet) and $MD_{0.6}$ (Figure 5a,b) that was less sensitive to altered parameter values than the DCMS group. The CMS group resulted in the largest MI and the lowest $MD_{0.6}$, both for the CL and the SL. On average, the addition of the second curve resulted in lower MIs and larger $MD_{0.6}$ (Figure 5a,b), as analyzed both throughout the channel and for a layer adjacent to the possible sensor location, CL and SL, respectively. This was mainly due to the decrease in mixing performance observed on W_4 (Figure 4d), and the increase in $MD_{0.6}$ from W_1 and W_4 (Figure 4a,d). The CMS group was found to yield a statistically significant different variance (smaller) of the MI at the outlet than the DCMS (at level of significance $\alpha = 0.05$).

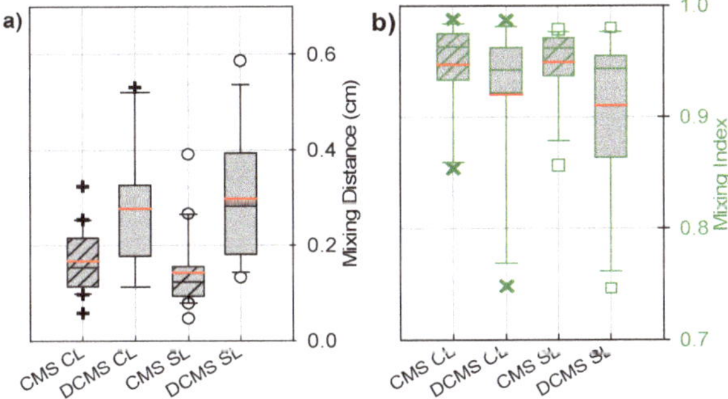

Figure 5. Boxplots depicting (**a**) a range of mixing distances required to achieve a mixing index of 0.6 (black) and (**b**) a range of mixing indices at the outlet (green) for the channel layers (CL) (black + and green ×) and surface layer (SL) (black ○ and green □) for the CMS (striped box) and DCMS (gray box) groups. The red and black/green lines within the boxes depict the mean and median values, respectively. The box corresponds to the interquartile range (50% of the data), whilst the lines (┬┴) represent 1.5× the interquartile range. The CMS data were obtained from our previous study [56].

The mixing efficiencies reported here for the asymmetric double curved mixing structures are also within a range obtained for other passive microfluidic mixers utilizing similar mixing lengths and Re numbers. Hossain et al. optimized the SHM placed on the top and bottom wall of a microchannel and achieved a simulated MI range of 0.68 to 0.93 for various design variables [42]. Kim et al. designed a barrier-embedded micromixer that achieved an MI of 0.40 at Re = 120 for a mixing length of 21 mm, although this was reduced to 0.23 for Re = 1 [57]. Comparing the present results also to include 3D dimensional mixers, the maximum MI reached was often over 0.9 [10], such as the serpentine crossing channel resulting in an MI ranging from 0.9–0.99 over an Re range of 0.01–120 [58–60]. Nevertheless, the 3D mixing designs such as the serpentine crossing channel do not readily support combination with biosensors integrated in the designs. Thus, the overall aim of those designs is directed to homogenization of the liquid at the outlet and does not support the feasibility for sensor integration, compared to the designs realized in the present study.

4. Conclusions

The results on the mixing index (MI) and the initial rate of mixing for a newly fabricated passive ceiling mixing structure group show a large dependence on the geometry of the structure. The addition of a second curved structure to the previously studied curved mixing structure (CMS) group [56] resulting in the novel double curved mixing structure (DCMS) had a drastic effect on the flow pattern induced by the mixers and the resulting efficiencies. The MI for the mixers were determined using confocal microscopy, where the homogenization process between water and fluorescein solution were analyzed for the specific regions of the channel volume, as well as for the sensor layer. This provided a description for the flow and progression throughout the channel in the XY and YZ planes of the channel.

The initial mixing rate was determined from the required mixing distance (MD) necessary for a design to achieve an MI = 0.6 ($MD_{0.6}$), where MI = 1 would be complete homogenization. Varying the geometric parameter channel height (H_C), mixing structure depth (D), spacing (S) and height (H_{PM}) resulted in a range of mixing efficiencies, where the three most important parameters to consider for the DCMS design were H_{PM}, D and S.

The most efficient DCMS, W_1 DCMS 3 resulted in an MI of 0.99, whilst the least efficient, W_4 DCMS 4, was 0.75, leaving a range from 0.75 to 0.99 for the MI at the outlet (after 1 cm of mixing). We compared these results to the CMS group reported previously [56], which showed an MI range from 0.99 (W_2 CMS 1) to 0.85 (W_1 CMS 2). The more extended overall range of MI at the outlet of the DCMS (0.75–0.99) as compared to that reported previously for the CMS (0.85–0.99) using similar spacings and pitches, indicating that the MI at the outlet are more sensitive to parameter values in the DCMS design than the CMS design. A similar relationship was also experienced at the surface of the channels, where the $MD_{0.6}$ range was 0.11–0.53 cm and 0.06–0.32 cm for DCMS and CMS respectively (Table A1).

Decreasing the height of the channel and the passive mixing structures resulted in an increase in the MI after 1 cm, as well as a reduction in $MD_{0.6}$ for the DCMS group. This could be due to the reduction in potential dead volumes experienced at the larger depth and height of the structure, or an increased efficiency at a larger velocity, as at the same flow rate the velocity of the fluid increases for smaller channels. The higher performance of the CMS could be due to the structure being more capable of inducing chaotic advection compared to the DCMS and the HBS. The DCMS group is therefore more sensitive to variation in its geometric parameters compared to the CMS group.

Here we have provided a large parametric study for the design of double curved passive mixing structure, where an efficient design can be achieved over a variation of channel heights. This provides a guide to design microchannels with efficient mixing.

Author Contributions: Conceptualization, I.H.O. and B.T.S.; methodology, I.H.O. and B.T.S.; software, A.Z.; validation, I.H.O. and B.T.S.; formal analysis, I.H.O.; investigation, I.H.O.; writing—original draft preparation, I.H.O.; writing—review and editing, I.H.O., A.Z. and B.T.S.; visualization, I.H.O.; supervision, B.T.S. All authors have read and agreed to the published version of the manuscript.

Funding: This work was supported by the Research Council of Norway within the Lab-on-a-chip Biophotonic Sensor Platform for Diagnostic, contract 248869/O70. The project is part of Center for Digital Life Norway and is also supported by the Research Council of Norway, grant 248810 The Research Council of Norway is acknowledged for the support to the Norwegian Micro- and Nano-Fabrication Facility, NorFab, project number 245963/F50.

Data Availability Statement: Relevant data are included in the manuscript and its Appendix A.

Conflicts of Interest: The authors declare no conflict of interest.

Appendix A

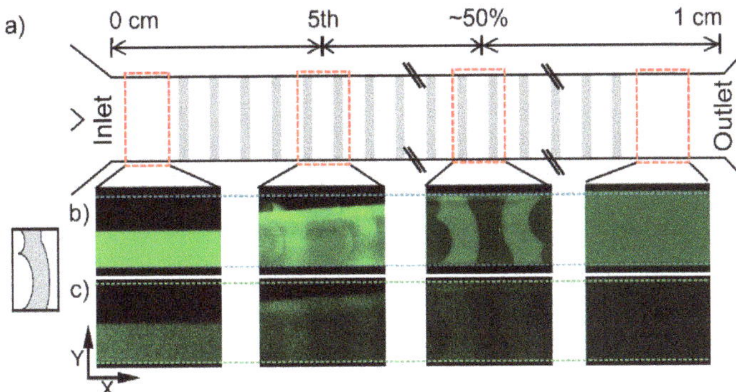

Figure A1. Schematic illustration of a mixing channel with arbitrary mixers (black dashed rectangles) for location purposes from the inlet (left) to the outlet (right). (**a**) Schematic illustration of a microfluidic mixing channel containing DCMS.Confocal XY view averaged intensity for (**b**) all Z levels and (**c**) the sensor layer. XYZT-stacks were acquired after each structure until the 10th structure, subsequent stacks were recorded after each structure. For mixing channels with fewer than 50 structures, stacks were recorded after every fifth structure.

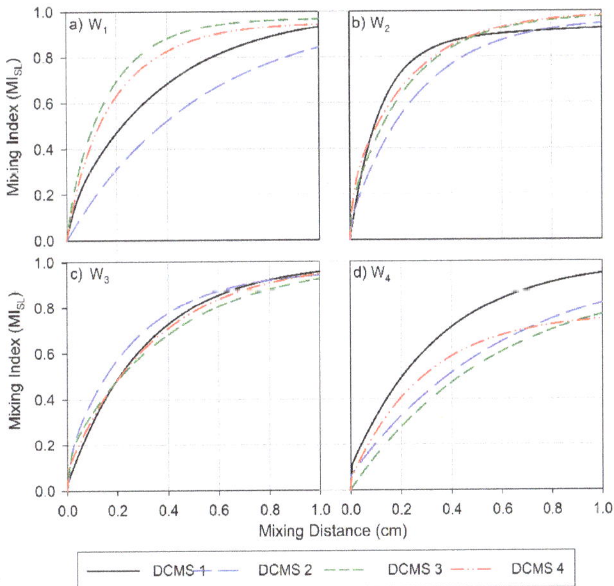

Figure A2. Mixing index at the surface layer versus the channel length for (**a**) Wafer 1, (**b**) Wafer 2, (**c**) Wafer 3 and (**d**) Wafer 4. Data obtained for channels with parameter design variables (PDV, Table 1) are depicted as follow: PDV 1 is a black solid line, PDV 2 is a blue long-dashed line, PDV 3 is a green three short-dashed line and PDV 4 is a red dashed-dot-dot line.

Table A1. Range in MI found at the outlet (1 cm) and the mixing distance (MD) required to obtain a mixing index of 0.6. The max and min values (and which design these represent) are presented for the channel layers (CL) and the surface layer (SL), as well as the range. Represented are the two design families compared in this article, CMS and DCMS, as well as a previously studied herringbone-type structure (HBS) presented in our previous article [56].

MI at Outlet	Max_{CL}		Min_{CL}		$Range_{CL}$	Max_{SL}		Min_{SL}		$Range_{SL}$
CMS	0.99	W_2 CMS 1	0.85	W_1 CMS 2	0.14	0.98	W_2 CMS 1	0.86	W_1 CMS 1	0.12
HBS	0.98	W_2 HBS 3	0.74	W_1 HBS 2	0.24	0.98	W_2 HBS 3	0.72	W_1 HBS 2	0.26
DCMS	0.99	W_1 DCMS 3	0.75	W_4 DCMS 4	0.24	*0.98*	W_2 DCMS 4	0.75	W_4 DCMS 4	0.23
MD (cm) at $MI_{0.6}$										
CMS	0.32	W_1 CMS 1	0.06	W_2 CMS 2	0.26	0.39	W_1 CMS 1	0.05	W_2 CMS 2	0.34
HBS	0.66	W_3 HBS 2	0.11	W_2 HBS 3	0.55	0.66	W_3 HBS 2	0.06	W_4 HBS 3	0.60
DCMS	0.53	W_4 DCMS 3	0.11	W_2 DCMS 1	0.42	0.59	W_4 DCMS 3	0.13	W_2 DCMS 1	0.46

References

1. Lee, C.Y.; Fu, L.M. Recent advances and applications of micromixers. *Sens. Actuators B Chem.* **2018**, *259*, 677–702. [CrossRef]
2. Gervais, L.; De Rooij, N.; Delamarche, E. Microfluidic chips for point-of-care immunodiagnostics. *Adv. Mater.* **2011**, *23*, 51–76. [CrossRef]
3. Hunt, H.K.; Armani, A.M. Label-free biological and chemical sensors. *Nanoscale* **2010**, *2*, 1544–1559. [CrossRef]
4. Marschewski, J.; Jung, S.; Ruch, P.; Prasad, N.; Mazzotti, S.; Michel, B.; Poulikakos, D. Mixing with herringbone-inspired microstructures: Overcoming the diffusion limit in co-laminar microfluidic devices. *Lab Chip* **2015**, *15*, 1923–1933. [CrossRef]
5. Frommelt, T.; Kostur, M.; Wenzel-Schäfer, M.; Talkner, P.; Hänggi, P.; Wixforth, A. Microfluidic Mixing via Acoustically Driven Chaotic Advection. *Phys. Rev. Lett.* **2008**, *100*, 034502. [CrossRef]
6. Shang, X.; Huang, X.; Yang, C. Vortex generation and control in a microfluidic chamber with actuations. *Phys. Fluids* **2016**, *28*, 122001. [CrossRef]
7. Ould El Moctar, A.; Aubry, N.; Batton, J. Electro-hydrodynamic micro-fluidic mixer. *Lab Chip* **2003**, *3*, 273–280.
8. West, J.; Karamata, B.; Lillis, B.; Gleeson, J.P.; Alderman, J.; Collins, J.K.; Lane, W.; Mathewson, A.; Berney, H. Application of magnetohydrodynamic actuation to continuous flow chemistry. *Lab Chip* **2002**, *2*, 224–230. [CrossRef]
9. Lee, C.Y.; Chang, C.L.; Wang, Y.N.; Fu, L.M. Microfluidic mixing: A review. *Int. J. Mol. Sci.* **2011**, *12*, 3263–3287. [CrossRef] [PubMed]
10. Cai, G.; Xue, L.; Zhang, H.; Lin, J. A Review on Micromixers. *Micromachines* **2017**, *8*, 274. [CrossRef]
11. Wang, X.; Liu, Z.; Wang, B.; Song, Q. Investigations into planar splitting and recombining micromixers with asymmetric structures. *J. Micromech. Microeng.* **2020**, *30*, 015006. [CrossRef]
12. Tsai, C.-H.D.; Lin, X.-Y. Experimental Study on Microfluidic Mixing with Different Zigzag Angles. *Micromachines* **2019**, *10*, 583. [CrossRef] [PubMed]
13. Chen, X.; Li, T.; Zeng, H.; Hu, Z.; Fu, B. Numerical and experimental investigation on micromixers with serpentine microchannels. *Int. J. Heat Mass Transf.* **2016**, *98*, 131–140. [CrossRef]
14. Vatankhah, P.; Shamloo, A. Parametric study on mixing process in an in-plane spiral micromixer utilizing chaotic advection. *Anal. Chim. Acta* **2018**, *1022*, 96–105. [CrossRef]
15. Rasouli, M.; Abouei Mehrizi, A.; Goharimanesh, M.; Lashkaripour, A.; Razavi, B.S. Multi-criteria optimization of curved and baffle-embedded micromixers for bio-applications. *Chem. Eng. Process. Process Intensif.* **2018**, *132*, 175–186. [CrossRef]
16. Borgohain, P.; Arumughan, J.; Dalal, A.; Natarajan, G. Design and performance of a three-dimensional micromixer with curved ribs. *Chem. Eng. Res. Des.* **2018**, *136*, 761–775. [CrossRef]
17. Lin, Y.C.; Chung, Y.C.; Wu, C.Y. Mixing enhancement of the passive microfluidic mixer with J-shaped baffles in the tee channel. *Biomed. Microdevices* **2007**, *9*, 215–221. [CrossRef]
18. Wang, L.; Ma, S.; Wang, X.; Bi, H.; Han, X. Mixing enhancement of a passive microfluidic mixer containing triangle baffles. *Asia-Pac. J. Chem. Eng.* **2014**, *9*, 877–885. [CrossRef]
19. Alam, A.; Afzal, A.; Kim, K.Y. Mixing performance of a planar micromixer with circular obstructions in a curved microchannel. *Chem. Eng. Res. Des.* **2014**, *92*, 423–434. [CrossRef]
20. Borgohain, P.; Dalal, A.; Natarajan, G.; Gadgil, H.P. Numerical assessment of mixing performances in cross-T microchannel with curved ribs. *Microsyst. Technol.* **2018**, *24*, 1949–1963. [CrossRef]
21. Meijer, H.E.H.; Singh, M.K.; Kang, T.G.; Den Toonder, J.M.J.; Anderson, P.D. Passive and active mixing in microfluidic devices. *Macromol. Symp.* **2009**, *279*, 201–209. [CrossRef]
22. Kang, T.G.; Singh, M.K.; Kwon, T.H.; Anderson, P.D. Chaotic mixing using periodic and aperiodic sequences of mixing protocols in a micromixer. *Microfluid. Nanofluid.* **2008**, *4*, 589–599. [CrossRef]
23. Floyd-Smith, T.M.; Golden, J.P.; Howell, P.B.; Ligler, F.S. Characterization of passive microfluidic mixers fabricated using soft lithography. *Microfluid. Nanofluid.* **2006**, *2*, 180–183. [CrossRef]

24. Mott, D.R.; Howell, P.B.; Golden, J.P.; Kaplan, C.R.; Ligler, F.S.; Oran, E.S. Toolbox for the design of optimized microfluidic components. *Lab Chip* **2006**, *6*, 540–549. [CrossRef]
25. Kee, S.P.; Gavriilidis, A. Design and characterisation of the staggered herringbone mixer. *Chem. Eng. J.* **2008**, *142*, 109–121. [CrossRef]
26. Mott, D.R.; Howell, P.B.; Obenschain, K.S.; Oran, E.S. The Numerical Toolbox: An approach for modeling and optimizing microfluidic components. *Mech. Res. Commun.* **2009**, *36*, 104–109. [CrossRef]
27. Fodor, P.S.; Itomlenskis, M.; Kaufman, M. Assessment of mixing in passive microchannels with fractal surface patterning. *EPJ Appl. Phys.* **2009**, *47*, 31301. [CrossRef]
28. Choudhary, R.; Bhakat, T.; Singh, R.K.; Ghubade, A.; Mandal, S.; Ghosh, A.; Rammohan, A.; Sharma, A.; Bhattacharya, S. Bilayer staggered herringbone micro-mixers with symmetric and asymmetric geometries. *Microfluid. Nanofluid.* **2011**, *10*, 271–286. [CrossRef]
29. Jain, M.; Rao, A.; Nandakumar, K. Numerical study on shape optimization of groove micromixers. *Microfluid. Nanofluid.* **2013**, *15*, 689–699. [CrossRef]
30. Kwak, T.J.; Nam, Y.G.; Najera, M.A.; Lee, S.W.; Strickler, J.R.; Chang, W.-J. Convex Grooves in Staggered Herringbone Mixer Improve Mixing Efficiency of Laminar Flow in Microchannel ed J Sznitman. *PLoS ONE* **2016**, *11*, e0166068. [CrossRef]
31. Nguyen, N.T.; Wu, Z. Micromixers—A review. *J. Micromech. Microeng.* **2005**, *15*, R1–R16. [CrossRef]
32. Haghighinia, A.; Movahedirad, S.; Rezaei, A.K.; Mostoufi, N. On-chip mixing of liquids with high-performance embedded barrier structure. *Int. J. Heat Mass Transf.* **2020**, *158*, 119967. [CrossRef]
33. Lee, C.Y.; Wang, W.T.; Liu, C.C.; Fu, L.M. Passive mixers in microfluidic systems: A review. *Chem. Eng. J.* **2016**, *288*, 146–160. [CrossRef]
34. Wang, H.; Iovenitti, P.; Harvey, E.; Masood, S. Numerical investigation of mixing in microchannels with patterned grooves. *J. Micromech. Microeng.* **2003**, *13*, 801–808. [CrossRef]
35. Yang, J.T.; Huang, K.J.; Lin, Y.C. Geometric effects on fluid mixing in passive grooved micromixers. *Lab Chip* **2005**, *5*, 1140–1147. [CrossRef]
36. Howell, P.B.; Mott, D.R.; Fertig, S.; Kaplan, C.R.; Golden, J.P.; Oran, E.S.; Ligler, F.S. A microfluidic mixer with grooves placed on the top and bottom of the channel. *Lab Chip* **2005**, *5*, 524–530. [CrossRef]
37. Stroock, A.D.; Dertinger, S.K.; Whitesides, G.M.; Ajdari, A. Patterning flows using grooved surfaces. *Anal. Chem.* **2002**, *74*, 5306–5312. [CrossRef]
38. Stroock, A.D.; Dertinger, S.K.W.; Ajdari, A.; Mezić, I.; Stone, H.A.; Whitesides, G.M. Chaotic mixer for microchannels. *Science* **2002**, *295*, 647–651. [CrossRef] [PubMed]
39. Cortes-Quiroz, C.A.; Zangeneh, M.; Goto, A. On multi-objective optimization of geometry of staggered herringbone micromixer. *Microfluid. Nanofluid.* **2009**, *7*, 29–43. [CrossRef]
40. Vasilakis, N.; Moschou, D.; Prodromakis, T. Computationally efficient concentration-based model for accurate evaluation of T-junction inlet staggered herringbone micromixers. *Micro Nano Lett.* **2016**, *11*, 236–239. [CrossRef]
41. Aubin, J.; Fletcher, D.F.; Xuereb, C. Design of micromixers using CFD modelling. *Chem. Eng. Sci.* **2005**, *60*, 2503–2516. [CrossRef]
42. Hossain, S.; Husain, A.; Kim, K.Y. Shape optimization of a micromixer with staggered-herringbone grooves patterned on opposite walls. *Chem. Eng. J.* **2010**, *162*, 730–737. [CrossRef]
43. Baik, S.J.; Cho, J.Y.; Choi, S.B.; Lee, J.S. Numerical investigation of the effects of geometric parameters on transverse motion with slanted-groove micro-mixers. *J. Mech. Sci. Technol.* **2016**, *30*, 3729–3739. [CrossRef]
44. Cortes-Quiroz, C.A.; Azarbadegan, A.; Zangeneh, M.; Goto, A. Analysis and multi-criteria design optimization of geometric characteristics of grooved micromixer. *Chem. Eng. J.* **2010**, *160*, 852–864. [CrossRef]
45. Bigham, S.; Nasr Isfahani, R.; Moghaddam, S. Direct molecular diffusion and micro-mixing for rapid dewatering of LiBr solution. *Appl. Therm. Eng.* **2014**, *64*, 371–375. [CrossRef]
46. Lopez, M.; Graham, M.D. Enhancement of mixing and adsorption in microfluidic devices by shear-induced diffusion and topography-induced secondary flow. *Phys. Fluids* **2008**, *20*, 053304. [CrossRef]
47. Arlett, J.L.; Myers, E.B.; Roukes, M.L. Comparative advantages of mechanical biosensors. *Nat. Nanotechnol.* **2011**, *6*, 203–215. [CrossRef]
48. Luchansky, M.S.; Bailey, R.C. Silicon photonic microring resonators for quantitative cytokine detection and T-cell secretion analysis. *Anal. Chem.* **2010**, *82*, 1975–1981. [CrossRef]
49. Wu, G.; Datar, R.H.; Hansen, K.M.; Thundat, T.; Cote, R.J.; Majumdar, A. Bioassay of prostate-specific antigen (PSA) using microcantilevers. *Nat. Biotechnol.* **2001**, *19*, 856–860. [CrossRef]
50. Gomez-Aranzadi, M.; Arana, S.; Mujika, M.; Hansford, D. Integrated Microstructures to Improve Surface-Sample Interaction in Planar Biosensors. *IEEE Sens. J.* **2015**, *15*, 1216–1223. [CrossRef]
51. Lynn, N.S.; Šípová, H.; Adam, P.; Homola, J. Enhancement of affinity-based biosensors: Effect of sensing chamber geometry on sensitivity. *Lab Chip* **2013**, *13*, 1413–1421. [CrossRef] [PubMed]
52. Lynn, N.S.; Homola, J. Biosensor enhancement using grooved micromixers: Part I, numerical studies. *Anal. Chem.* **2015**, *87*, 5516–5523. [CrossRef] [PubMed]
53. Lynn, N.S.; Špringer, T.; Slabý, J.; Špačková, B.; Gráfová, M.; Ermini, M.L.; Homola, J. Analyte transport to micro-and nano-plasmonic structures. *Lab Chip* **2019**, *19*, 4117–4127. [CrossRef] [PubMed]

54. Lynn, N.S.; Martínez-López, J.I.; Bocková, M.; Adam, P.; Coello, V.; Siller, H.R.; Homola, J. Biosensing enhancement using passive mixing structures for microarray-based sensors. *Biosens. Bioelectron.* **2014**, *54*, 506–514. [CrossRef]
55. Lynn, N.S.; Bocková, M.; Adam, P.; Homola, J. Biosensor enhancement using grooved micromixers: Part II, experimental studies. *Anal. Chem.* **2015**, *87*, 5524–5530. [CrossRef] [PubMed]
56. Oevreeide, I.H.; Zoellner, A.; Mielnik, M.; Stokke, B.T. Curved passive mixing structures: A robust design to obtain efficient mixing and mass transfer in microfluidic channels. *J. Micromech. Microeng.* **2021**, *31*, 015006. [CrossRef]
57. Kim, D.S.; Lee, S.W.; Kwon, T.H.; Lee, S.S. A barrier embedded chaotic micromixer. *J. Micromech. Microeng.* **2004**, *14*, 798–805. [CrossRef]
58. Raza, W.; Hossain, S.; Kim, K.Y. A review of passive micromixers with a comparative analysis. *Micromachines* **2020**, *11*, 455. [CrossRef]
59. Hossain, S.; Kim, K.Y. Optimization of a Micromixer with Two-Layer Serpentine Crossing Channels at Multiple Reynolds Numbers. *Chem. Eng. Technol.* **2017**, *40*, 2212–2220. [CrossRef]
60. Liu, R.H.; Stremler, M.A.; Sharp, K.V.; Olsen, M.G.; Santiago, J.G.; Adrian, R.J.; Aref, H.; Beebe, D.J. Passive mixing in a three-dimensional serpentine microchannel. *J. Microelectromech Syst.* **2000**, *9*, 190–197. [CrossRef]

Article

Numerical and Experimental Study of Cross-Sectional Effects on the Mixing Performance of the Spiral Microfluidics

Omid Rouhi [1,2,†], Sajad Razavi Bazaz [1,†], Hamid Niazmand [2], Fateme Mirakhorli [1], Sima Mas-hafi [3,4], Hoseyn A. Amiri [3,4], Morteza Miansari [3,4] and Majid Ebrahimi Warkiani [1,5,*]

1 School of Biomedical Engineering, University of Technology Sydney, Sydney, NSW 2007, Australia; omid.rouhi1989@gmail.com (O.R.); sajad.razavibazaz@student.uts.edu.au (S.R.B.); fateme.mirakhorli@student.uts.edu.au (F.M.)

2 Department of Mechanical Engineering, Ferdowsi University of Mashhad, Mashhad 91779-48974, Iran; niazmand@um.ac.ir

3 Micro+Nanosystems & Applied Biophysics Laboratory, Department of Mechanical Engineering, Babol Noshirvani University of Technology, P.O. Box 484, Babol 47148-71167, Iran; sima.mashafi@gmail.com (S.M.-h.); aamirihoseyn@gmail.com (H.A.A.); mmiansari@nit.ac.ir (M.M.)

4 Department of Cancer Medicine, Cell Science Research Center, Royan Institute for Stem Cell Biology and Technology, Isar 11, Babol 47138-18983, Iran

5 Institute of Molecular Medicine, Sechenov University, 119991 Moscow, Russia

* Correspondence: majid.warkiani@uts.edu.au

† These authors contributed equally as the first author.

Abstract: Mixing at the microscale is of great importance for various applications ranging from biological and chemical synthesis to drug delivery. Among the numerous types of micromixers that have been developed, planar passive spiral micromixers have gained considerable interest due to their ease of fabrication and integration into complex miniaturized systems. However, less attention has been paid to non-planar spiral micromixers with various cross-sections and the effects of these cross-sections on the total performance of the micromixer. Here, mixing performance in a spiral micromixer with different channel cross-sections is evaluated experimentally and numerically in the *Re* range of 0.001 to 50. The accuracy of the 3D-finite element model was first verified at different flow rates by tracking the mixing index across the loops, which were directly proportional to the spiral radius and were hence also proportional to the Dean flow. It is shown that higher flow rates induce stronger vortices compared to lower flow rates; thus, fewer loops are required for efficient mixing. The numerical study revealed that a large-angle outward trapezoidal cross-section provides the highest mixing performance, reaching efficiencies of up to 95%. Moreover, the velocity/vorticity along the channel length was analyzed and discussed to evaluate channel mixing performance. A relatively low pressure drop (<130 kPa) makes these passive spiral micromixers ideal candidates for various lab-on-chip applications.

Keywords: 3D printing; spiral micromixers; dean flow; trapezoidal cross-section; mixing index; convection and diffusion

Citation: Rouhi, O.; Razavi Bazaz, S.; Niazmand, H.; Mirakhorli, F.; Mas-hafi, S.; A. Amiri, H.; Miansari, M.; Ebrahimi Warkiani, M. Numerical and Experimental Study of Cross-Sectional Effects on the Mixing Performance of the Spiral Microfluidics. *Micromachines* **2021**, *12*, 1470. https://doi.org/10.3390/mi12121470

Academic Editor: Kwang-Yong Kim

Received: 29 October 2021
Accepted: 23 November 2021
Published: 29 November 2021

1. Introduction

With the progress in device miniaturization technology, micro-platforms have drawn significant attention in various realms, including environmental science, chemistry, and therapeutics [1]. Among them, microfluidics, which manipulate fluid flow inside a microchannel, has progressed considerably due to its privileges such as enhanced control over the experiment, reduced reagent consumption, and the ability to fabricate compact and portable devices [2]. Accompanied by immense improvements in such platforms, their application has flourished drastically in several fields, including in drug delivery [3], cell isolation and lysis [4,5], polymerization [6], DNA amplification [7], and crystallization [8]. In many microfluidic devices, the sample should be mixed with a reagent prior to experimentation. However, the flow regime in most of these microfluidic devices is laminar, where slow mixing occurs by pure diffusion, making efficient mixing challenging. To address this concern, diverse micromixer designs have been proposed and optimized. Thus, once the purpose of an experiment has been identified, the optimum design and fabrication method should be selected based on the operational requirements and the available equipment of the practiced application.

Generally, micromixers are divided into two categories based on the external energy usage aspect: active and passive [9]. Active micromixers employ various acoustic, electric, magnetic, or optic forces to enhance the mixing efficiency in a short period of time. However, they have some serious drawbacks, such as detrimental effects on the biological reagents or troublesome integration with other microfluidic devices since they are inherently slow, and matching their flow rates with other microfluidic components is challenging [10]. On the other hand, passive micromixers rely on optimizing the channel geometry, creating chaotic advection, or enhancing the molecular diffusion to boost the mixing efficiency. In this regard, their simplicity, low-cost fabrication, and ease of integration with other lab-on-a-chip platforms provide better functionality for practical applications. Hence, passive micromixers have become of interest and are considered to be a viable option for commercialization [11].

In passive micromixers, diffusion and advection are the two main mechanisms that can be used to control mixing inside a micromixer. Molecular diffusion is inherently very slow. There are limited methods that can be used to increase the efficiency of this method: lamination or split and recombine (SAR), although they are not as productive as advection-based mixing mechanisms. On the other hand, advection has more impact. To enhance the mixing, different types of micromixers use chaotic advection concepts, such as staggered herringbones [12], serpentines [13,14], and spirals [15]. The idea of combining different mixing units has also been tested [16]. Due to the centrifugal force inside of a curved channel, the maximum velocity point in the parabolic profile of laminar flow is drawn towards the outer wall, creating a velocity gradient and consequently an adverse pressure gradient along the channel cross-section. These changes cause two counter-rotating vortices that are perpendicular to the mainstream that are called Dean vortices. The Dean number indicates how strong those vortical patterns are and is proportional to the Reynolds number (Re), hydraulic diameter (D_h), and radius of curvature (R) [17,18].

In addition to the significant impact of high mixing efficiency in a sample preparation step during an experiment, incubation time is of great importance in numerous applications, including during protein coating and crystallization [19–21]. This can be explained by the fact that for a complete reaction, two or more samples and reagents need to contact each other for a specific amount of time, and the only way to attain this is to increase the mixing length. Nevertheless, there are limitations in increasing the mixing length, such as extending the fluid path, increasing the pressure drop, and enlarging the channel footprint, which conflicts with miniaturization and portability. Therefore, there is a need to have a maximum mixing length in a minimum footprint, i.e., a compact, efficient micromixer [22]. In this regard, the spiral microchannels are proper candidates. This design not only satisfies the requirements of a compact micromixer but also has a relatively high mixing index and straightforward fabrication process.

Howell et al. [23] were the first to propose spiral-shaped mixer devices. Although the channel dimensions were big, they studied the vortex formation at different Reynolds numbers, channel aspect ratios, and initial radii. Expectedly, later studies have developed this structure for micromixers and have applied different techniques with various modifications to enhance the mixing efficiency. Moreover, a combination of varying mixing effects such as SAR and expansion with Dean vortices have been studied [24]. These effects were further characterized using confocal microscopy techniques, which presented a cross-sectional view of transverse Dean flow inside of the curved channel. Various geometrical features can influence the mixing quality. The angle between two inlets, for example, is an essential parameter in mixing processes. Zhang et al. [25] investigated this effect in the mixing performance of spiral micromixers both numerically and experimentally. The mixing results showed that the mixing index (MI) improved by increasing the inlet mixing angle, with the best angles being obtained between 90–180 degrees. As microchannel cross-sections are of great diversity, numerical methods are applied as an efficient comparative tool for various operational conditions. Kumar et al. [26] computationally investigated the fluid dynamics of a cross-section for a curved tube in contrast with a straight tube. Based on Reynolds number and Schmidt number (*Sc*), the study showed that mixing is enhanced at a moderate *Re* (~10) for a higher *Sc*, while this behavior does not apply to a *Re* in the order of 0.1. Another research study on the effect of the cross-section aspect ratio in spiral micromixers showed that mixing performance in high-aspect ratio micromixers is better for a wide range of *Re* ranging from 1 to 468 [27]. Furthermore, a parametric study was carried out on the mixing process in spiral micromixers [28]. This study discussed various key variables on mixing efficiency, such as initial radius and cross-section shape, where they compared the impact of inlet velocity between straight and spiral micromixers. They reported that the mixing performance increases when the initial radius decreases while the straight micromixer acts oppositely. Very recently, Wang and colleagues studied the mixing enhancement of various spiral structures, including Archimedean, Logarithmic, Hyperbolic, Golden, and Fibonacci spirals, and concluded that the Archimedean spiral has the best performance among the other types [29]. However, so far, the effects of channel cross-section on the total performance of spiral micromixers have not been investigated.

Alongside all of the parameters and functionalities of micromixers, finding the proper fabrication method is vital for device miniaturization [30]. The softlithography method has been widely used due to the channel transparency for live monitoring, biocompatibility, and gas permeability [31,32]. However, in this method, master mould fabrication requires several laborious processes such as photolithography or micro-milling, which are time-consuming and are mainly applied to the manufacture of planar devices. To address these issues, additive manufacturing or 3D printing has evolved as a promising method for fabricating complex devices. In this method, parts with intricate geometries can be manufactured with high precision [33,34].

The present work concentrates on analyzing the mixing qualities of several spiral micromixers with unconventional cross-sections. Using numerical and experimental results, we nominated the most efficient cross-section for flow field and performance analysis in terms of the mixing index and the pressure drop at the outlet. The experimentally validated numerical model was conducted using the 3D finite element method to solve the convection–diffusion equation as well as the Navier–Stokes momentum equation. In addition, the mixing qualities and pressure drops were evaluated across a wide range of Reynolds numbers. Then, the effect of the velocity field and Dean flow on the mixing mechanism was investigated in detail for several flow conditions. Finally, the vortical impact was explored as the fluid poured into each loop, leading to compact, efficient micromixer structures being obtained according to the required application and flow rates. The selected designs were compared with each other using the mentioned performance parameters. Consequently, a general map of the mixing index was obtained for the most appropriate design by changing the *Re* along the loops.

2. Materials and Methods

In the following, the proposed spiral micromixer structures, design, and fabrication process are described. After selecting the micromixer designs, their efficiency is evaluated at the same *Re* by employing numerical simulation. The underlying theories and governing equations for the current physics are discussed, and analysis tools are used to achieve the optimal design from a range of geometrical/operational features.

2.1. Spiral Designs and Fabrication Methods

A spiral microchannel with trapezoidal cross-section was designed by our group [35] for CTC isolation and had a width of 600 μm, a 80 μm height in the inner wall, and a

130 μm height for the outer wall, which was previously modified here in terms of the height ratio, cross-section, and the position of the walls in order to investigate the effect of each factor on the mixing index. Here, two major changes have been made to the original cross-sectional design: First, the channel height ratio, being the larger height relative to the smaller one, has changed compared to the original design. Thus, the 80 μm wall was reduced to 60 μm and 40 μm, while the larger other wall helped to maintain a constant hydraulic diameter and volume (enabling valid comparison based on *Re*). Since the inner wall (80 μm) was modified to be shorter than the outer wall in the original pattern, the second step was to flip the channel cross-section, i.e., switching the position of the sidewalls. In the end, a rectangular cross-section was drawn in order to gain insight into the effect of the cross-section shape. Therefore, in all of the geometries, the key variable is channel cross-section. While the fluid flows through a curve, the inner/outer near-wall velocity decreases/increases due to the centrifugal force and the pressure gradient. The pressure variation enforces a secondary flow that is perpendicular to the main flow, which benefits the mixing process in high inertial fluid states. Hence, as illustrated in Figure 1, a total of seven cross-sections comprising outward and inward (mirrored) trapezoidal and rectangular geometries were studied to determine the most efficient structure.

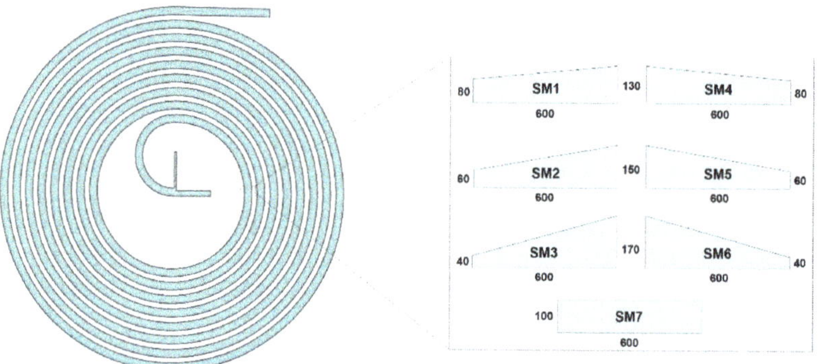

Figure 1. Schematic of the microchannel and the proposed cross-sections along with the applied boundary conditions. SM1 to SM3 in the left column are trapezoidal cross-sections before flipping when the outer wall is shorter, while SM4 to SM6 on the right column refer to mirrored ones with shorter inner walls. The cross-section hydraulic diameter and total fluid volume capacity of the micromixer are kept equal in all cases. All dimensions are in micrometers.

All of the devices were fabricated using a high-resolution DLP 3D printer (Miicraft, Hsinchu, Taiwan) with a 32 × 57 × 120 mm^3 printing volume and 30 μm XY resolution. The geometry was printed on one side and was then attached to a substrate (PMMA sheet) using a double-sided adhesive. The desired micromixer device was initially drafted using a commercially available CAD drawing software (SolidWorks 2019, Waltham, MA, USA) and was then exported in STL format. This STL format is suitable for 3D printer language. The Miicraft software (Version 6.1.0.t4, Miicraft, Hsinchu, Taiwan) was used to slice the file in the Z direction. The slicing in the Z direction (slice thickness) can be adjusted from 5 to 200 μm with an increment of 5 μm [36,37].

The complexity of the channel assists the user in choosing the proper slicing option. A ramp or steps in geometries leads to a smaller slice thickness, while orthogonal or planar channel structures result in a higher slice thickness. Afterward, the file, which was previously sliced, was sent to the 3D printer, which has a UV wavelength of 358–405 nm UV. The UV light was projected from the bottom of the resin bath, which had already been filled with BV-007 resin. The UV light passed from a transparent Teflon film and cured the resin on the picker of the 3D printer. As soon as one layer formed on the picker, the stepper motor relocated one layer and started printing another layer. This process continued until the whole part was printed successfully. Once the part was printed, it was removed from the picker and was washed carefully with isopropanol three times. As a post-curing step, the channel was placed in a UV chamber with a wavelength of 405 $\pm$ 5 nm. Finally, the tubes were mounted on the parts using tweezers. The inlet and outlet holes were printed in a way where they could keep the tubes firmly together to eliminate any chance of leakage [38]. The parts printed via this method were of high quality; we previously characterized the surface characteristics of the fabricated parts and measured and reported the surface roughness (*Sa*) of the parts that were printed via this method. Its value was less than 300 nm, indicating that surface roughness does not disturb channel performance and that it is appropriate for various microfluidic applications [39].

2.2. Numerical Model

Newtonian fluid flow is considered to be steady, incompressible, and laminar; hence, general governing equations for fluid mixing include the continuity of mass and momentum [40], which are shown as:

$$\nabla \cdot \boldsymbol{V} = 0 \tag{1}$$

$$\frac{\partial \boldsymbol{V}}{\partial t} + \rho(\boldsymbol{V} \cdot \nabla)\boldsymbol{V} = -\nabla P + \mu \nabla^2 \boldsymbol{V} \tag{2}$$

where $\boldsymbol{V}$, P, ρ, and μ are the fluid velocity field, pressure, density (998 kg/m^3), and dynamic viscosity (8.9 $\times$ 10^{-4} Pa·s), respectively. The boundary conditions include constant velocity inlets ($V_{in,1} = U_1$ and $V_{in,2} = U_2$), the pressure outlet ($P_{out} = 0$), and the no-slip walls for all cases. Since the hydraulic diameter and therefore the flow volume does not change in the different investigated structures, the inlet velocities differ throughout the *Re* range. The convection–diffusion equation governs the species concentrations (c) for fluids with a diffusion coefficient of D (2 $\times$ 10^{-9} m^2/s), which is similar to other works [26–28,35,40–42] and can be determined as:

$$\frac{\partial c}{\partial t} + (\boldsymbol{V} \cdot \nabla)c = \frac{1}{ReSc}\nabla^2 c \tag{3}$$

where Re is the Reynolds number of the channel and where Sc is the Schmidt number, which defines as the ratio of momentum to mass diffusivity. The concentrations are considered as 0 and $c_0 = 1$ mol/m^3 for the inlet boundary condition, where the average molar fraction of $\bar{c} = 0.5$ mol/m^3 is referred to as complete mixing. Consequently, the coupled governing equations were solved in COMSOL Multiphysics 5.3a (COMSOL, Burlington, MA, USA) by applying the proper boundary conditions. Mixing quality was evaluated by the mixing index criterion that was defined based on the deviation from $\bar{c}$ [43]:

$$MI = 1 - \sqrt{\frac{1}{n}\sum_{i=1}^{n}\left(\frac{c_i - \bar{c}}{\bar{c}}\right)^2} \tag{4}$$

where MI, n, c_i, and $\bar{c}$ are the mixing index, number of sample points, concentration of the species, and average concentration, respectively.

2.3. Sample Preparation

In order to visualize the mixing process through bright field imaging, food coloring (Queen Fine Foods, Alderley, Queensland, Australia) was used. To prepare the sample, 1 mL from each dye was dissolved in 49 mL of MACS buffer (Miltenyi Biotec, Bergisch Gladbach, Germany) containing phosphate-buffered saline (PBS) and 2 mM ethylenediaminetetraacetic acid (EDTA) supplemented with 0.5% BSA, and 0.09% sodium azide.

2.4. Experimental Results

An inverted microscope (IX73, Olympus, Tokyo, Japan) equipped with a CCD camera (DP80, Olympus, Tokyo, Japan) was used for brightfield imaging. Samples (green and red) were filled into the 50 mL plastic Syringe (BD, Franklin Lakes, NJ, USA) and were injected into the micromixer using a syringe pump (Chemyx Fusion 200, Stafford, TX, USA).

3. Results and Discussions

3.1. Grid Study

Grid independence is of significant importance and directly affects the accuracy of the results. The quality of the results obtained with a small number of elements is not high since only a small number of discrete points within the domain are evaluated and solved. On the other hand, simulation with an increased number of elements results in more accurate data, increasing computational time. Consequently, a trade-off between the number of elements and calculation time is required. Accordingly, an initial simulation was performed with three grid numbers to evaluate the effect of the element size on the mixing index in a trapezoidal spiral micromixer. Figure 2A shows a big gap between the results for the 2×10^6 and 3×10^6 mesh numbers. Subsequently, another simulation was performed with 5×10^6 elements, and the difference between the two results is negligible. Henceforth, the latter was selected for the simulations, and its details, such as tetrahedral elements and boundary layer cells, are demonstrated in Figure 2B.

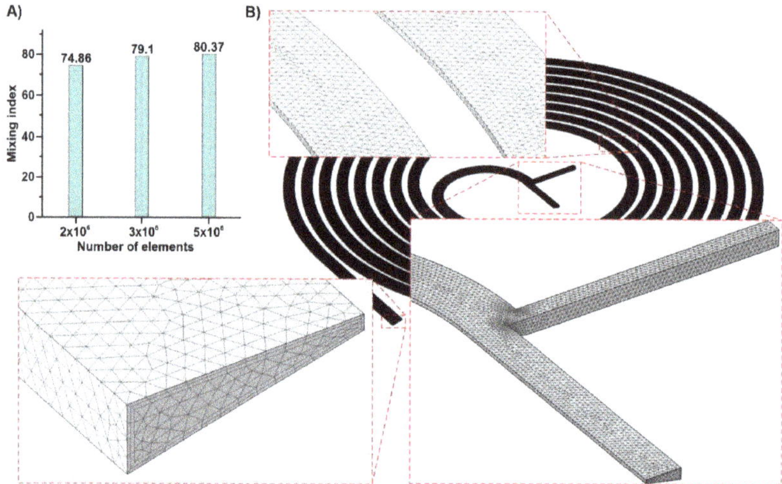

Figure 2. The information for the applied grids. (**A**) The grid independence study; (**B**) The selected grid is shown in detail in several regions of the channel.

3.2. Spiral Micromixers with Various Cross-Sections

Spiral microfluidic devices have shown great promise in the area of microfluidic because of their flexibility when parameters such as cross-section, initial radius, and the number of loops change. Additionally, changing the above parameters can flexibly affect enormous interactive forces and regimes inside a spiral microfluidic device. This feasibility leads to various applications for these types of devices such as mixing, focusing, and separation, all of which being applications where the channel cross-section plays the most critical role in their functionality [44,45]. In this regard, Figure 1 shows the changes that were applied in order to determine the effect of the cross-sectional parameters in a spiral micromixer based on the initial geometry.

The numerical model was implemented on a variety of spiral micromixers that had been discussed earlier to capture the slightest concentration gradient and to identify the most efficient structure. The final mixing index at $Re = 5$, which is shown in Figure 3A, is generally known as having the least efficient Reynolds number, highlights that SM3 offers the best overall mixing qualities. Therefore, the higher the height ratio is, the better the mixing index becomes, and this is due to the broadened centrifugal force and the secondary flows. As a result, when the fluid moves from the shorter wall to the larger wall, the higher wall height ratio creates a greater surface area near the longer wall, allowing sufficient space for secondary flows, increasing the chance of mixing. Upon that, the cross-section benefits from the forces acting on in- or outward direction of the fluid. Moreover, due to the significantly smaller height ratios, the mixers with an 80 μm height (SM1 and SM4) perform somewhat similarly to the rectangular micromixer (SM7), exhibiting the least predictable mixing performances.

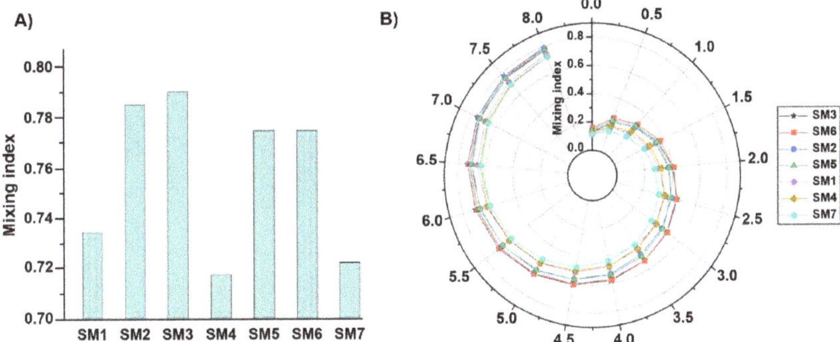

Figure 3. Evaluation of the mixing performance at Re = 5 for the proposed micromixers based on (**A**) final and (**B**) along-the-loop mixing index. The impact of height ratios on mixing quality is more significant than the outward/inward position of the shorter wall.

In Figure 3B, the mixing index growth along the spirals suggests that the performance of these mixers rises gradually as the fluids flow through each loop. The difference between the designs with the high height ratio (SM3 and SM6) and other geometries is evident from the initial loops, which is remarkable until the sixth loop. In contrast, poor performance was observed for cases with a low height ratio, which is evident from the beginning of the spiral. Nevertheless, another trend that was observed shows that the outward channels outperform the inward channels. Overall, the height ratio is far more influential than the inward/outward positioning of the walls. Final evaluations confirm SM3 as the most efficient structures. Thus, it was nominated as the micromixer structure of choice.

3.3. Numerical Validation

Figure 4 depicts the comparison of the mixing trends between the numerical method and experimental runs of the nominate channel (SM3). In this experiment, the samples were loaded into BD plastic syringes with Luer-Lok tips, were mounted on a syringe pump (Chemyx Fusion 200, Stafford, TX, USA), and were then loaded into the channels via Tygon tubing. We used tubing with an inner diameter of 0.02 and an outer diameter of 0.06 inches. For the visualization of the mixing process in the bright field, food coloring (Queen Fine Foods, Alderley, Queensland, Australia) were used. The inlet flow rates were set to 12.5 and 9.5 μL/min for Re = 1, 125 and 95 μL/min for Re = 10, and 250 and 190 μL/min for Re = 20 for each inlet. The results for the three nominal Reynolds numbers show similar mixing behavior before, at, and after the critical Re = 5. The snapshots were captured after a sufficient amount of time, meaning that the flow represents a steady state. This agreement between both methodologies is satisfactory in the initial loops wherein the widthwise concentration gradient is relatively more intense. Moreover, the mixing that takes place in the earlier stages occurs at the two-fluid interface, producing a narrow area. This mixing interface is shifted towards the center of the spiral with the Re growth due to the involvement of the Dean vortices. Accordingly, an efficiently mixed flow is obtained with fewer loops and when using a greater velocity. To quantitatively validate numerical simulations, Fiji, an image processing software, was used to extract the values of the mixing index from the experimental snapshots. To this end, a line that was parallel to the channel width at the inlet and outlet of the microchannel was drawn, and the grayscale values of the image were extracted. These values have been normalized using Equation (5).

$$I_i = \frac{I_i^* - I_{min}^*}{I_{max}^* - I_{min}^*} \quad (5)$$

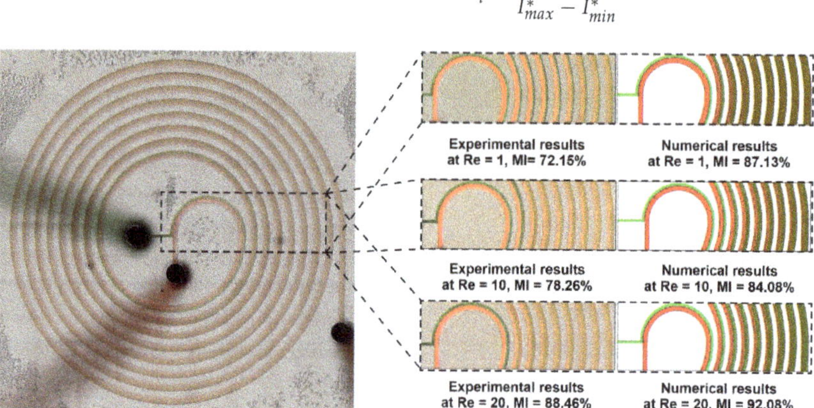

Figure 4. Numerical model validation and assessment with experiments at *Re* = 1, 10, and 20 using SM3. An overall view of mixing performance obtained via experimental observation (**left**) and numerical simulation (**right**). Fluids in the channel are shown in two distinct colors: green and red. The numerical errors seem to be insignificant, and the model agrees well with experiments in different situations.

Here, I_i^* is the actual intensity, and I_{min}^* and I_{max}^* are the minimum and maximum intensity values at the channel inlet (unmixed fluid), respectively. Using these values, the experimental mixing index can be calculated using Equation (6).

$$MI_{experimental} = 1 - \frac{\sqrt{\frac{1}{N}\sum_{i=1}^{N}\left(\frac{I_i - \bar{I}}{\bar{I}}\right)^2}}{\sqrt{\frac{1}{N}\sum_{i=1}^{N}\left(\frac{I_{min,max} - \bar{I}}{\bar{I}}\right)^2}} \quad (6)$$

where the pixel numbers in the inlet or outlet image along a line parallel to the channel width are represented by N, $I_{min,max}$ is either 0 or 1, and $\bar{I}$ is 0.5. Using the above formulas, the $MI_{experimental}$ for *Re* = 1, 10, and 20 was extracted and calculated and is shown in Figure 4. The difference between the values is mainly related to the fact that the mixing index in numerical simulations is evaluated along the channel cross-section, while in experimental snapshots, it is evaluated from the top view. In all, both the qualitative and quantitative comparison of the mixing process reveals that the numerical simulations agree well with experimental ones, proving that computational fluid dynamics is a proper approach for the investigation of the mixing process that occurs within micromixers.

3.4. Nominated Designs: A Comparison

As observed in Figure 5, the nominated micromixer was analyzed in a wide range of *Re* = 0.001–50 to highlight the performance advantages in different working conditions. More cases were analyzed between *Re* = 1–10 because there were more fluctuations and sensitivity during this period. There is a negative balance spot between the diffusion and convection regimes at *Re* = 5. Thus, Figure 5A confirms *Re* = 5 as the critical condition with the least efficient mixing process, which still demonstrates a suitable amount of mixing at 80%. Before this *Re* and hence at lower flow velocities, there is more time for the fluid molecules to disperse and diffuse onto the interface. In general, the available time reduces by increasing the *Re*, and the vortices/secondary flows are not initially strong enough to compensate for the time reduction. As a result, the mixing index is poor until the cross-sectional flow can overcome this weakness after *Re* = 5, leading to a rapid increase in the mixing quality. In conclusion, starting from a *Re* = 0.001 with M.I. = 88%, the mixing quality first decreases, and the critical *Re* increases afterwards. Nevertheless, in the present case, the mixing efficiency usually saturates after a flow rate is assigned to *Re* = 30. Furthermore, the flows with the highest flow rate create the best mixing, with mixing efficiencies of up to 95% being demonstrated at *Re* = 50. However, this mixing quality also causes the highest pressure drop, which is depicted in Figure 5A. Although the micromixer performs appropriately in the studied velocity range, the optimum efficiency is achieved at moderate Reynolds numbers (*Re* = 20–40) without reaching the 100 kPa pressure-drop limit.

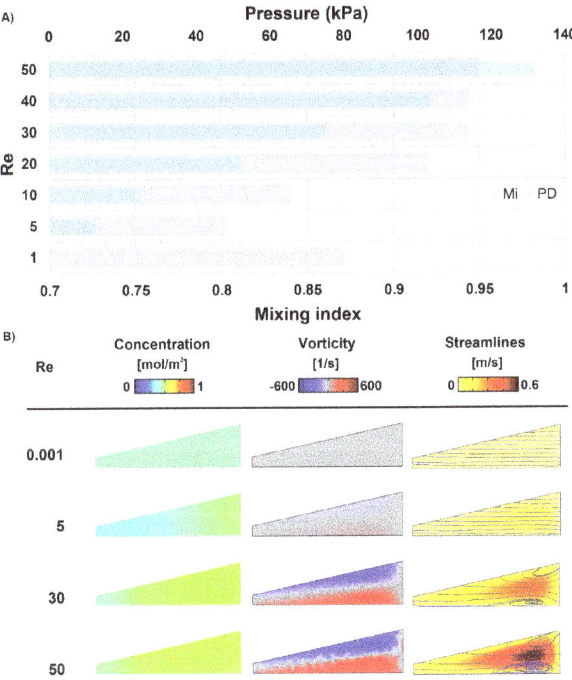

Figure 5. Mixing parameters evaluated at the final loop of SM3 in *Re* = 0.001–50. (**A**) Mixing index variations and pressure drop changes achieved by altering flow rates; (**B**) Concentration distribution, vorticity intensity, and fluid streamlines at the final loop of the nominated design for the selected flow conditions.

The influence of the *Re* is further confirmed by observing the concentration distribution, vorticity contours, and streamlines at the outlet in Figure 5B. The outlets have a uniform concentration distribution at *Re* < 5, where mixing is mainly guided by molecular diffusion; hence, the order of magnitude is small for the vorticity, leading to insignificant changes in the streamlines of the laminar flow. Therefore, both fluids are evenly mixed throughout the channels due to the low speed and sufficient time for molecular mixing. This uniformity, however, does not appear at a moderate *Re* of more than five since the molecular diffusion becomes less dominant and because the transition to the leading vortex effect occurs. By moving further away from that region, the mixing index starts to be much higher, and the concentration is not heterogenous all over the cross-section. Finally, the outer wall regions achieved better mixing due to the strengthened vortices at *Re* > 30 (*Re* = 30 is when the vortical pattern emerges) given the adequate total length/mixing time. However, the no-slip condition produces dead zones for mass transfer and for the stirring effect around the inner/smaller wall. On the other hand, the streamlines, which reveal that the exact fluid flow paths in the cross-section continue to increase as the *Re* increases. This results in enhanced mixing, as these secondary flows create wider/stronger vortices.

Velocity analysis was also performed to highlight the mechanism by which the vortices and secondary flows enhance the mixing process. Figure 6A shows the Dean number map for the current design based on the Reynolds number at a loop number. It is known that the mixing efficiency improves as the *Re* rises since stronger vortices emerge, which is indicated by *De* as a manifestation of the flow characteristics. Moreover, as previously demonstrated, by increasing the velocity, the Dean number changes dramatically and is able to go through the loops; this therefore increase the changes of the fluids being properly mixed. The vortex magnitude is illustrated in Figure 6B at *Re* = 50, which shows large decreases in the strength of the vortices and in the *De* number as the fluid moves along the spiral. Thus, the primary loops offer better secondary flow effects. This is a result of the inverse relationship between the Dean flow and curvature radius. Therefore, long channels do not aid the secondary flow influence in the spiral channels, and/or fewer loops are necessary since the mixing is less dependent on the number of loops in this secondary flow state.

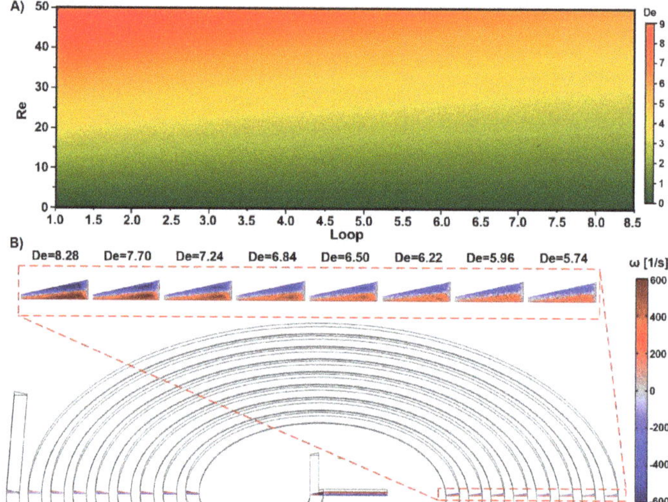

Figure 6. Illustration of mixing dynamics and mechanism by means of Dean flow and vortices in SM3. (**A**) The map of the Dean number is provided at Reynolds number and channel loops; (**B**) The out-of-plane component of the vorticity field at *Re* = 50 through the loops. Vortices are reduced throughout the spiral channel, and the initial loops provide better secondary flows since the dean flow grows in contrast to the curve radius.

Another aspect to be considered is the required loops in each flow condition. As the flow goes through the loops of the spiral micromixer, two distinct mixing patterns were detected in the mixing index growth, one is diffusion based and one is advection based, which are depicted in Figure 7A,B. The assessment of this trend is summarized and mapped for all of the applied flow rates in Figure 7C. Contrary to $1 < Re < 10$ flows, it can be seen that the very low and high *Re* conditions achieved proper mixing in a much smaller number of loops. In these cases, the mixing remains unchanged after only a few loops. Thus, those loops are required within these ranges, and the structure can be deliberately modified according to the application requirements that substantially reduce the input pressure usage.

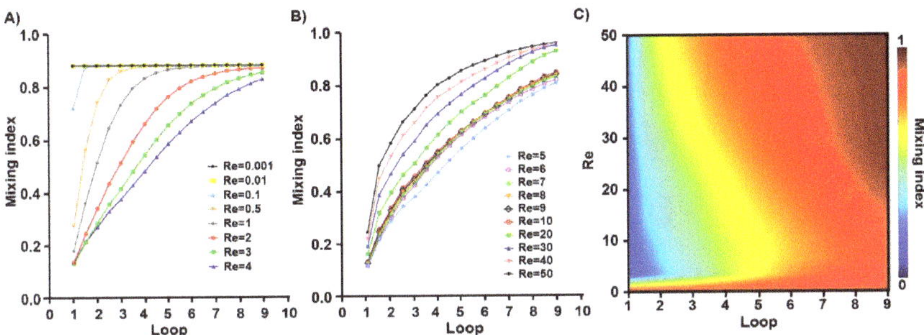

Figure 7. The mixing trend index for SM3 loops and Reynolds numbers. The mixing index growth over the length (**A**) before $Re = 5$; (**B**) after $Re = 5$; (**C**) the mixing contour along the spiral loops is shown for all of the applied flow rates. Very low and high velocities show proper mixing in fewer loops. No further mixing occurs after the initial loops in high *Re*, benefiting the fabrication and application requirements.

4. Conclusions

In the present work, seven different spiral geometries were proposed and examined by experiments and using a 3D-finite element model. The nominated channel was further analyzed after finding the most efficient cross-section based on mixing performance, pressure drop, and velocity variations. The mixer with the large-angle outward trapezoidal cross-section achieved the best mixing performance and achieved the highest height aspect ratio that was studied (4.25). The mixing index of the nominated device reduced the diffusion-dominant mixing mechanism from 88% at $Re = 0.001$ to 80% at $Re = 5$. The turning point at $Re = 5$ was associated with the transition of the mixing mechanism from a diffusion to a convection mechanism. Once the convection mechanism became dominant, the mixing index reached 95% at $Re = 50$, corresponding to a pressure drop of 130 kPa. Further investigation of the loop number effect in terms of velocity and vortex emergence, which influences the mixing performance, indicated that the desired mixing rate can be achieved at a higher *Re* and by using a lower number of spiral loops. This can significantly decrease the device footprint. This study may open up new avenues for the further understanding of spiral micromixers and can be used for various applications ranging from particle synthesis to crystallization, where proper mixing and enough incubation time is necessary.

Author Contributions: Conceptualization, M.E.W.; data curation, O.R. and S.R.B.; formal analysis, O.R. and S.R.B.; funding acquisition, M.E.W.; investigation, O.R. and S.R.B.; methodology, O.R. and S.R.B.; project administration, M.E.W.; resources, M.E.W.; software, F.M., S.M.-h. and H.A.A.; supervision, H.N., M.M. and M.E.W.; visualization, F.M.; writing—original draft, O.R., S.R.B., F.M., S.M.-h., H.A.A. and M.M.; writing—review and editing, O.R., S.R.B., H.N., F.M., S.M.-h., H.A.A., M.M. and M.E.W. All authors have read and agreed to the published version of the manuscript.

Funding: This research was funded by the Australian Research Council through Discovery Project Grants (DP170103704 and DP180103003) and the National Health and Medical Research Council through the Career Development Fellowship (APP1143377).

Data Availability Statement: The data presented in this study are available upon request from the corresponding author.

Acknowledgments: M.E.W. would like to acknowledge the support from the Australian Research Council through Discovery Project Grants (DP170103704 and DP180103003) and the National Health and Medical Research Council through the Career Development Fellowship (APP1143377).

Conflicts of Interest: The authors declare no conflict of interest.

References

1. Franke, T.A.; Wixforth, A. Microfluidics for Miniaturized Laboratories on a Chip. *ChemPhysChem* **2008**, *9*, 2140–2156. [CrossRef] [PubMed]
2. Whitesides, G.M. The origins and the future of microfluidics. *Nature* **2006**, *442*, 368–373. [CrossRef] [PubMed]
3. Damiati, S.; Kompella, U.B.; Damiati, S.A.; Kodzius, R. Microfluidic Devices for Drug Delivery Systems and Drug Screening. *Genes* **2018**, *9*, 103. [CrossRef] [PubMed]
4. Han, J.Y.; Wiederoder, M.; DeVoe, D.L. Isolation of intact bacteria from blood by selective cell lysis in a microfluidic porous silica monolith. *Microsyst. Nanoeng.* **2019**, *5*, 30. [CrossRef]
5. Rzhevskiy, A.S.; Razavi Bazaz, S.; Ding, L.; Kapitannikova, A.; Sayyadi, N.; Campbell, D.; Walsh, B.; Gillatt, D.; Ebrahimi Warkiani, M.; Zvyagin, A.V. Rapid and Label-Free Isolation of Tumour Cells from the Urine of Patients with Localised Prostate Cancer Using Inertial Microfluidics. *Cancers* **2019**, *12*, 81. [CrossRef]
6. Nowbahar, A.; Mansard, V.; Mecca, J.M.; Paul, M.; Arrowood, T.; Squires, T.M. Measuring Interfacial Polymerization Kinetics Using Microfluidic Interferometry. *J. Am. Chem. Soc.* **2018**, *140*, 3173–3176. [CrossRef]
7. Zhang, C.; Xu, J.; Ma, W.; Zheng, W. PCR microfluidic devices for DNA amplification. *Biotechnol. Adv.* **2006**, *24*, 243–284. [CrossRef] [PubMed]
8. Shi, H.-h.; Xiao, Y.; Ferguson, S.; Huang, X.; Wang, N.; Hao, H.-X. Progress of crystallization in microfluidic devices. *Lab A Chip* **2017**, *17*, 2167–2185. [CrossRef] [PubMed]
9. DeMello, A.J. Control and detection of chemical reactions in microfluidic systems. *Nature* **2006**, *442*, 394–402. [CrossRef] [PubMed]
10. Lang, Q.; Ren, Y.; Hobson, D.; Tao, Y.; Hou, L.; Jia, Y.; Hu, Q.; Liu, J.; Zhao, X.; Jiang, H. In-plane microvortices micromixer-based AC electrothermal for testing drug induced death of tumor cells. *Biomicrofluidics* **2016**, *10*, 064102. [CrossRef] [PubMed]
11. Minteer, S.D. *Microfluidic Techniques: Reviews and Protocols*; Humana Press: Totowa, NJ, USA, 2005.
12. Channon, R.B.; Menger, R.F.; Wang, W.; Carrão, D.B.; Vallabhuneni, S.; Kota, A.K.; Henry, C.S. Design and application of a self-pumping microfluidic staggered herringbone mixer. *Microfluid Nanofluid.* **2021**, *25*, 31. [CrossRef]
13. Rampalli, S.; Dundi, T.M.; Chandrasekhar, S.; Raju, V.R.K.; Chandramohan, V.P. Numerical Evaluation of Liquid Mixing in a Serpentine Square Convergent-divergent Passive Micromixer. *Chem. Prod. Process. Model.* **2020**, 20190071. [CrossRef]
14. Bazaz, S.R.; Hazeri, A.; Mehrizi, A.A. Increasing Efficiency of Micromixing Within a Biomicrofluidic Device Using Acceleration, Deceleration Technique. In Proceedings of the 2017 24th National and 2nd International Iranian Conference on Biomedical Engineering (ICBME), Tehran, Iran, 30 November–1 December 2017; pp. 315–327.
15. Tripathi, E.; Patowari, P.K.; Pati, S. Numerical investigation of mixing performance in spiral micromixers based on Dean flows and chaotic advection. *Chem. Eng. Process.Process. Intensif.* **2021**, *169*, 108609. [CrossRef]
16. Bazaz, S.R.; Mehrizi, A.A.; Ghorbani, S.; Vasilescu, S.; Asadnia, M.; Warkiani, M.E. A hybrid micromixer with planar mixing units. *RSC Adv.* **2018**, *8*, 33103–33120. [CrossRef]
17. Dean, W.R.; Hurst, J.M.J.M. Note on the motion of fluid in a curved pipe. *Mathematika* **1959**, *6*, 77–85. [CrossRef]
18. Dean, W.R. The streamline motion of fluid in a curved pipe. *Phil. Mag.* **1928**, *5*, 673–693. [CrossRef]
19. Nhien, N.T.T.; Huy, N.T.; Uyen, D.T.; Deharo, E.; Hoa, P.T.L.; Hirayama, K.; Harada, S.; Kamei, K. Effect of Inducers, Incubation Time and Heme Concentration on IC(50) Value Variation in Anti-heme Crystallization Assay. *Trop. Med. Health* **2011**, *39*, 119–126. [CrossRef] [PubMed]
20. Debayle, M.; Balloul, E.; Dembele, F.; Xu, X.; Hanafi, M.; Ribot, F.; Monzel, C.; Coppey, M.; Fragola, A.; Dahan, M.; et al. Zwitterionic polymer ligands: An ideal surface coating to totally suppress protein-nanoparticle corona formation? *Biomaterials* **2019**, *219*, 119357. [CrossRef] [PubMed]
21. Mahmoodi, Z.; Mohammadnejad, J.; Razavi Bazaz, S.; Abouei Mehrizi, A.; Ghiass, M.A.; Saidijam, M.; Dinarvand, R.; Ebrahimi Warkiani, M.; Soleimani, M. A simple coating method of PDMS microchip with PTFE for synthesis of dexamethasone-encapsulated PLGA nanoparticles. *Drug Deliv. Transl. Res.* **2019**, *9*, 707–720. [CrossRef] [PubMed]
22. Nguyen, N.-T.; Wereley, S.T.; Shaegh, S.A.M. *Fundamentals and Applications of Microfluidics*; Artech House: Norwood, MA, USA, 2019.
23. Howell, J.P.B.; Mott, D.R.; Golden, J.P.; Ligler, F.S. Design and evaluation of a Dean vortex-based micromixer. *Lab A Chip* **2004**, *4*, 663–669. [CrossRef] [PubMed]
24. Sudarsan, A.P.; Ugaz, V.M. Fluid mixing in planar spiral microchannels. *Lab A Chip* **2006**, *6*, 74–82. [CrossRef]
25. Zhang, W.; Wang, X.; Feng, X.; Yang, C.; Mao, Z.-S. Investigation of Mixing Performance in Passive Micromixers. *Ind. Eng. Chem. Res.* **2016**, *55*, 10036–10043. [CrossRef]
26. Kumar, V.; Aggarwal, M.; Nigam, K.D.P. Mixing in curved tubes. *Chem. Eng. Sci.* **2006**, *61*, 5742–5753. [CrossRef]
27. Duryodhan, V.S.; Chatterjee, R.; Govind Singh, S.; Agrawal, A. Mixing in planar spiral microchannel. *Exp. Therm. Fluid Sci.* **2017**, *89*, 119–127. [CrossRef]
28. Vatankhah, P.; Shamloo, A. Parametric study on mixing process in an in-plane spiral micromixer utilizing chaotic advection. *Anal. Chim. Acta* **2018**, *1022*, 96–105. [CrossRef] [PubMed]
29. Wang, X.; Liu, Z.; Wang, B.; Cai, Y.; Wan, Y. Vortices degradation and periodical variation in spiral micromixers with various spiral structures. *Int. J. Heat Mass Transf.* **2022**, *183*, 122168. [CrossRef]
30. Niculescu, A.-G.; Chircov, C.; Bîrcă, A.C.; Grumezescu, A.M. Fabrication and Applications of Microfluidic Devices: A Review. *Int. J. Mol. Sci.* **2021**, *22*, 2011. [CrossRef] [PubMed]

31. Xia, Y.; Whitesides, G.M. Soft Lithography. *Annu. Rev. Mater. Sci.* **1998**, *37*, 550–575. [CrossRef]
32. Gale, B.K.; Jafek, A.R.; Lambert, C.J.; Goenner, B.L.; Moghimifam, H.; Nze, U.C.; Kamarapu, S.K. A Review of Current Methods in Microfluidic Device Fabrication and Future Commercialization Prospects. *Inventions* **2018**, *3*, 60. [CrossRef]
33. Razavi Bazaz, S.; Hazeri, A.H.; Rouhi, O.; Mehrizi, A.A.; Jin, D.; Warkiani, M.E. Volume-preserving strategies to improve the mixing efficiency of serpentine micromixers. *J. Micromech. Microeng.* **2020**, *30*, 115022. [CrossRef]
34. Chai, M.; Razavi Bazaz, S.; Daiyan, R.; Razmjou, A.; Ebrahimi Warkiani, M.; Amal, R.; Chen, V. Biocatalytic micromixer coated with enzyme-MOF thin film for CO2 conversion to formic acid. *Chem. Eng. J.* **2021**, *426*, 130856. [CrossRef]
35. Warkiani, M.E.; Guan, G.; Luan, K.B.; Lee, W.C.; Bhagat, A.A.S.; Kant Chaudhuri, P.; Tan, D.S.-W.; Lim, W.T.; Lee, S.C.; Chen, P.C.Y.; et al. Slanted spiral microfluidics for the ultra-fast, label-free isolation of circulating tumor cells. *Lab A Chip* **2014**, *14*, 128–137. [CrossRef] [PubMed]
36. Razavi Bazaz, S.; Kashaninejad, N.; Azadi, S.; Patel, K.; Asadnia, M.; Jin, D.; Ebrahimi Warkiani, M. Rapid Softlithography Using 3D-Printed Molds. *Adv. Mater. Technol.* **2019**, *4*, 1900425. [CrossRef]
37. Shrestha, J.; Ghadiri, M.; Shanmugavel, M.; Razavi Bazaz, S.; Vasilescu, S.; Ding, L.; Ebrahimi Warkiani, M. A rapidly prototyped lung-on-a-chip model using 3D-printed molds. *Organs Chip* **2019**, *1*, 100001. [CrossRef]
38. Vasilescu, S.A.; Bazaz, S.R.; Jin, D.; Shimoni, O.; Warkiani, M.E. 3D printing enables the rapid prototyping of modular microfluidic devices for particle conjugation. *Appl. Mater. Today* **2020**, *20*, 100726. [CrossRef]
39. Razavi Bazaz, S.; Rouhi, O.; Raoufi, M.A.; Ejeian, F.; Asadnia, M.; Jin, D.; Ebrahimi Warkiani, M. 3D Printing of Inertial Microfluidic Devices. *Sci. Rep.* **2020**, *10*, 5929. [CrossRef] [PubMed]
40. Sidharthan, S.; Gandhi, A.D.; Balashanmugam, N.; Jeyaraj, P. Numerical Approach for Fabrication of Micromixers Using Microstereolithography. *Procedia Mater. Sci.* **2014**, *5*, 527–534. [CrossRef]
41. White, F.M. *Fluid Mechanics*; McGraw-Hill Education: New York, NY, USA, 2003.
42. Balasubramaniam, L.; Arayanarakool, R.; Marshall, S.D.; Li, B.; Lee, P.S.; Chen, P.C.Y. Impact of cross-sectional geometry on mixing performance of spiral microfluidic channels characterized by swirling strength of Dean-vortices. *J. Micromech. Microeng.* **2017**, *27*, 095016. [CrossRef]
43. Razavi Bazaz, S.; Amiri, H.A.; Vasilescu, S.; Abouei Mehrizi, A.; Jin, D.; Miansari, M.; Ebrahimi Warkiani, M. Obstacle-free planar hybrid micromixer with low pressure drop. *Microfluid Nanofluid.* **2020**, *24*, 61. [CrossRef]
44. Mihandoust, A.; Bazaz, S.R.; Maleki-Jirsaraei, N.; Alizadeh, M.; Taylor, R.A.; Warkiani, M.E. High-Throughput Particle Concentration Using Complex Cross-Section Microchannels. *Micromachines* **2020**, *11*, 440. [CrossRef] [PubMed]
45. Herrmann, N.; Neubauer, P.; Birkholz, M. Spiral microfluidic devices for cell separation and sorting in bioprocesses. *Biomicrofluidics* **2019**, *13*, 061501. [CrossRef] [PubMed]

 micromachines

Article

Mixing Performance of a Passive Micro-Mixer with Mixing Units Stacked in Cross Flow Direction

Makhsuda Juraeva and Dong-Jin Kang *

School of Mechanical Engineering, Yeungnam University, Gyoungsan 712-749, Korea; mjuraeva@ynu.ac.kr
* Correspondence: djkang@yu.ac.kr; Tel.: +82-(0)53-810-2463

Abstract: A new passive micro-mixer with mixing units stacked in the cross flow direction was proposed, and its performance was evaluated numerically. The present micro-mixer consisted of eight mixing units. Each mixing unit had four baffles, and they were arranged alternatively in the cross flow and transverse direction. The mixing units were stacked in four different ways: one step, two step, four step, and eight step stacking. A numerical study was carried out for the Reynolds numbers from 0.5 to 50. The corresponding volume flow rate ranged from 6.33 μL/min to 633 μL/min. The mixing performance was analyzed in terms of the degree of mixing (*DOM*) and relative mixing energy cost (*MEC*). The numerical results showed a noticeable enhancement of the mixing performance compared with other micromixers. The mixing enhancement was achieved by two flow characteristics: baffle wall impingement by a stream of high concentration and swirl motion within the mixing unit. The baffle wall impingement by a stream of high concentration was observed throughout all Reynolds numbers. The swirl motion inside the mixing unit was observed in the cross flow direction, and became significant as the Reynolds number increased to larger than about five. The eight step stacking showed the best performance for Reynolds numbers larger than about two, while the two step stacking was better for Reynolds numbers less than about two.

Keywords: passive micro-mixer; mixing unit; cross flow direction; baffle impingement; swirl motion; mixing performance

Citation: Juraeva, M.; Kang, D.-J. Mixing Performance of a Passive Micro-Mixer with Mixing Units Stacked in Cross Flow Direction. *Micromachines* **2021**, *12*, 1530. https://doi.org/10.3390/mi12121530

Academic Editor: Kwang-Yong Kim

Received: 25 October 2021
Accepted: 6 December 2021
Published: 9 December 2021

1. Introduction

Mixing at micro scales is an essential design task in the bio- and microfluidic systems such as micro-total analysis system (μ-TAS), lab on a chip, and micro-reactors. As the associated dimensions of a microfluidic system are small, the molecular diffusion is a major mechanism of mixing, the corresponding Reynolds number is also very small and the flow is laminar. Meanwhile, many biochemical processes require rapid and complete mixing, and mixing enhancement is essential in designing these systems [1].

A variety of micro-mixers have been devised to enhance the mixing in a micro-fluidic system [2]. They are named either active or passive micro-mixers, depending on the usage of an external energy source. As active micro-mixers utilize an external energy source to achieve the mixing enhancement, the structure of a micro-mixer becomes more complicated than that of passive micro-mixers. In addition, it is more costly compared with a passive micro-mixer. Examples of external energy sources used for active micro-mixers are acoustic [3], magnetic [4], electric [5], thermal [6], electro [7,8], and pressure fluctuating [9]. On the contrary, passive micro-mixers use micro-mixer geometry to agitate and generate secondary flow. They are expected to promote chaotic advection of fluids leading to the mixing enhancement. Accordingly, the mixing enhancement by a passive micro-mixer could be limited compared with that by a similar, active micro-mixer. However, passive micro-mixers are simpler to fabricate and easier to integrate into microfluidic systems; therefore, they are widely used in microfluidic systems.

Many different types of passive micro-mixers have been designed and investigated to enhance their mixing performance [10]. Some examples include recessed grooves in the channel wall [11], herringbone-type walls [12], baffles [13,14], channel wall twisting [15], junction contraction [16], split and recombine (SAR) [17,18], serpentine microchannel [19,20], converging and diverging microchannel [21], spiral channel [22,23], and periodic geometric features [24].

Among various geometric components, baffles are one of the most widely adopted components in designing a passive micro-mixer. Many researchers have shown that micro-mixers based on baffles perform effectively over a wide range of Reynolds numbers [13,25]. For example, Tsai et al. [25] used radial baffles in a curved microchannel and showed that radial baffles induced vortices in multiple directions. Kang [13] placed rectangular baffles in a cyclic order along the channel wall and enhanced the mixing performance of a T-shaped micromixer. Santana et al. [14] used triangular baffles to obtain a high mixing index for various micro-channel dimensions. Sotowa et al. [26] showed that the indentations and baffles attached to the micromixer wall enhance the mixing by means of secondary flow in deep micro-channel reactors. Chen et al. [27] distributed baffles on both side walls of a micromixer based on the Koch fractal principle. They showed that the baffles distributed on the both side of the microchannel wall performs better than the mixing performance of the baffles distributed on the one side. Chung et al. [28] implemented planar baffles and showed that the mixing performance was enhanced at both of diffusion dominant and convection dominant regime of mixing. Raza et al. [29] improved the mixing performance of a SAR micromixer by embedding baffles just after each SAR unit. The enhancement was shown as achievable over a Reynolds number range from 0.1 to 80.

Recently, several complicated three-dimensional (3D) micro-mixers showed an improved mixing performance compared with those of two-dimensional (2D) micro-mixers of similar size [30–33]. However, 2D planar micromixers have an advantage of simplicity in fabrication compared with that of complex 3D micromixers. Various design techniques were attempted to promote 3D effects: baffles, channel wall twisting, spiral channel, and split and recombine. For example, Kang [13] showed that a cyclic configuration of baffles generates a rotational flow in the cross section of a micromixer. The same idea was adopted in the present micromixer to promote 3D effects, and four baffles were attached to the four side walls of the micromixer in a cyclic order. The overall layout of a passive micromixer is also an important design concept [33]. For example, Tripathi [23] studied three different layout of a passive micromixer based on spiral microchannel and showed that the mixing performance is quite dependent on the micromixer layout. The eight mixing units were stacked in the cross flow direction; this kind of stacking has not been studied yet. The stacking was designed as four different layouts: one step, two step, four step, and eight step stacking. The combined effects of a cyclic baffle arrangement and mixing unit stacking in the cross flow direction was the main design concept of the present micromixer.

Most of micromixers used in biological and chemical applications usually operate in the range of milliseconds of mixing time, and the corresponding Reynolds number is less than about 100 [34–36]; the corresponding volume flow rate is an order of mL/h. In this range, the mixing mechanism can be divided into three regimes: molecular dominance, transition, and advection dominance [10,18,33]. The effects of micromixer layout design as well as a passive device such as baffles are known to be significant in all of the three regimes. Accordingly, a numerical study was carried out for the Reynolds numbers from 0.5 to 50, covering all of the three regimes. The corresponding volume flow rate ranged from 6.33 μL/min to 633 μL/min.

A numerical simulation has several advantages to study the fluid dynamic and mixing features involved in a micro-mixer, including easy visualization of the mixing process and the associated flow characteristics such as streamlines and vortex formation. Accordingly, a numerical approach is widely accepted in studying mixing enhancement mechanism of a micro-mixer. For example, Rhoades et al. [37] used the commercial software COMSOL Multiphysics 5.1 (COMSOL, Inc., Burlington, MA, USA) to study the mixing performance of grooved serpentine microchannels. Volpe et al. [38] studied the flow dynamics of a continuous size-based sorter microfluidic device by using the lattice Boltzmann method (LBM).

Kang [13] used the commercial software ANSYS® Fluent 19.2 (ANSYS, Inc., Canonsburg, PS, USA) [39] to simulate the mixing performance of a passive micromixer with baffles and quantitatively evaluated the mixing performance.

In this paper, the commercial software ANSYS® Fluent 19.2 was used to simulate the mixing performance. The mixing performance was estimated in terms of the degree of mixing (*DOM*) and corresponding mixing energy cost (*MEC*). The numerical simulation was carried out for Reynolds numbers ranging from 0.5 to 50, corresponding to volume flow rates ranging from 6.33 μL/min to 633 μL/min.

2. Passive Micro-Mixer with Mixing Units Stacked in the Cross Flow Direction

Figure 1 shows a mixing unit with four hexahedron baffles inside. The four baffles were arranged in a cyclic order. The first baffle was attached to the lower wall, and the third baffle was attached to the upper wall. The second and fourth baffles were attached to the front and back walls, respectively. Each baffle was 30 μm thick, and blocked the flow passage by half in each direction. Two consecutive baffles were separated by 60 μm, and each mixing unit was 600 μm long. Figure 2 depicts the five different layouts of a micromixer simulated in this paper. Each micromixer had eight mixing units, and each mixing unit consisted of four baffles. Figure 2a shows the baseline layout of eight mixing units, and Figure 2b–e shows mixing units stacked in the cross stream direction (y direction); the mixing units were overlapped 150 μm in the main stream direction (x direction). The inlets and outlet had a rectangular cross-section of 300 μm × 120 μm. The two inlet branches were 1000 μm long, and the outlet branch was 600 μm long.

For the sake of simplicity, we assumed that the same aqueous solution flows into the two inlets: Inlet 1 and Inlet 2. The properties of the aqueous solution are similar to those used in BioMEMS systems. The kinematic viscosity and mass diffusion coefficient of the aqueous solution are $v = 10^{-6}$ m^2/s and $D = 10^{-9}$ m^2/s, respectively [18,33]. The corresponding Schmidt (Sc) number is 10^3; it is defined as the ratio of the kinetic viscosity and the mass diffusion coefficient. The Reynolds number is defined as $Re = \frac{U_{mean} d_h}{v}$, where U_{mean}, d_h and v indicate the mean velocity at the outlet, the hydraulic diameter of outlet, and the kinematic viscosity of the fluid, respectively.

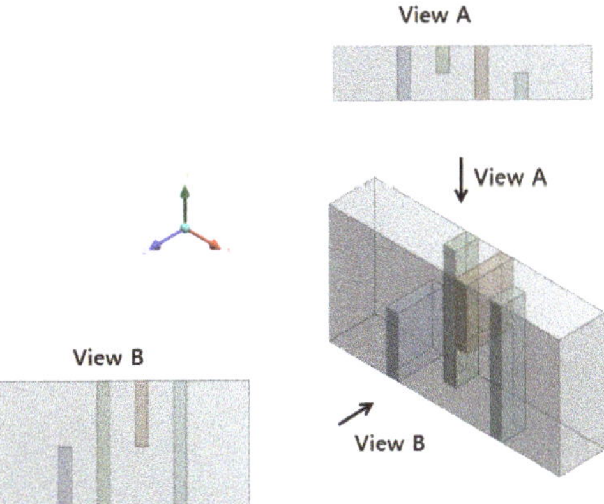

Figure 1. Schematic diagram of the four-baffle arrangement within a mixing unit.

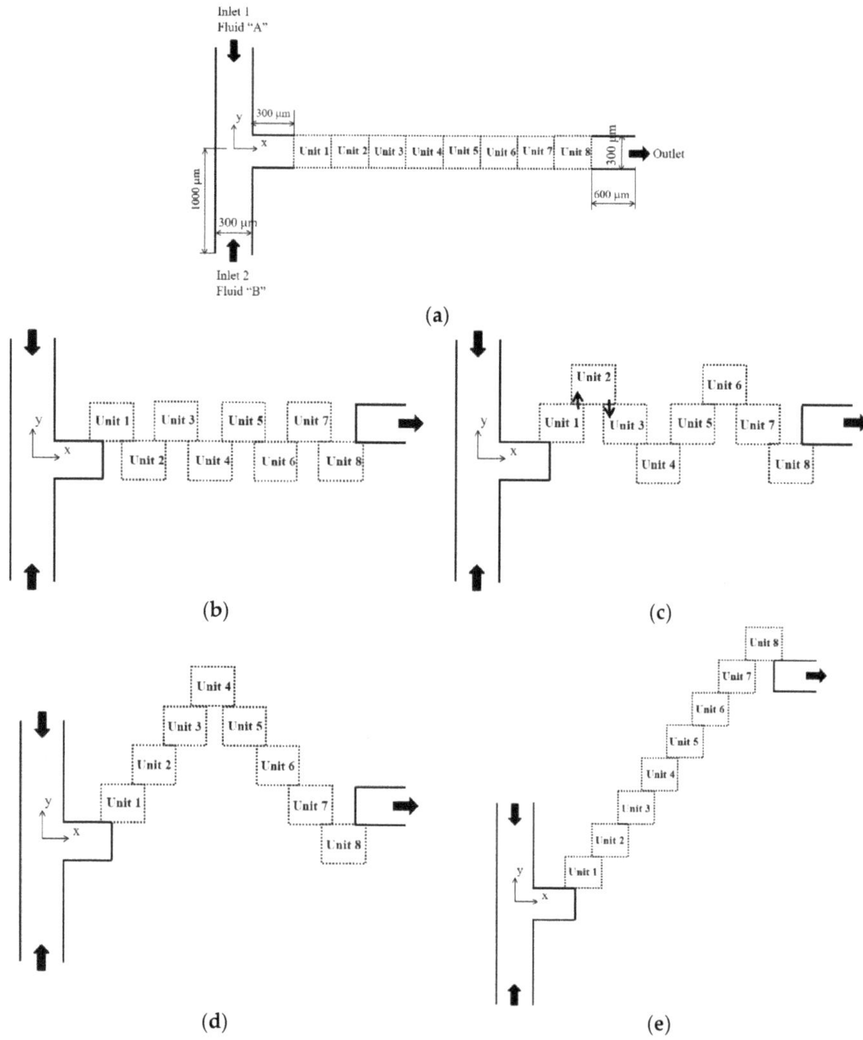

Figure 2. Present passive micro-mixers with mixing units stacked in the cross flow direction. (**a**) Baseline layout. (**b**) One step stacking. (**c**) Two step stacking. (**d**) Four step stacking. (**e**) Eight step stacking.

3. Governing Equations and Computation Procedure

Transport phenomena in micromixers can be described theoretically at two different levels: the molecular level and the continuum level. The two different levels are related to the typical length scale involved. The continuum model can describe most transport phenomena in micromixers with a length scale ranging from micrometers to centimeters [40]. The present micromixers were in this range of length scale.

In this study, the fluid entering the two inlets were water and dyed water. They were assumed to have the same physical properties of water at a temperature of 20 °C; the fluid A at Inlet 1 was the dyed water and the fluid B at Inlet 2 was water. The flow inside the micromixer was assumed to be steady, incompressible, laminar, and Newtonian. Therefore, the fluid flow in the micromixer was simulated solving the incompressible Navier–Stokes and the continuity equations:

$$\left(\vec{u}\cdot\nabla\right)\vec{u} = -\frac{1}{\rho}\nabla p + \nu\nabla^2\vec{u} \tag{1}$$

$$\nabla\cdot\vec{u} = 0 \tag{2}$$

where ρ, $\vec{u}$, p, and ν are the density, the velocity vector, the pressure, and the kinematic viscosity of the fluid, respectively. The density and viscosity were specified as 998 kg/m^3 and 0.001 kg/(ms), respectively.

The flow field thus obtained was used to simulate the mixing process throughout the micromixer. In this study, the mixing was assumed to be governed by two fluid dynamic mechanisms: molecular diffusion and advection. A scalar advection-diffusion transport equation was used to simulate the mixing process [33,37]:

$$\left(\vec{u}\cdot\nabla\right)\phi = D\nabla^2\phi \tag{3}$$

where D and ϕ are the diffusion coefficient and the dye concentration, respectively. The dyed water concentration is $\phi = 1$ at Inlet 1, and $\phi = 0$ at Inlet 2. In general, a concentration gradient is a density gradient which can induce a convective flow. The flow velocity due to concentration gradients can vary from 0.1 to 10 μms^{-1} depending on properties such as the channel dimensions, fluid viscosity, fluid type, gradient magnitude [41]; in this paper, the flow velocity in the micromixer was in the range of 0.001 to 0.14 ms^{-1}. Therefore, the convective flow induced by any concentration gradient was neglected in this paper.

We used the commercial software ANSYS® Fluent to solve the governing Equations (1)–(3), and it is based on the finite volume method. In general, a certain amount of numerical diffusion is introduced in discretizing the convective terms, but it can be limited by using a high-order discretization scheme. The non-linear convective terms in Equations (1) and (3) were approximated using the QUICK scheme (quadratic upstream interpolation for convective kinematics), and its theoretical accuracy is third order. The velocity was assumed uniform at the two inlets, while it was assumed fully developed at the outlet. Along all the other walls, the no-slip boundary was applied.

In order to obtain a fully converged solution, every computation was continued until oscillation of the residual of all equations was negligibly small. Some researchers used an adaptive mesh technique to reduce the oscillation [42]. In this study, the residuals of continuity, momentum, and concentration equations oscillated in the range smaller than 10^{-11}, 10^{-14}, and 10^{-8}, respectively; they were sufficiently small to obtain reliable numerical solutions.

Accurate numerical simulation of the mixing in micromixers is still a challenging problem, especially for high Peclet numbers. Some studies do not deal with computational issues related to the mesh dependence of numerical solution. In this paper, a detailed study was conducted in the section of validation of numerical solution. According to Okuducu et al. [43], the accuracy of numerical solutions is also dependent on the type of cells; structured hexahedral cells show the most reliable numerical solution, in comparison with tetrahedral and prism cells. In this paper, all cells for every micromixer layout were structured and hexahedral.

The mixing performance of present micro-mixer was evaluated using the degree of mixing (*DOM*) and mixing energy cost (*MEC*). We calculated the *DOM* in the following form:

$$DOM = 1 - \frac{1}{\xi}\sqrt{\sum_{i=1}^{n}\frac{(\phi_i - \xi)^2}{n}} \tag{4}$$

where φ_i and n are the mass fraction of fluid A in the ith cell and total number of cells, respectively, and ξ is 0.5 when the two fluids are completely mixed. The *MEC* is calculated in the following form, and measures the effectiveness of a micro-mixer [44,45]:

$$MEC = \frac{\Delta p/\rho u_{mean}^2}{DOMX100} \tag{5}$$

where u_{mean} is the average velocity at the outlet, and Δp is the pressure difference between the inlet and the outlet.

4. Validation of Numerical Study

For high Sc number simulations, numerical diffusion is known to deteriorate the accuracy of simulated results, in general [46–49]. To minimize the numerical diffusion problem, several approaches can be chosen. These include a particle-based simulation such as the Monte Carlo method [46] or a grid-based method with a small cell Peclet number. Here, the cell Peclet number is $Pe = \frac{U_{cell}l_{cell}}{D}$ where U_{cell} and l_{cell} are the local flow velocity and cell size, respectively. However, these approaches are computationally too expensive to adopt in a study like this paper. Most numerical studies prefer a practical approach to obtain numerical solutions with a reasonable degree of accuracy. To achieve that, a detailed study of grid independence including the grid convergence index (GCI) test is usually adopted [18,29,31]. In this paper, a similar procedure was followed.

To quantitatively validate the present numerical approach, we simulated a passive micro-mixer experimented by Tsai et al. [25]. Figure 3 shows a schematic diagram of the micro-mixer. The two inlets had a rectangular cross section of width 45 μm by depth 130 μm. The micromixer had four baffles of width 45 μm by height 97.5 μm. The fluid was assumed to have the properties of density 997 kg/m^3, viscosity 0.00097 kg/(ms), and diffusion coefficient 3.6 × 10^{-10} m^2/s, as reported by Tsai et al. [25]. The numerical simulation was carried out for three different Reynolds numbers of *Re* = 1, 9, and 81 and the results were compared with the corresponding experimental data.

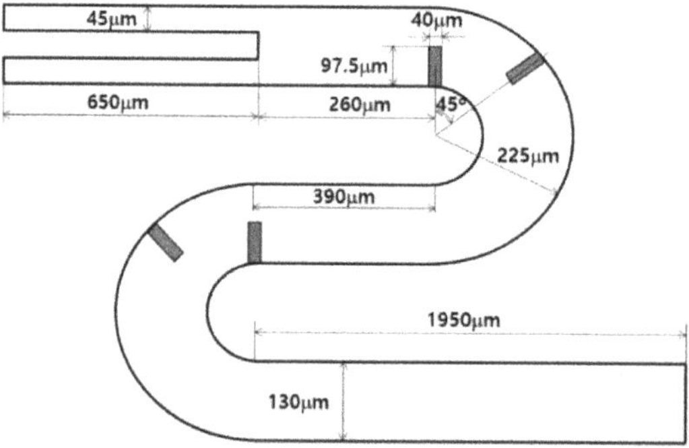

Figure 3. Diagram of the micromixer experimented by Tsai et al. [25].

Hexahedral cells were used to mesh the computational domain sketched in Figure 3; they are all structured. Before detailed simulations, a preliminary study was carried out to check the grid independence of numerical solutions for the Reynolds number of 9. Figure 4a shows the dependence of the numerical solution for *Re* = 1, and about 1 million of the computational cells were enough to obtain a numerical solution with a reasonable accuracy. Here, DOM_T stands for the degree of mixing defined by Tsai et al. [25] in the following way:

$$DOM_T = 1 - \frac{\sigma_D}{\sigma_{D,o}} \tag{6}$$

and

$$\sigma = \sqrt{\frac{1}{n}\sum_{i=1}^{n}(\phi_i - \phi_{ave})^2} \tag{7}$$

where σ_D is the standard deviation of ϕ on a cross section normal to the flow, $\sigma_{D,o}$ is the standard deviation at the inlet, and ϕ_{ave} is the average value of ϕ at a sampled cross section.

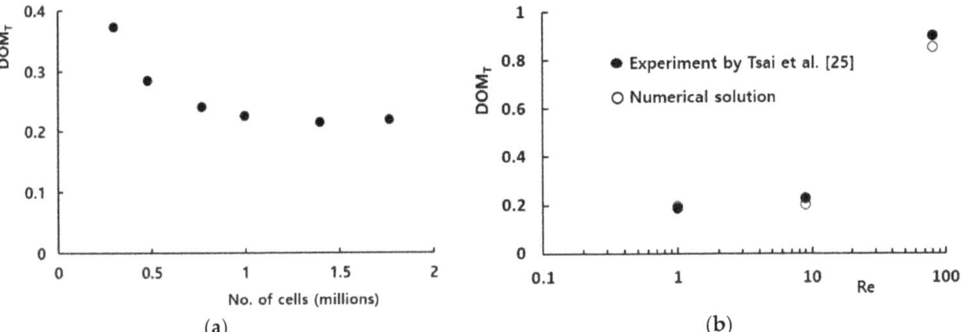

Figure 4. Validation of numerical solution. (**a**) Grid dependence of the numerical solution. (**b**) *DOM* vs. *Re*.

Figure 4b compares the simulation results with the corresponding experimental data by Tsai et al. [25] for Reynolds numbers from 1 to 81. The numerical solution and experimental data showed similar behavior as the Reynolds number increased, and the discrepancy was acceptable. The discrepancy between the experimental data and numerical solution was less than 4%, and became smaller as the Reynolds number decreased, and as the Peclet number decreased. The discrepancy is attributed to several factors such as the numerical diffusion, experimental uncertainty, etc.

Prior to carrying out the final simulations, an additional set of preliminary simulations was carried out to determine an appropriate mesh size for the simulation of the present micro-mixer. We used the present micro-mixer of Figure 2d for this study, and simulated for the Reynolds number of $Re = 3$. The size of every hexahedral cell was varied from 2.5 μm to 5 μm. Figure 5 shows the dependence of the calculated *DOM* on the edge size. The deviation of 3.5 μm solution from that of 4 μm was 0.4%. Therefore, 3.5 μm was small enough to obtain grid independent solutions.

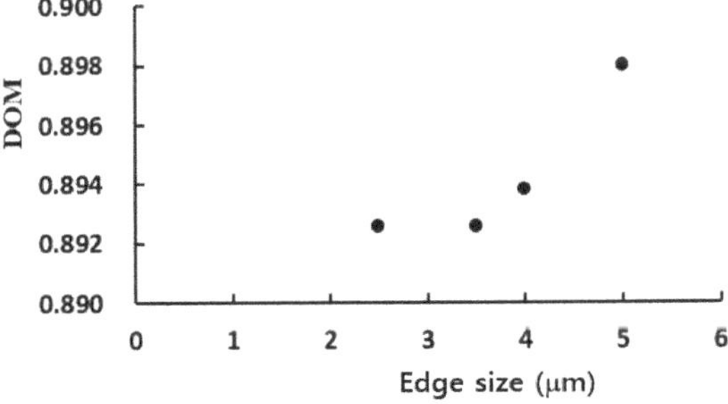

Figure 5. Dependence of the numerical solution on the edge size.

Using the preliminary simulation results, the uncertainty of grid convergence was evaluated using the grid convergence index (*GCI*) [50,51]. According to the Richardson extrapolation methodology, the *GCI* is defined as follows:

$$GCI = F_s \frac{|\varepsilon|}{r^p - 1} \tag{8}$$

$$\varepsilon = \frac{f_{coarse} - f_{fine}}{f_{fine}} \tag{9}$$

where F_s, r, and p are the safety factor of the method, grid refinement ratio, and the order of accuracy of the numerical method, respectively. f_{coarse} and f_{fine} are the numerical results obtained with a coarse grid and fine grid, respectively. We specified F_s as 1.25 according to the suggestion by Roache [50]. For the GCI evaluation, the edge size was varied from 2.5 μm to 5 μm, 2.5 μm, 3.5 μm, 4 μm, and 5 μm. Table 1 summarizes the result of the *GCI* test. The *GCI* based on the *DOM* at the outlet was 1% and 0.57% for 4 μm and 3.5 μm, respectively. Therefore, the edge size of 3.5 μm was small enough to obtain numerical solutions with reasonable accuracy.

Table 1. Results of the GCI test.

Edge Size (μm)	*DOM*	ε	r	*GCI*
2.5	0.89253	0.00001	1.4	0.00001
3.5	0.89254	0.0014	1.14286	0.00572
4	0.89379	0.00466	1.25	0.01036
5	0.89796			

5. Results

A new passive micro-mixer with mixing units stacked in the cross flow direction was proposed, its mixing performance was simulated for Reynolds numbers ranging from 0.1 to 50. The present micro-mixer consisted of eight mixing units, and each mixing unit contained four baffles; the baffles were arranged in three different ways. The mixing units were stacked in four different ways to study their effect on the mixing performance.

The velocity at the two inlets was uniform in the range from 0.0293 mm/s to 146 mm/s, and the corresponding volume flow rates were from 6.33 to 633 μL/min. The mixing performance was evaluated in terms of the *DOM* at the outlet and the corresponding *MEC*.

Figure 6 shows the effects of mixing unit stacking in the cross flow direction on the mixing performance in terms of the *DOM* and the corresponding *MEC*. All of the cross flow stacking layouts resulted in a noticeable variation from the baseline layout, and the variation was prominent in the Reynolds number range from 1 to 30. For the Reynolds numbers less than about one, the two step stacking showed the best performance in terms of *DOM*. For example, the *DOM* of the two step stacking at *Re* = 1 showed a 9% increase from that of the baseline layout. On the contrary, the eight step stacking showed the best *DOM* for the Reynolds numbers larger than about two. For example, the *DOM* of the eight step stacking at *Re* = 20 was enhanced 14% from that of the baseline layout. However, the eight step stacking showed the least effective performance for the Reynolds numbers less than about one, as can be seen in the *MEC* distribution; it required the largest pressure load.

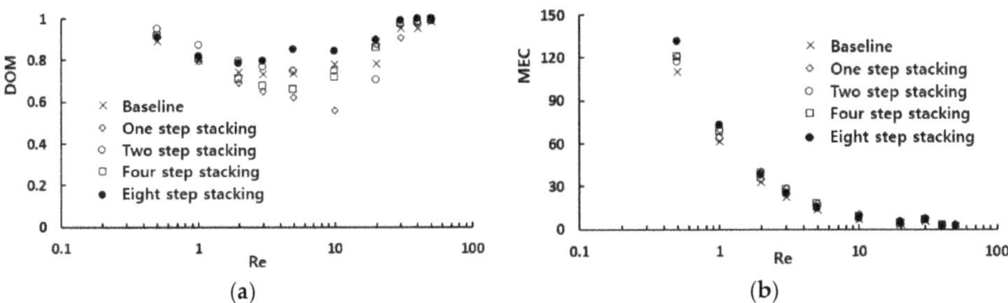

Figure 6. Effects of mixing unit stacking on the mixing performance. (**a**) *DOM* vs. *Re*. (**b**) *MEC* vs. *Re*.

We examined how the mixing unit stacking affected the mixing performance of a passive micro-mixer. Figure 7 shows the distribution of mass concentration of the fluid A at the plane of channel half width $z = 60$ μm for $Re = 1$. The concentration contours showed that a stream of high concentration of fluid A (reddish yellow line in the figure) formed along the center of the flow passage. Its appearance from the second mixing unit suggests that the mixing in the first mixing unit was actively achieved by the baffles. One interesting thing to note is that the stream of high concentration impinged on the baffle in the mixing units, especially in the two step stacking layout. Considering that the convective mixing effects for $Re = 1$ were quite limited, the flow characteristics of baffle impingement is an interesting phenomenon in the diffusion dominant flow regime. The baffle impingement seemed most intensive for the case of two step stacking. To quantitatively check this explanation, the *DOM* increment in each mixing unit was compared for the baseline and two step stacking layouts. Figure 8 shows the *DOM* increment in each mixing unit for $Re = 1$. Here, the *DOM* increment means the difference of *DOM* between at the exit and entry of each mixing unit. For example, the flow entry and exit in the second mixing unit of the two step layout are indicated as the arrows pointing upward and downward (see Figure 2c), respectively. The entry and exit *DOMs* were calculated at the corresponding cross section normal to the flow. The increment of the two step stacking in the third and fourth mixing units was noticeably larger than that of the baseline layout. For example, the *DOM* increment in the third mixing unit of two step stacking was about 64% larger than the baseline layout.

Figure 9 shows the distribution of mass concentration of the fluid A at the plane of channel half width $z = 60$ μm for $Re = 20$. The concentration contours showed a very chaotic distribution in the first and second mixing units. However, the stream of fluid A (red stream in the figure) and fluid B (blue stream in the figure) started to segregate from each other in the third mixing unit for the baseline layout. On the other hand, the eight step stacking layout showed a quite different flow pattern. The stream of fluid B impinged on the baffle in the third and fourth mixing units while the stream of fluid A impinged on the baffle in the second, fifth, and sixth mixing units. This suggests that the stream of high concentration changes their position as it flows down along the mixing unit. Figure 10 plots the distribution of mass concentration of the fluid A at the cross sections before and after each mixing unit for the eight step stacking layout, with contours in the *zx* plane (in the cross flow direction). The streams of fluid A and B showed a swirl motion as they passed through the mixing units. Since the swirl motion was in the y-direction (*zx* plane), the stacking of the mixing units in the cross flow direction contributed to the mixing enhancement discussed in Figure 6. This suggests that the mixing enhancement in the convection dominant flow regime can be easily obtained by a simple stacking method of the mixing units in the cross flow direction.

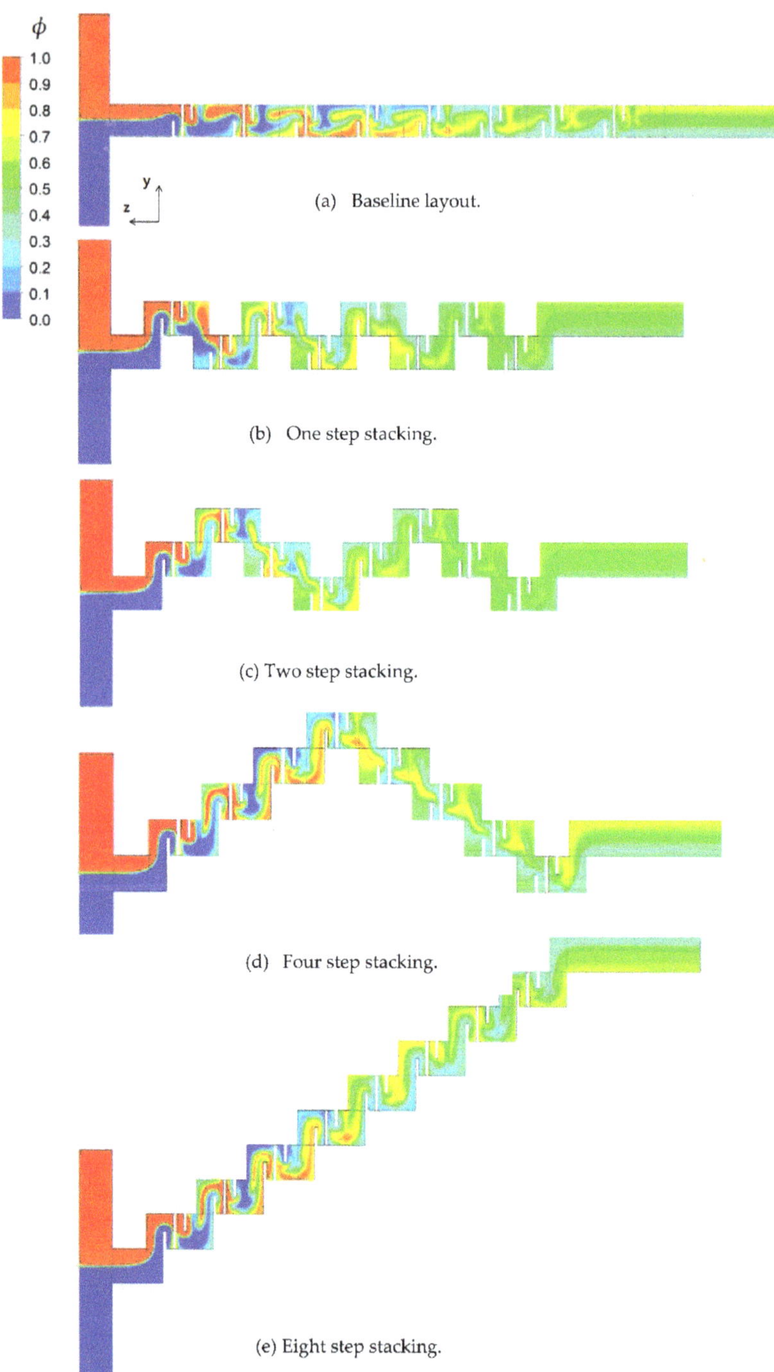

Figure 7. Concentration distribution for $Re = 1$ at the plane of half width $z = 60$ μm.

0.2
0.15
0.1
0.05
0
Δ (DOM)
Baseline
Two step stacking
1 2 3 4 5 6 7 8
Mixing unit

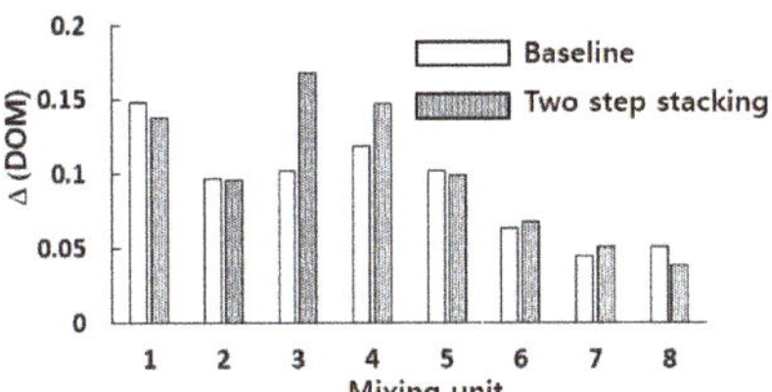

Figure 8. Comparison of *DOM* increment in each mixing unit for $Re = 1$.

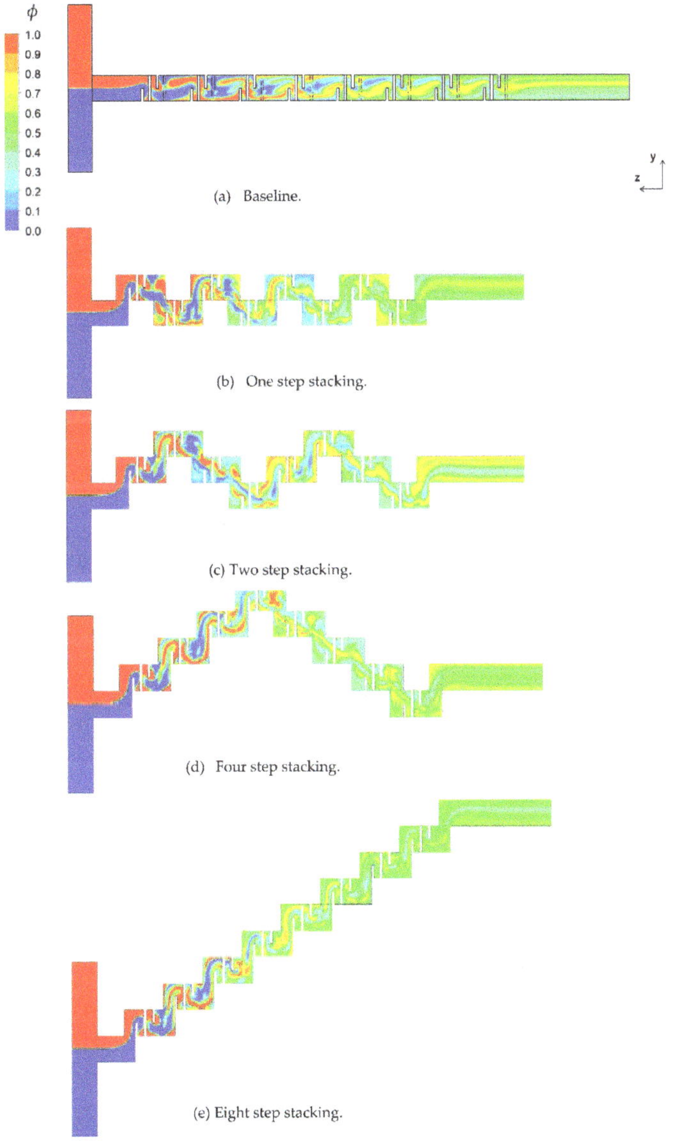
ϕ
1.0
0.9
0.8
0.7
0.6
0.5
0.4
0.3
0.2
0.1
0.0
y
z
(a) Baseline.
(b) One step stacking.
(c) Two step stacking.
(d) Four step stacking.
(e) Eight step stacking.

Figure 9. Concentration distribution for $Re = 20$ at the plane of half width $z = 60$ μm.

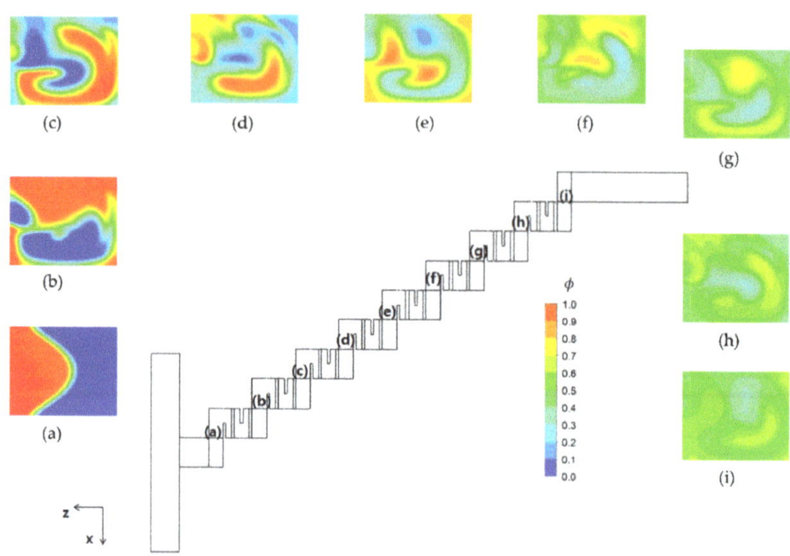

Figure 10. Concentration distribution at the cross section before and after the mixing unit for $Re = 20$.

Figure 11 shows the velocity vector field at the junction of third and fourth mixing units (at section (d) in Figure 10) for $Re = 20$, 5, and 2. The velocity vector field in Figure 11a shows two counter rotating vortices. A larger and stronger vortex flow was observed on the lower right corner, and the other vortex was on the upper left corner. Comparing this with Figure 10d, the vortex flow pattern caused a stream of high concentration of fluid A (a red stream on the lower right corner) swirling in the clockwise direction along the mixing unit walls. However, this swirl motion became weak as the Reynolds number decreased, as can be seen in Figure 11b,c. A weak vortex seemed to form near the lower right corner for $Re = 5$, while no explicit pattern of vortex was observed for $Re = 2$. This suggests that the mixing unit stacking in the cross flow direction generated a swirl motion inside the mixing unit, even if the swirl motion was insignificant for Reynolds numbers less than about five The swirl motion inside the mixing unit is the second flow characteristic attributed to the mixing enhancement described in the Figure 8.

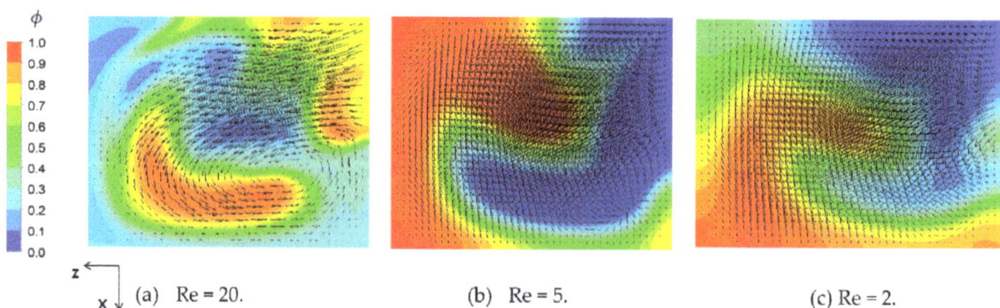

(a) Re = 20. (b) Re = 5. (c) Re = 2.

Figure 11. Velocity vector field at the junction of the third and fourth mixing units.

Figure 12 compares the streamlines starting from the two inlets and concentration contours at a plane within each mixing unit for the baseline and eight step stacking layouts for $Re = 20$. The streamline plots were obtained at the cross section of half width $z = 60$ μm. On the right upper side, an enlarged view of the streamlines passing through from the third to fifth mixing units are shown. The concentration contours were obtained at the plane of the first baffle within each mixing unit along the micromixer. The streamlines from the inlet 1 (red) and streamlines from the inlet 2 (blue) changed their relative position more frequently for the eight step stacking layout as they passed down the mixing units; this flow pattern was more clearly seen in the enlarged view shown on the right upper side. The rapid change of their relative position led to faster mixing at the cross section, as can be seen in the centration contours within the mixing units. This flow pattern is also explained by the velocity vector field shown on the right lower side. The eight step stacking layout showed a vortex flow on the right lower corner, and the vector field was wavier than the baseline layout. This vortex flow in the x-direction is another flow characteristic attributed to the stacking of mixing unit in the cross flow direction.

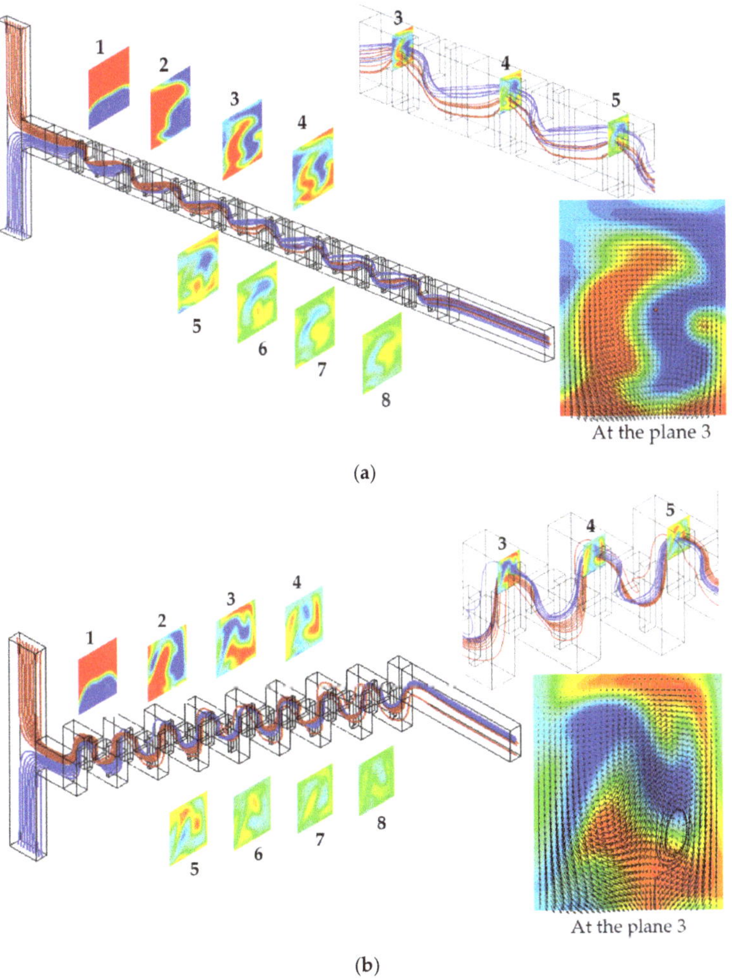

Figure 12. Streamlines starting from the two inlets and concentration contours at the plane of the first baffle within each mixing unit for $Re = 20$. (**a**) Baseline. (**b**) Eight step stacking.

Figure 13 compares the simulated mixing performance in terms of the *DOM* and required pressure load with those from other passive micromixers. In general, the mixing performance of micromixers increased with the diffusion coefficient. The value of diffusion coefficient depends on the fluid used in actual applications. To neglect the effects of the diffusion coefficient on the mixing performance, several passive micromixers based on the same size of the diffusion coefficient were selected for comparison; the diffusion coefficient was 1.2×10^{-9} m^2/s. Tripathi et al. [23] numerically investigated the mixing performance of spiral micromixers and observed transverse flow due to the centrifugal effect. A pair of counter rotating vortices was found to enhance the mixing performance in the spiral micromixer. Raza et al. [29] proposed a SAR mixing unit combined with baffles in a curved channel and simulated its mixing performance. They showed that the mixing performance was better than their earlier version of SAR micro-mixer, and attributed the mixing enhancement to the combined effects of collision, Dean vortex, and secondary flows. Makhsuda et al. [52] proposed an optimal combination of the cross channel split and recombine (CC-SAR) and the mixing cell with baffles (MC-B). They showed that the mixing performance of the optimized micromixer was noticeably improved compared with a micromixer based on the CC-SAR alone. Hossain et al. [53] studied a micromixer with unbalanced three split rhombic sub-channels. According to their results, a rhombic angle of 90° provides the best mixing performance. The mixing performance was enhanced by the two pairs of counter rotating vortices in the cross section. The present micromixer showed a meaningful enhancement of the mixing performance for Reynolds numbers less than about 20, as compared with other micromixers. The enhancement was caused by the two flow characteristics: the baffle impingement by a stream of high concentration and a swirl motion inside the mixing unit. The baffle impingement by a stream of high concentration was observed throughout all Reynolds numbers, while the swirl motion inside the mixing unit became significant only for Reynolds numbers larger than about five. Figure 13b shows that the required pressure load of the present micromixer was comparable with those of other micromixers; the mixing enhancement was achieved without an additional pressure load. This confirms that the mixing unit stacking in the cross flow direction is a practical approach to enhance the mixing performance, especially for Reynolds numbers less than about 20.

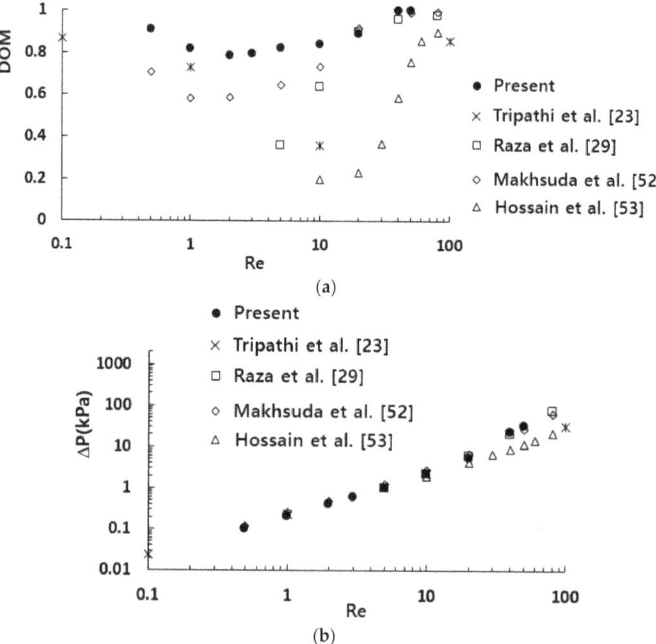

Figure 13. Comparison of the simulated mixing performance with other micromixers. (**a**) *DOM* vs. *Re*. (**b**) Pressure load vs. *Re*.

6. Conclusions

A new passive micro-mixer with the mixing units stacked in the cross flow direction was proposed and estimated. It consisted of eight mixing units, and each mixing unit contained four baffles; the baffles were staggered alternatively in the cross flow and transverse directions. The mixing performance of the present micro-mixer was evaluated numerically by examining the *DOM* and *MEC*. The numerical simulation was carried out for Reynolds numbers ranging from 0.5 to 50.

The numerical solutions showed that the mixing unit stacking in the cross flow direction noticeably enhanced the mixing performance for Reynolds numbers less than about 20, as compared with other micromixers. The stacking layout would be different according to the Reynolds number. For Reynolds numbers less than about two, the two step stacking in the cross flow direction showed the best mixing performance. On the other hand, the eight step stacking showed the best mixing performance for Reynolds numbers larger than about two.

The mixing enhancement due to the mixing unit stacking in the cross flow direction was caused by two different flow characteristics. One is the baffle impingement by a stream of high concentration. A stream of high concentration was observed to impinge on the baffle wall throughout all Reynolds numbers. It was observed even in the limited convection flow regime such as $Re = 0.5$. The other one is the swirl motion inside the mixing unit in the cross flow direction, and it was significant for Reynolds numbers larger than about five.

Since both halves of the present micro-mixer in the transverse direction are planar, they can be simply stacked to construct the present micromixer. The present micromixer performed better than other passive micro-mixers for low Reynolds numbers less than about 20. Therefore, it is expected to be useful as a necessary part of lab-on-a-chip devices and micro-total analysis systems.

Author Contributions: Simulation, M.J.; formal analysis, D.-J.K.; visualization, M.J.; writing—original draft, D.-J.K. All authors have read and agreed to the published version of the manuscript.

Funding: This work was supported by BOKUK© 2021 Research Grant.

Conflicts of Interest: The authors declare no conflict of interest.

Nomenclature

D	diffusion coefficient (m^2/s)
DOM	degree of mixing
f_{coarse}	numerical solution obtained with coarse grid
f_{fine}	numerical solution obtained with fine grid
GCI	grid convergence index
DOM_T	mixing index
MEC	mixing energy cost
Δp	pressure load (Pa)
r	grid refinement ratio
Re	Reynolds number
U_{mean}	average velocity at the outlet (m/s)
x, y, z	Cartesian coordinates
Greek Symbols	
ε	relative error
μ	fluid viscosity (kg/(ms))
ν	fluid kinematic viscosity (m^2/s)
ϕ	mass fraction of the fluid A
φ	fluid density (kg/m^3)
σ	standard deviation

References

1. Liu, L.; Cao, W.; Wu, J.; Wen, W.; Chang, D.C.; Sheng, P. Design and integration of an all-in -one bio-micro-fluidic chip. *Biomicrofluidics* **2008**, *2*, 034103. [CrossRef]
2. Cai, G.; Xue, L.; Zhang, H.; Lin, J. A Review on Micro-mixers. *Micromachines* **2017**, *8*, 274. [CrossRef]
3. Huang, P.H.; Xie, Y.; Ahmed, D.; Rufo, J.; Nama, N.; Chen, Y.; Chan, C.Y.; Huang, T.J. An acoustofluidic micro-mixer based on oscillating sidewall sharp-edges. *Lab Chip* **2013**, *13*, 3847–3852. [CrossRef]
4. Eickenberg, B.; Wittbracht, F.; Stohmann, P.; Schubert, J.R.; Brill, C.; Weddemann, A.; Hutten, A. Continuous flow particle guiding based on dipolar coupled magnetic superstructures in rotating magnetic fields. *Lab Chip* **2013**, *13*, 920. [CrossRef]
5. Wu, Y.; Ren, Y.; Tao, Y.; Hu, Q.; Jiang, H. A novel micro-mixer based on the alternating current flow field effect transistor. *Lab Chip* **2016**, *17*, 186–197. [CrossRef] [PubMed]
6. Huang, K.R.; Chang, J.S.; Chao, S.D.; Wung, T.S.; Wu, K.C. Study of active micro-mixer driven by electrothermal force. *Jpn. J. Appl. Phys.* **2012**, *51*, 047002. [CrossRef]
7. Mondal, B.; Metha, S.K.; Pati, S.; Patowari, P.K. Numerical analysis of electroosmotic mixing in a heterogeneous charged micromixer with obstacles. *Chem. Eng. Proc.-Process Intensif.* **2021**, *168*, 108585. [CrossRef]
8. Gunti, K.; Bhattacharya, A.; Chakraborty, S. Analysis of micromixing of non-Newtonian fluids driven by altering current electrothermal flow. *J. Non-Newton. Fluid Mech.* **2017**, *247*, 123–131.
9. Wang, X.; Ma, X.; An, L.; Kong, X.; Xu, Z.; Wang, J. A pneumatic micro-mixer facilitating fluid mixing at a wide range of flow rate for the preparation of quantum dots. *Sci. China Chem.* **2012**, *56*, 799–805. [CrossRef]
10. Raza, W.; Hossain, S.; Kim, K.Y. A review of passive micromixers with a comparative analysis. *Micromachines* **2020**, *11*, 455. [CrossRef]
11. Stroock, A.D.; Dertinger, S.K.; Ajdari, A.; Mezic, I.; Stone, H.A.; Whitesides, G.M. Chaotic mixer for micro-channels. Patterning flows using grooved surfaces. *Science* **2002**, *295*, 647–651. [CrossRef]
12. Somashekar, V.; Olse, M.; Stremler, M.A. Flow structure in a wide microchannel with surface grooves. *Mech. Res. Comm.* **2009**, *36*, 125–129. [CrossRef]
13. Kang, D.J. Effects of baffle configuration on mixing in a T-shaped microchannel. *Micromachines* **2015**, *6*, 765–777. [CrossRef]
14. Santana, H.S.; Silva, J.L.; Taranto, O.P. Optimization of micro-mixer with triangular baffles for chemical process in millidevices. *Sens. Actuators B Chem.* **2019**, *281*, 191–203. [CrossRef]
15. Kang, D.J. Effects of channel wall twisting on the mixing in a T-shaped micro-channel. *Micromachines* **2019**, *11*, 26. [CrossRef]
16. Kang, D.J.; Song, C.H.; Song, D.J. Junction contraction for a T-shaped microchannel to enhance mixing. *Mech. Res. Comm.* **2012**, *33*, 739–746.
17. Ansari, M.A.; Kim, K.Y.; Anwar, K.; Kim, S.M. A novel passive micro-mixer based on unbalanced splits and collisions of fluid stream. *J. Micromech. Microeng.* **2010**, *20*, 055007. [CrossRef]
18. Makhsuda, J.; Kang, D.J. Mixing Performance of a Cross-Channel Split-and-Recombine Micro-Mixer Combined with Mixing Cell. *Micromachines* **2020**, *11*, 685.
19. Clark, J.; Kaufman, M.; Fodor, P. Mixing enhancement in serpentine micro-mixers with a non-rectangular cross-section. *Micromachines* **2018**, *9*, 107. [CrossRef]
20. Mondal, B.; Mehta, S.K.; Patowari, P.K.; Pati, S. Numerical study of mixing in wavy micromixers: Comparison between raccoon and serpentine mixer. *Chem. Eng. Proc.-Process Intensif.* **2019**, *136*, 44–61. [CrossRef]
21. Afzal, A.; Kim, K.-Y. Convergent–divergent micromixer coupled with pulsatile flow. *Sens. Actuators B Chem.* **2015**, *211*, 198–205. [CrossRef]
22. Schönfeld, F.; Hardt, S. Simulation of helical flows in microchannels. *AIChE J.* **2004**, *50*, 771–778. [CrossRef]
23. Tripathi, E.; Patowari, P.K.; Pati, S. Numerical investigation of mixing performance in spiral micromixers based on Dean flows and chaotic advection. *Chem. Eng. Proc.-Process Intensif.* **2021**, *169*, 108609. [CrossRef]
24. Fang, Y.; Ye, Y.; Shen, R.; Guo, P.Z.; Hu, Y.; Wu, L. Mixing enhancement by simple periodic geometric features in microchannels. *Chem. Eng. J.* **2012**, *187*, 306–310. [CrossRef]
25. Tsai, R.; Wu, C. An efficient micro-mixer based on multidirectional vortices due to baffles and channel curvature. *Biomicrofluidics* **2011**, *5*, 014103. [CrossRef]
26. Sotowa, K.I.; Yamamoto, A.; Nakagawa, K.; Sugiyama, S. Indentations and baffles for improving mixing rate in deep microchannel reactors. *Chem. Eng. J.* **2011**, *167*, 490–495. [CrossRef]
27. Chen, X.; Tian, Y. Passive micro-mixer with baffles distributed on both sides of micro-channels based on the Koch fractal principle. *J. Chem. Tech. Biotechnol.* **2020**, *95*, 806–812. [CrossRef]
28. Chung, C.K.; Shiha, T.R.; Changa, C.K.; Lai, C.W.; Wub, B.H. Design and experiments of a short-mixing-length baffled microreactor and its application to microfluidic synthesis of nanoparticles. *Chem. Eng. J.* **2011**, *168*, 790–798. [CrossRef]
29. Raza, W.; Kim, K.Y. Asymmetrical split-and-recombine micro-mixer with baffles. *Micromachines* **2019**, *10*, 844. [CrossRef]
30. Lim, T.W.; Son, Y.; Jeong, Y.J.; Yang, D.Y.; Kong, H.J.; Lee, K.S.; Kim, D.Y. Three dimensionally crossing manifold micro-mixer for fast mixing in a short channel length. *Lab Chip* **2011**, *11*, 100–103. [CrossRef] [PubMed]
31. Borgohain, P.; Arumughan, J.; Dalal, A.; Natarajan, G. Design and performance of a three-dimensional micromixer with curved ribs. *Chem. Eng. Res. Des.* **2018**, *136*, 761–775. [CrossRef]

32. Liu, K.; Yang, Q.; Chen, F.; Zhao, Y.; Meng, X.; Shan, C.; Li, Y. Design and analysis of the cross-linked dual helical micro-mixer for rapid mixing at low Reynolds numbers. *Microfluid. Nanofluid.* **2015**, *19*, 169–180. [CrossRef]
33. Kashid, M.; Renken, A.; Kiwi-Minsker, L. Mixing efficiency and energy consumption for five generic microchannel designs. *Chem. Eng. J.* **2011**, *167*, 436–443. [CrossRef]
34. Hessel, V.; Löwe, H.; Schönfeld, F. Micromixers-a review on passive and active mixing principles. *Chem. Eng. Sci.* **2006**, *60*, 2479–2501. [CrossRef]
35. Liao, Y.; Mechulam, Y.; Lassale-Kaiser, B. A millisecond passive micromixer with low flow rate, low sample consumption and easy fabrication. *Sci. Rep.* **2021**, *11*, 20119. [CrossRef]
36. Cook, K.J.; Fan, Y.F.; Hassan, I. Mixing evaluation of a passive scaled-up serpentine micromixer with slanted grooves. *ASME J. Fluids Eng.* **2013**, *135*, 081102. [CrossRef]
37. Rhoades, T.; Kothapalli, C.R.; Fodor, P. Mixing optimization in grooved serpentine microchannels. *Micromachines* **2020**, *11*, 61. [CrossRef]
38. Volpe, A.; Paie, P.; Ancona, A.; Osellame, R.; Lugara, P.M. A computational approach to the characterization of a microfluidic device for continuous size-based inertial sorting. *J. Phys. D Appl. Phys.* **2017**, *50*, 255601. [CrossRef]
39. *Ansys®Fluent 19.2 Computer Software*, ANSYS Inc.: Canonsburg, PA, USA, 2018.
40. Nguyen, N.T. *Micromixers*, 2nd ed.; Elsevier Science Ltd.: Amsterdam, The Netherlands, 2012; pp. 9–72.
41. Gu, Y.; Hegde, C.; Bishop, K.J.M. Measurement and mitigation of free convection in microfluidic gradient generators. *Lab Chip* **2018**, *18*, 3371–3378. [CrossRef]
42. Kefala, I.; Papadopoulos, V.E.; Karpou, G.; Kokkoris, G.; Papadakis, G.; Tserepi, A. A labyrinth split and merge micromixer for bioanalytical applications. *Microfluid. Nanofluid* **2015**, *19*, 1047–1059. [CrossRef]
43. Okuducu, B.; Aral, M.M. Performance analysis and numerical evaluation of mixing in 3-D T-shape passive micromixers. *Micromachines* **2018**, *9*, 210. [CrossRef] [PubMed]
44. Ortega-Casanova, J. Enhancing mixing at a very low Reynolds number by a heaving square cylinder. *J. Fluids Struct.* **2016**, *65*, 1–20. [CrossRef]
45. Valentin, K.; Ekaterina, S.B.; Wladimir, R. Numerical and experimental investigations of a micromixer with chicane mixing geometry. *Appl. Sci.* **2018**, *8*, 2458.
46. Soleymani, A.; Kolehmainen, E.; Turunen, I. Numerical and experimental investigations of liquid mixing in T-type micro-mixers. *Chem. Eng. Sci.* **2008**, *135S*, S219–S228. [CrossRef]
47. Matsunaga, T.; Lee, H.; Nishino, K. An approach for accurate simulation of liquid mixing in a T-shaped micro-mixer. *Lab Chip* **2013**, *13*, 1515–1521. [CrossRef]
48. Vikhansky, A. Quantification of reactive mixing in laminar micro-flows. *Phys. Fluids* **2004**, *16*, 4738–4741. [CrossRef]
49. Mengeaud, V.; Josserand, J.; Girault, H. Mixing processes in a zigzag microchannel: Finite element simulations and optical study. *Anal. Chem.* **2002**, *74*, 4279–4286. [CrossRef]
50. Roache, P. Perspective: A method for uniform reporting of grid refinement studies. *ASCE J. Fluids Eng.* **1994**, *116*, 307–357. [CrossRef]
51. Roache, P. *Verification and Validation in Computational Science and Engineering*; Hermosa: Albuquerque, NM, USA, 1998.
52. Makhsuda, J.; Kang, D.J. Optimal combination of mixing units using the design of experiments method. *Micromachines* **2021**, *11*, 685.
53. Hossain, S.; Kim, K.Y. Mixing analysis of passive micro-mixer with unbalanced three split rhombic sub-channels. *Micromachines* **2014**, *5*, 913–928. [CrossRef]

 *micromachines*

Article

Evaluation of Hydrodynamic and Thermal Behaviour of Non-Newtonian-Nanofluid Mixing in a Chaotic Micromixer

Naas Toufik Tayeb [1], Shakhawat Hossain [2,*], Abid Hossain Khan [3], Telha Mostefa [4] and Kwang-Yong Kim [5,*]

1 Gas Turbine Joint Research Team, University of Djelfa, Djelfa 17000, Algeria; toufiknaas@gmail.com
2 Department of Industrial and Production Engineering, Jashore University of Science and Technology, Jashore 7408, Bangladesh
3 Institute of Nuclear Power Engineering, Bangladesh University of Engineering and Technology, Dhaka 1000, Bangladesh; khanabidhossain@gmail.com
4 Mechanical Engineering Department, Ziane Achour University of Djelfa, Djelfa 17000, Algeria; telhamostefa@gmail.com
5 Department of Mechanical Engineering, Inha University, 100 Inha-ro, Michuhol-gu, Incheon 22212, Korea
* Correspondence: shakhawat.ipe@just.edu.bd (S.H.); kykim@inha.ac.kr (K.-Y.K.); Tel.: +82-32-872-3096 (K.-Y.K.); Fax: +82-32-868-1716 (K.-Y.K.)

Abstract: Three-dimensional numerical investigations of a novel passive micromixer were carried out to analyze the hydrodynamic and thermal behaviors of Nano-Non-Newtonian fluids. Mass and heat transfer characteristics of two heated fluids have been investigated to understand the quantitative and qualitative fluid faction distributions with temperature homogenization. The effect of fluid behavior and different Al_2O_3 nanoparticles concentrations on the pressure drop and thermal mixing performances were studied for different Reynolds number (from 0.1 to 25). The performance improvement simulation was conducted in intervals of various Nanoparticles concentrations ($\varphi = 0$ to 5%) with Power-law index (n) using CFD. The proposed micromixer displayed a mixing energy cost of 50–60 comparable to that achieved for a recent micromixer (2021y) in terms of fluid homogenization. The analysis exhibited that for high nanofluid concentrations, having a strong chaotic flow enhances significantly the hydrodynamic and thermal performances for all Reynolds numbers. The visualization of vortex core region of mass fraction and path lines presents that the proposed design exhibits a rapid thermal mixing rate that tends to 0.99%, and a mass fraction mixing rate of more than 0.93% with very low pressure losses, thus the proposed micromixer can be utilized to enhance homogenization in different Nano-Non-Newtonian mechanism with minimum energy.

Keywords: chaotic micromixer; Nano-Non-Newtonian fluid; mass mixing index; thermal mixing index; low generalized Reynolds number; minimal mixing energy cost

Citation: Tayeb, N.T.; Hossain, S.; Khan, A.H.; Mostefa, T.; Kim, K.-Y. Evaluation of Hydrodynamic and Thermal Behaviour of Non-Newtonian-Nanofluid Mixing in a Chaotic Micromixer. *Micromachines* **2022**, *13*, 933. https://doi.org/10.3390/mi13060933

Academic Editor: Shizhi Qian

Received: 27 April 2022
Accepted: 2 June 2022
Published: 11 June 2022

1. Introduction

Heat and mass transfer produced by different fluid concentrations is generally applied in high mixing processes that affect the dynamic transport of chemical species within physical chaotic advection. Mixing of non-Newtonian fluids process of micromixers occurs in multiple applications of various devices and industrial applications that have significant utility in bioengineering fields [1], chemical engineering [2], extraction, polymerization techniques and cleaning systems [3–7], etc. To enhance the kinematic behaviour [8] and fluid homogenization performance, many researchers contribute various solutions using chaotic advection inside microchannels [9,10] while reducing pressure drops [11–14].

Xia et al. [15] investigated experimentally and numerically different passive micromixers with multi-layer crossing channels. Their proposed micromixer can give a fast mixing (more than 95%) at minimum Reynolds number (not exceeding 40), for example, a high mixing index of 0.96 was found for very low flow regimes. The same geometry was selected by Hossain et al. [16] to carry out a parametric study. To develop the mixing efficiency, the effect of the geometrical parameter was analyzed at (not exceeding 40) low Reynolds numbers.

The mixing enhancement of split and recombination micromixers using Newtonian fluid flow with low regimes was investigated numerically by Hossain et al. [17]. They evaluated the homogenization performance and the pressure-drop characteristics compared to the TLCCM micromixer which was selected by Xia et al. [15], in a range of Reynolds numbers between 0.2 and 120. Raza et al. [18,19] developed the same problem to create an unbalanced split and recombination micromixer (1.5 mm). They measured numerically the mixing efficiency within various cross-sections via low flow regime. They found that the optimal configuration has 88% mixing efficiency. A parametric study using five geometric parameters of the micromixer was performed by Naas et al. [8]. They presented the effect of a low Reynolds number on mixing efficiency. Their proposed short micromixer gives mixing rates greater than 99% throughout the Reynolds number range in a short length of three-dimensional micromixer, which is rarely achieved in early micromixer configurations.

The effects of thermal condition of non-Newtonian fluids rheology were investigated by Naas et al. for C-shape chaotic geometry [20,21] and Multi-layer micromixers [22]. They selected a highly efficient heat mixing system and better thermodynamic performance, for rheology index ranging from 1 to 0.49. The results showed that the decrease in heat transfer flow of shear-thinning flows was higher than that of Newtonian fluids.

In order to enhance the heat transfer efficiency, several researchers carried out numerically [23–30] and experimentally [31,32] the thermo-physical enhancement of various applied thermal engineering systems. They used effective nanoparticles such as: Al_2O_3 [23,24], H2O/SWCNT [29], CuO [28–30]. The authors proved experimentally that the nanofluids have significantly more powerful thermal conductivities than the same flow with zero nano concentration.

Inside tubes, Xuan and Li [33] investigated experimentally the thermal performance of nanofluid swirling under wall heat transfer. Compared with pure water, the heat transfer coefficient was augmented more than 39% with the Cu nanoparticles concentration of 2.0%.

Within rectangular microchannels, convection laminar regimes with nano-shear thinning flow have been carried out by Esmaeilnejad et al. [34]. Their obtained results presented that, with Peclet number of 700 and 4% particle concentration, the thermal coefficient will decease around 27.2% and the pressure drop will increase roughly 50.7%. Evangelos et al. [35] studied the mixing performances of heated water with the effect of an electromagnetic field. For different mixers, Pouya [36] investigated numerically, the heat transfer enhancement and mixing quality of a hybrid nanofluid. They found that for high Reynolds numbers and a constant frequency, the mixing rate increased over time.

In other works [37], the authors created a chaotic geometry called C-shape to enhance the heat transfer characteristics of non-Newtonian nanofluid. They found that the development of chaotic advection causes more enhancement to thermal rate in comparison with pumping power.

In recent years, Antar and Kacem [24] analyzed the enhancement of forced convection flow within a straight pipe under the effect of nano-particles of Al_2O_3subjected to a fixed wall of heat. They examined the effects of both fluid behavior index and nanofluid concentrations on the gradient temperature and heat transfer rates. Their obtained results indicated that, for various Peclet and Brinkman numbers, the heat transfer rate increases when nanoparticles concentration.

Due to the diminutive measurement of micromixers, it's difficult to enhance the molecular diffusion operation. While the improvement in dynamic behaviors is limited, the injection of nanofluid concentrations is a suitable manner in which to enhance mass transfer and thermal mixing efficiency within a passive micromixer. Therefore, our contribution aims to improve high mixing performances for Nano-non-Newtonian fluids in terms of mass and heat transfer inside the optimum multi-layer micromixer as it was recently used by Naas et al. [8] beneath a very low Reynolds number. Various nanoparticles concentrations with different values of fluid behavior index were proposed to investigate the chaotic flow formation and thermal mixing performances within the suggested micromixer. In order to get important energy efficiency, the homogenization of the fluid indexes and mixing energy cost will be appraised.

2. Micromixer Structure and Problem Statement

A novel micromixer was suggested in this study, namely a Two-Layer Crossing Modified geometry (TLCM) which was used recently and firstly by Naas et al. [8] to achieve High-Mixing Processing under the effect of nanofluid concentrations for power-law non-Newtonian fluid [24], see Figure 1. The micromixer is constructed of a couple of twisted channels; the higher and lower channels are arranged with a periodic chamber. The following mixing components occur in reconstructed forms of several grooves. Dimensions detail is presented in Table 1, where D is the chamber diameter, d is the grooves diameter, d_{hyd} is the hydraulic diameter, l is the distance between the inlets and L* is the length of the geometry. The same nano-non-Newtonian fluid of Antar and Kacem [23,24] and Santra et al. [36] was proposed as working fluid, represented in Tables 2 and 3 respectively.

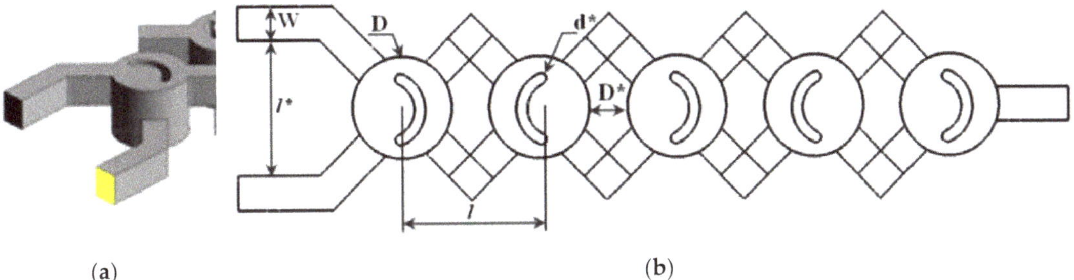

(a) (b)

Figure 1. (**a**): 3-D view design of the proposed micromixers, (**b**): geometric parameters.

Table 1. Dimensions information of the proposed micromixer.

W	0.2 mm
*l**	0.8 mm
D*	0.2 mm
d*	0.1 mm
d_{hyd}	0.22 mm
D	0.6 mm
L	4.5 mm

Table 2. Rheological parameters of Al_2O_3 Nano-non-Newtonian, reproduced with permission from [23,24].

φ %	m (N s^n m^{-2})	N
0.5	0.00187	0.88
1.0	0.00230	0.83
1.5	0.00283	0.78
2.0	0.00347	0.730
2.5	0.00426	0.680
3.0	0.00535	0.625
3.5	0.00641	0.580
4.0	0.00750	0.540
4.5	0.00876	0.500
5.0	0.01020	0.460

Table 3. Thermal parameters of Al_2O_3 Nano-non-Newtonian, reproduced with permission from [23,24].

(φ %)	ρ (Kg/m³)	C_p (J/Kg k)	k (w/m k)
0.5	1013.1	4165.9	0.6248
1.0	1027.9	4148.8	0.6367
1.5	1042.8	4131.7	0.6488
2.0	1057.6	4114.6	0.6610
2.5	1072.5	4097.5	0.6734
3.0	1087.4	4080.5	0.6859
3.5	1102.2	4063.4	0.6987
4.0	1117.1	4046.3	0.7116
4.5	1131.9	4029.2	0.7246
5.0	1146.8	4012.1	0.7379

Heated and cold Nano-non-Newtonian fluids flow from the inlets with a uniform velocity profile. The outsides are deemed adiabatic and the other boundaries include no-slip. Pressure outlet condition is awarded at the outflow section. A numerical CFD code Fluent© numerically solved the governing equations, which are given by the following expressions [22,38]:

$$div\vec{V} = 0 \tag{1}$$

where $\vec{V}$ is the velocity vector.

$$\vec{V} \cdot \overline{\overline{\nabla}}\vec{V} = -\frac{1}{\rho_{nf}}\vec{\nabla}P + div\tau \tag{2}$$

where σ (Pa) is shear stress and P is the pressure.

$$\rho_{nf}c_{nf}\vec{V} \cdot \vec{\nabla}T = \lambda_{nf}\Delta T \tag{3}$$

where ρ_{nf}, λ_{nf} and T are the density, the conductivity and the temperature of the nanofluid, respectively.

The constitutive connection among the shear rate $\dot{\gamma}$ (s^{-1}) and shear stress τ (Pa) can be characterized by a simple power-law equation:

$$\tau = m\dot{\gamma}^{n} \tag{4}$$

where, m (Pas^{-1}) is fluid consistency index and n is the fluid behavior index.

The apparent viscosity is:

$$\mu_{nf} = k\dot{\gamma}^{n-1} \tag{5}$$

The advised boundary conditions are:

- Uniform velocity profile imposed to the inlets flow, and the temperatures equal to T_{min} = 300 k for one inlet, and the other one T_{max} = 330 k
- No-slip conditions within the solid walls.
- Pressure outlet condition is considered at the outlet section flow.

2.1. Mass Transfer Characteristics of the Chaotic Flows

Metzner and Reed [39,40] defined the generalized Reynolds number (Re_g) for power-law fluids as follows:

$$Re_g = \frac{\rho_{nf}u^{2-n}d_{hyd}^{n}}{\left[8^{n-1}\left(b^* + \frac{a^*}{n}\right)^{n}m\right]} \tag{6}$$

a^* Geometric parameter (a^* = 0.2121)
b^* Geometric parameter (b^* = 0.6771)

ρ_{nf} Density $[kg \cdot m^{-3}]$
u Average speed $[m \cdot s^{-1}]$
d_{hyd} Hydraulic diameter [m]
m Consistency index (N s^n m^{-2})
n Rheological behavior index of the fluid

Mixing efficiency of two Nano-non-Newtonian fluids inside the micromixer is characterized by:

$$MM_i = 1 - \frac{\sigma}{\sigma_0} \tag{7}$$

σ is the standard deviation measured by post-CFD.
σ_0 is the maximum standard deviation.

$$\sigma^2 = \frac{1}{N}\sum_{i=1}^{N}\left(C_i - \overline{C}\right)^2 \tag{8}$$

For fully mixed fluids, the standard deviation rate was taken from the minimum values and maximum for unmixed fluids. N presents the total number of nodes. $\overline{C}$ is the average mass fraction. The maximum standard deviation σ^2 is measured by:

$$\sigma_0^2 = \left(1 - \overline{C}\right) \tag{9}$$

The mixing energy cost (MEC) is used to decide the effectiveness of the micromixer and is established by merging the pressure losses and the mixing degree, as follows [41]:

$$MEC = \frac{\Delta P \times Q}{MMi} \tag{10}$$

where ΔP and Q are the pressure drop and the flow rate (m^3/s) along the geometry, respectively.

2.2. Thermal Characteristics of the Chaotic Flows

Heat transfer coefficient, h, different inlet temperature is assumed as:

$$h = \frac{q''}{(T_b - T_w)} \tag{11}$$

where, T_b (k) is the mean bulk temperature fluid, T_w (k) is the perimeter average wall temperature and q'' (w/m^2) is the wall heat flux.

These two temperatures are defined as:

$$T_w(s) = \frac{1}{P}\int_P T_w dp \tag{12}$$

$$T_b(s) = \frac{1}{AU_i}\iint_A \vec{V}\cdot\vec{n}T\cdot dA \tag{13}$$

The mean heat transfer coefficient, h_{mean}, is defined as:

$$h_{mean} = \frac{1}{L}\int_0^L h(s)ds \tag{14}$$

The Thermal Mixing index (TM_i) utilized by Naas et al. [22] of two fluids (hot and cold) is given by the following equation:

$$TM_i = 1 - \frac{\iint_s \left|T - T_{avg}\right| ds}{(T_{max} - T_{min})s} \tag{15}$$

where, $T_{avg} = (T_{min} + T_{max})/2$. T_{min}= 300 K and $T_{max} = 330$.

K The values of TM_i range from 0 for the unmixed flow case, to 1 for fully mixed flow.

In order to confirm the efficiency of the scalar temperature T which was measured between two values; (T_a, T_b) at a given section, a probability density function PDF % (T) was calculated which was equal to the number of nodes within (T_a, T_b) divided by the total number of nodes [22].

All governing equations in the present work were solved in a laminar flow by using ANSYS Fluent 16© CFD software [42], which works by the (FVM) finite volumes method. The SIMPLEC scheme was chosen for pressure and velocity coupling. To determine the mass and momentum equations a second-order upwind scheme was selected. The computations were ensured and simulated to be converged at 10^{-7} of root mean square (RMS) residual values. Non-Newtonian power-law fluids were done as a working fluid for various Al_2O_3 nanoparticles concentrations.

3. Mesh Sensitivity Test

To check the sensitivity of the CFD results, a quantitative grid study was carried outby varying the total number of cells. Using an unstructured mesh with uniform tetrahedral cells, four mesh grids were analyzed ranging from 100,000 to 800,000.

The evolutions of the velocity along the X-axis of the outlet section with a Reynolds number equal to 10 are shown in Figure 2. It can be observed that the outlet velocity rates were sensitive to the mesh, except for the mesh densities with 700,000 and 800,000 nodes where no important difference is seen. As a consequence, the 700,000 nodes grid is selected as the favorable mesh for the analyses (Figures 2 and 3).

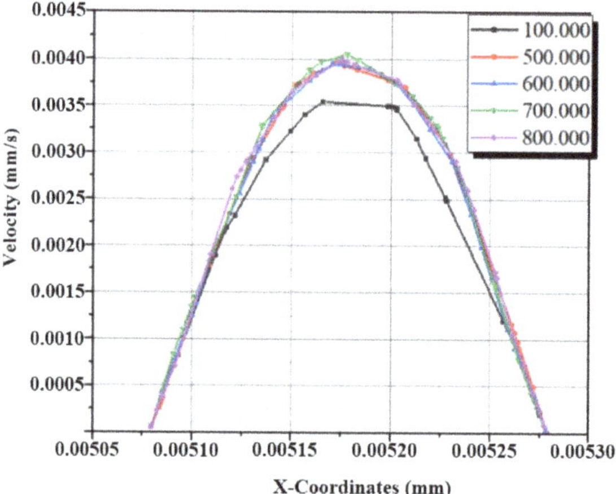

Figure 2. Evolutions of outlet velocities at the exit mid-line of X coordinates.

Figure 3. Structure of the generated mesh grid.

4. Results and Discussion

Kinematic and thermal behaviors of mixing Nano-non-Newtonian flows are investigated in detail within novel micromixers, which are compared with potential micromixers used recently in the literature. Heat, mass transfer and fluid mixing process are investigated for several low generalized Reynolds number ranging between 0.1 to 25.

4.1. CDF Validation Case

The numerical solution procedure of mass transfer performances has been reported and validated thoroughly by comparing the present results with the results of Xia et al. [15] and Hossain et al. [16], for Reynolds number equal to 0.2, quantitative and quantitative comparison for mixing efficiency index and mass transfer contours for different successive cross-sectional planes, as shown in Figure 4.

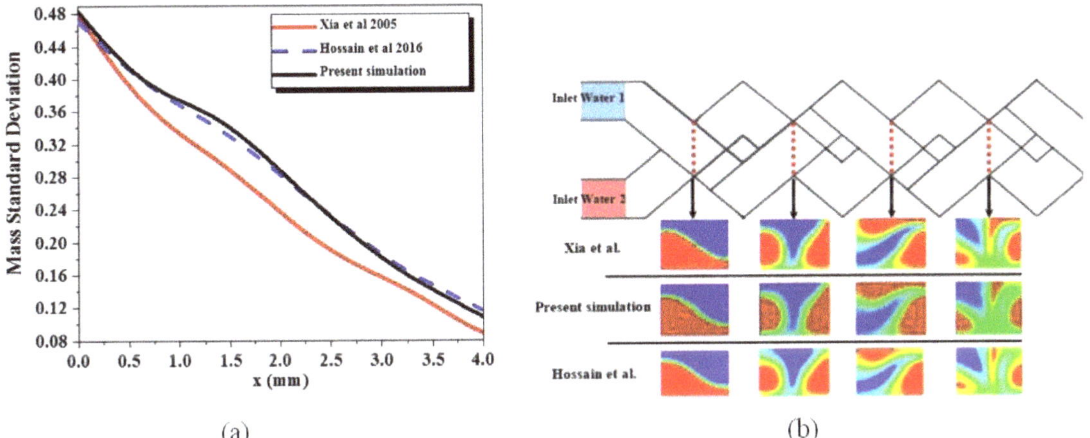

Figure 4. Quantitative validation for mixing efficiency index (**a**), and quantitative validation of local mass fraction contours for various planes (**b**) With the results of Xia et al., reproduced with permission from [15] and Hossain et al. reproduced with permission from [16].

An excellent agreement is seen to exist between the present numerical values and the literature values of mass mixing index. Based on these comparisons, it is perhaps reasonable to conclude that the present results are reliable to within ⊥0.4%. Deviations of this order are not at all uncommon in numerical studies and arise due to the differences in the flow schematics, problem formulations, grid and/or domain sizes, discretization schemes and numerical methods.

Moreover, a quantitative numerical validation was carried out with those obtained by Ning et al. (2016) [43], and the results present a heat transfer rate as a function of various Reynolds numbers for non-Newtonian cases. The comparison is satisfactory and revealed good agreements among results which are shown in Figure 5.

In addition, we added another validation with those obtained by Jibo et al. [44]. The problem was analyzed for multiphase fluids of mixing enhancement inside micromixers, see Figure 6, where the relative error with the numerical results is less than 1%.

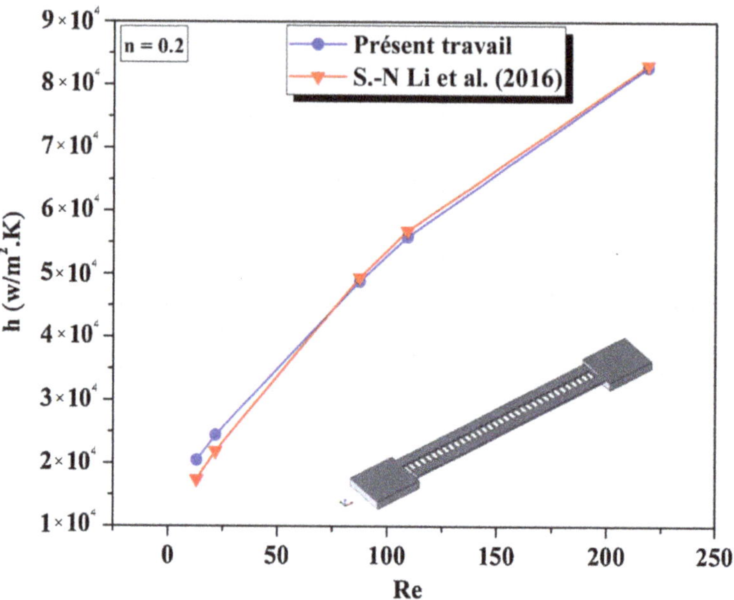

Figure 5. Heat transfer coefficient Vs Reynolds number for non-Newtonian case with Li et al. reproduced with permission from [43].

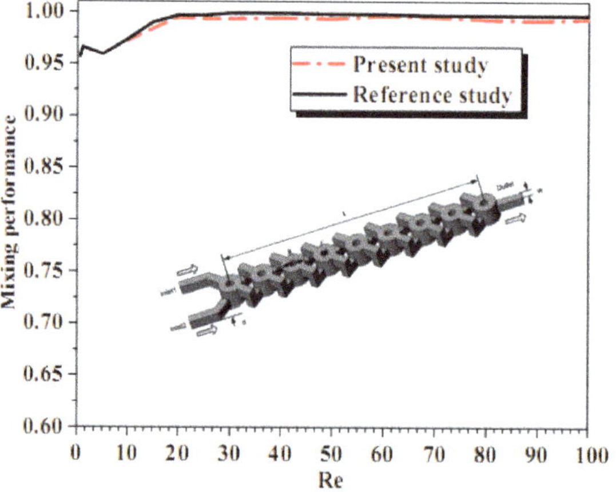

Figure 6. Evolution of mixing rates Vs Reynolds number with Jibo et al. [44].

4.2. Mass Transfer and Fluid Mixing Processing

Flow visualization of mass fraction between two nanofluids is shown in Figure 7. Different cases of fluid concentrations are subjected within the fluid pattern to understand the development of visual mixing inside the new configuration. We remark that the fluid mixing augments with the Reg for all fluid behavior index n and there is no exactly black region in the second unites. Therefore, the new micromixers have a quick mass transfer performance.

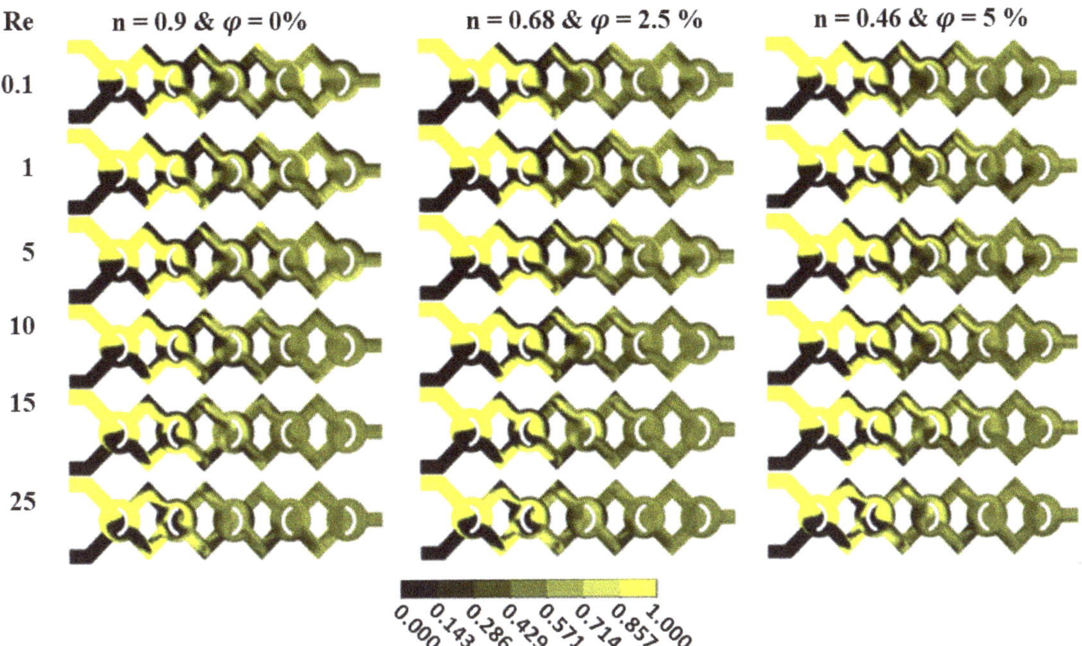

Figure 7. Mass fraction contours at various Reynolds numbers with different fluid concentrations (φ = 0.5 to 5%).

Figure 8 shows the evaluation of the mass mixing efficiency of two fluids as a function of Re_g for various cases of nanofluid concentrations (φ = 0.5 to 5%). As the flow homogenization was only performed by the nanoparticles, the prevalence of Reynolds number is very low (Re < 5) and the flow behaviors were not effective for improving mixing (molecular diffusion dominates). When the generalized Reynolds fluid number increased, the homogenization was more effective, and the mixing intensity developed rapidly. Note that in large Reynolds numbers, the nano-particles are more effective, and the mixing intensity develops quickly so the most select mixing state is reached when the concentration increases. Moreover, it is observed that the proposed micromixer shows a 2.22% enhancement of mixing intensity when the fluid behavior index decreases to 0.46, as compared to the case of n = 0.88.

The homogeneity achieved over midlines at the exit of the several cases of nano-fluid concentrations inside the micromixer at a given Re_g is demonstrated by the standard deviation like mass transfer profiles as shown in Figure 5. Significant changes in the values of the mass transfer for all cases of Nano-non-Newtonian fluids can be observed. For the case of φ = 5%, the results indicates that the best degree of mixing done at the exit of the micromixer.

To explain the structure of the fluid flow inside the proposed micromixer, a visualization of vectors and streamlines of the mass fraction and vortex core regions are shown in Figure 9. The kinematic behavior has an important role in enhancing homogenization for nanofluids. As can be seen in Figure 10, the micromixer has a single strong vortex region inside each corner which develops the mixing rate inwardly of the geometry, whereas it has a low-pressure drop near the outlet part. Furthermore, we can see that the flow is more chaotic and dynamic due to the structure and the curve of the configuration. Moreover, it is remarkable that the pathline inside the selected new micromixer produces a reversed flow pattern and strong secondary flows are created which can enhance more mass transfer efficiency and ensure excellent quality of homogenization.

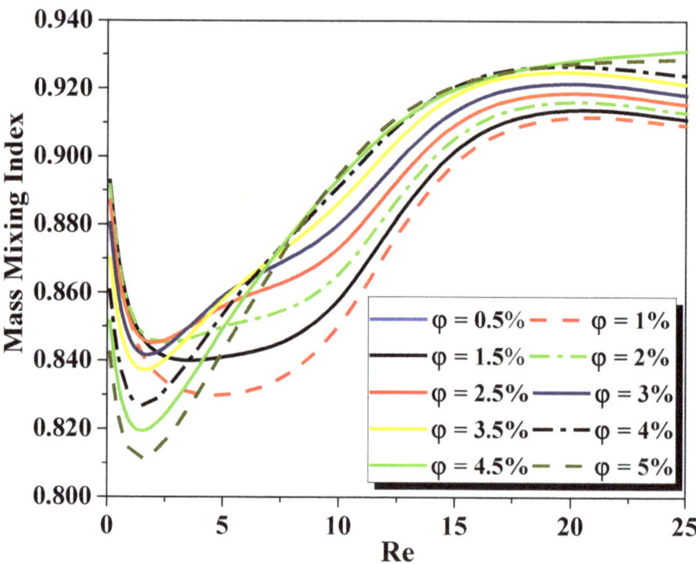

Figure 8. Development of mass mixing performance for different Reynolds numbers with variation cases of nanofluid concentration (φ = 0.5 to 5%).

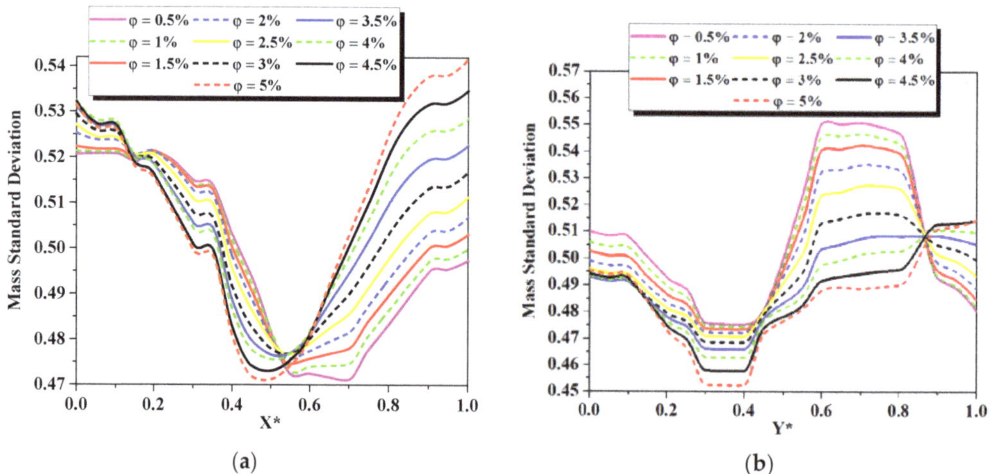

Figure 9. Evolution of the mass standard deviation along the (**a**): exit X-line and (**b**) exit Y-line of different cases of nano-fluid concentration (φ = 0.5 to 5%) at fixed Re_g = 25.

4.3. Heat Transfer and Thermal Mixing Processing

Thermal mixing fluids between hot and cold Nano-non Newtonian fluids are measured for various cases of nanofluid concentrations (φ = 0.5 to 5%). For this, the cold fluid is injected in one inlet at 300K and the heated fluid is injected in the other inlet at 330 K, see Figure 11. In this part, we investigate thermal mixing performances and mining energy coast in the present micromixer, and its performances will be compared to other strong micromixers. Figure 11 shows a top view of the temperature contours in three cases of nanofluid concentration (φ = 0, 2.5 and 5%) for Reg ranging from 0.1 to 25. For a given value of nanofluid concentration or fluid behavior index, the thermal homogenization quality is more vigorous when the Re_g is more important. The development of the generalized number highly improves the dynamic of the movement, the kinematic of the fluid nanoparticles varies considerably and the mixing intensity will be improved.

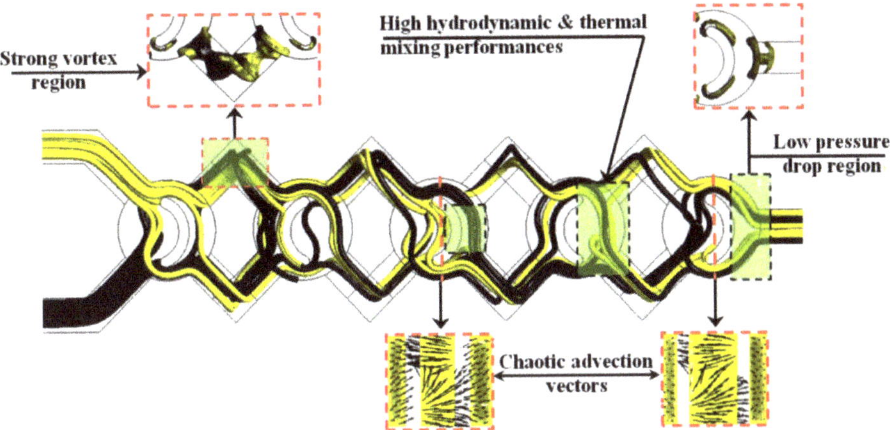

Figure 10. Streamlines and vectors of the mass fraction with vortex core region.

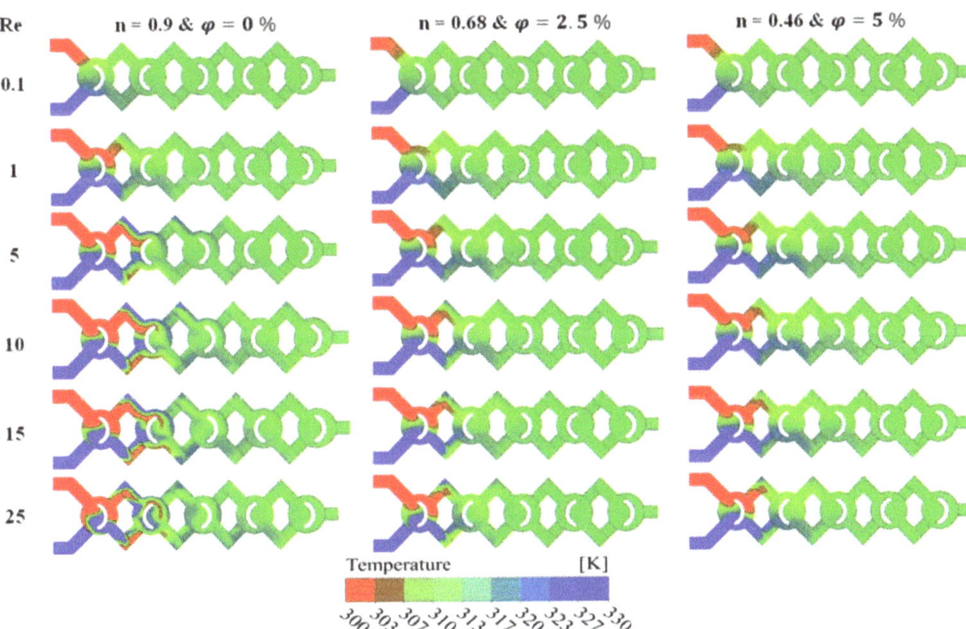

Figure 11. Qualitative representation of Temperature contours for different generalized Reynolds number at the horizontal middle section of each nano-fluid concentrations (φ = 0.5 to 5%).

The influence of the nanofluid concentration loading on the improved thermal efficiency is analyzed, see Figures 12–14. Figure 12 shows the evolution of heat transfer coefficient for several generalized Reynolds numbers with different nano-fluid concentrations (φ = 0.5 to 5%). As the nanofluid concentration ranges from 0.5% to 5%, the heat transfer augments from 122 W/m^2 K to 222 W/m^2 K for Re_g = 25. It can be resolved that other devices besides thermal conductivity increase can be efficient for thermal flow enhancement. Thermal nanofluid performance is under the effect of various parameters. It appears that nanofluid behavior may develop the temperature gradient within the flow and thus increase the heat of the nanofluid, see Figure 13. Figure 14 shows the evaluation of the thermal mixing efficiency of two fluids for different Re_g for various cases of nanofluid concentrations ($\varphi = 0.5$ to 5%).

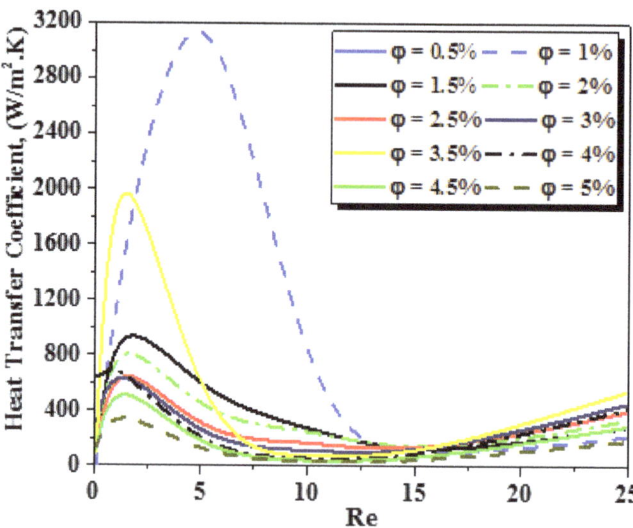

Figure 12. Evolutions of heat transfer coefficient for different generalized Reynolds numbers with various nanofluid concentrations (φ = 0.5 to 5%).

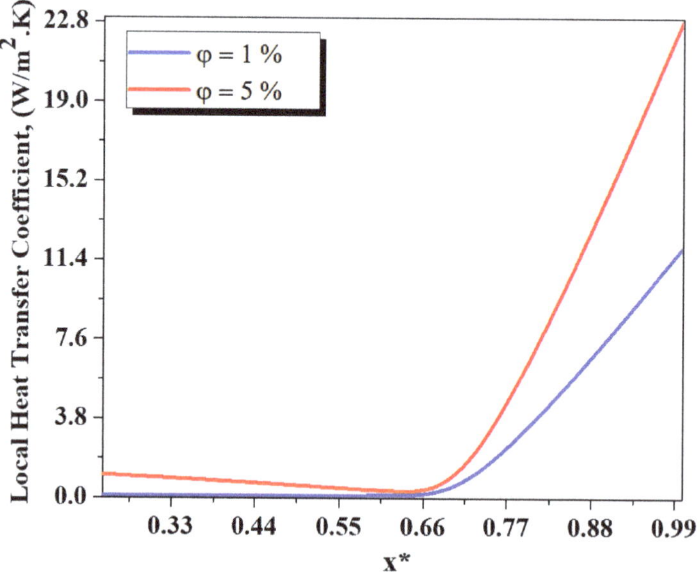

Figure 13. Evolutions of local heat transfer coefficient as a function along the geometry with φ = 1 and 5%.

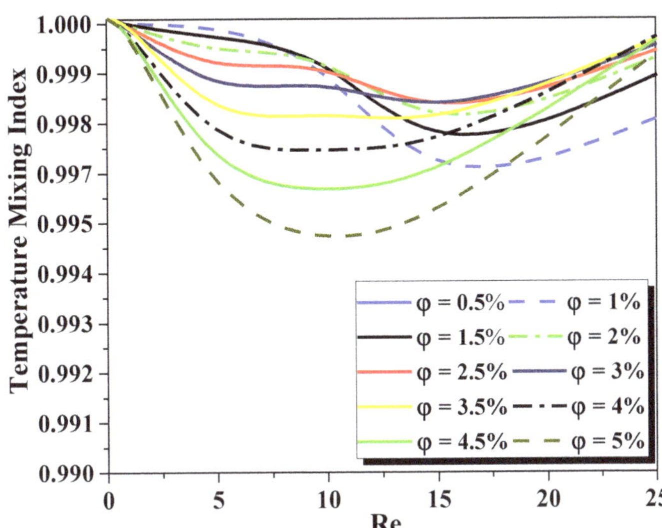

Figure 14. Improvement of thermal mixing performance for different generalized Reynolds numbers for variation cases of nanofluid concentrations (φ = 0.5 to 5%).

Due to the heat transfer being done only by conduction, it is indicated that the mixing quality is not affected by the fluid concentrations for low Re_g numbers. Therefore, the rheology fluid index n has no effect on the mode of energy transfer in the flow. When Re_g augments, the flow is chaotic which leads to greater mixing. This behavior enhances greatly the thermal efficiency as a function of Re_g number for all cases of nanofluid concentrations. For Re_g numbers equal to 25, the mixing degree is close to 1 and the quality of mixing is perfect. The degree of thermal mixing has an insignificant difference as a function of the fluid concentrations considered, so the fluid behavior index n has the same effect on the TMI as described above.

Table 4 presents the probability density function of temperature distribution for different cross-sections. As it is known, the greatest PDF is 100% for T = 315 K. Despite the fluid passing through the micromixer, the fluids are well mixed and tend to be homogenized (T = 315 K) under the effect of the chaotic behavior of the flow. For all cases of various nanofluid concentrations, the temperature distribution from the third plane (P3) is examined in a small variety of the order of two Kelvin, where the best of this arrangement is compared to the coveted mixing fluid temperature, 315 K.

Table 4. Local Temperature PDF ranging from 314 K to 316 K for different cross-sections.

φ (%)	P1 = 0.93 mm	P2 = 1.82 mm	P3 = 2.62 mm	P4 = 3.47 mm	Outlet
0.5	5%	31.45%	99.99%	99.99%	99.99%
2	7.83%	26.23%	99.99%	99.99%	99.99%
4	7.85%	24.43%	99.99%	99.99%	99.99%
5	12.24%	30.49%	99.99%	99.99%	99.99%

To choose the better Nano-non-Newtonian fluid concentration with high mixing efficiency and lower cost of mixing energy, Figure 15 presents Mixing Energy Cost under the effect of generalized Reynolds number ranging (Re_g = 0.1 to 25) for various nanofluid concentrations (φ = 0.5 to 5%). The flow of low Re_g has qualitatively the same behavior in terms of cost of the mixing energy, due to the fluid being more viscous and the secondary flows not yet active. When the Re_g overrun the value of 5, the difference enhances obviously for both the kinematic and thermal process. As can be seen, when the Reg increases, the

MEC become larger inside the micromixers because the fluid flow turns out more sheared and agitated. Moreover, the fluid flow concentration of 0.5% displays low rates of energy cost compared to the other cases.

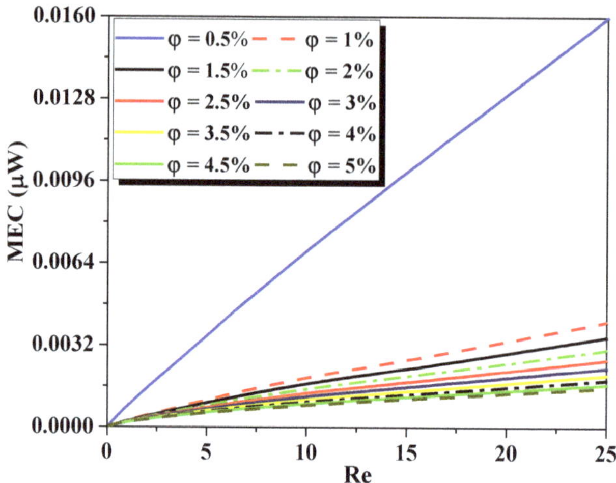

Figure 15. Development of Mixing Energy Cost for several generalized Reynolds numbers (Re_g = 0.1 to 25) for various nanofluid concentrations (φ = 0.5 to 5%).

The development of Mixing Energy Costis illustrated in Table 5 for various Reynolds numbers. For cases of Reynolds numbers, it can be clearly remarked that the present micromixer has the most economical mixing energy cost compared to the micromixer studied recently by Embarek et al. in [39]. In addition, to compare quantitatively the mixing index, pressure-losses, and the mixing energy cost, Table 6 shows different recent works (2017, 2019, 2021y) for these parameters at fixed Reynolds number (Re = 70). It is shown that the proposed micromixer has the lower pressure drops and the higher mixing index value of 99.99%, and the best mixing energy cost.

Table 5. Quantitative comparison of the Energy Mixing Cost for varied generalized Reynolds number (Re = 1, 5, 15 and 15) with that obtained by Embarek et al. reprinted with permission from [41] in 2021y.

Re	MEC (μW) of Embarek [39]	MEC (μW) of Present Work	Evolution Parentage
1	0.00920	0.0064	30%
5	0.23600	0.194	18%
15	2.57302	1.548	40%
30	12.99	7.43	45%

Table 6. Comparison of the pressure losses, mixing degree and energy mixing cost with different recent works at fixed Reynolds number (Re = 70).

Micromixer, Year	Unit	Length (mm)	ΔP (Pa)	MI	MEC (μW)
Split and recombination micromixer, 2012 [14]	6	10.66	16500	0.83	210.25
Convergent and divergent micromixer, 2017 [41]	6	11	14000	0.94	209.55
rectangular obstacles micromixer 2019 [45]	5	8.0	12542	0.98	179.75
triangular obstacles micromixer 2019 [45]	5	8.0	11485	0.93	175.38
teardrop obstacles micromixer 2019 [45]	5	8.0	10908	0.94	163.22
Elongation micromixer with 2021 [41]	4	3.5	8002	0.99	111.48
Present Two-layer Modified micromixer	5	4.58	3455	0.99	55.67

5. Conclusions

A numerical CFD work was performed to analyze the flow characteristics of Nano-non-Newtonian shear-thinning fluid inside a novel micromixer. This study outlines the hydrodynamic and thermal mixing performances for different generalized Reynolds numbers with nanofluid concentrations (φ = 0.5 to 5%). According to this study, the following points can be given:

- Effects of generalized Reynolds numbers on the hydrodynamic behavior of non-Newtonian flow were improved within the proposed micromixers.
- Strong secondary flows are created inside the micromixer to enhance the mixing quality for all cases of nanofluid concentration.
- The non-Newtonian fluid of φ = 0.5% exhibits low mass transfer compared to the preferable nanofluid concentration case (φ = 5%).
- As Reynolds number increases, the flow visualization of both heat and mass transfer revealed that the vortex created in the micromixer had more vigorous intensity.
- Higher rates of generalized Reynolds number have more effects to increase both mass and thermal homogenization rates.
- The heat transfer coefficient increases from 122 W/m^{2}K to 222 W/m^{2}K when φ rises from 0.5 to 5%.
- The cases of low fluid behavior index n have a more effective improvement in the mixing efficiency than the other cases
- For all of Reg, high intensity thermal and fluid mixing is obtained for high nanofluid concentration (φ = 5%).

Author Contributions: Conceptualization, Numerical Simulation, N.T.T.; Writing Original Draft, S.H.; Conceptualization, Methodology, Writing Original Draft. A.H.K.; Writing—review and editing, K.-Y.K.; Funding, Supervision, T.M.; Numerical Simulation, Writing Original Draft. All authors have read and agreed to the published version of the manuscript.

Funding: This research received no external funding.

Acknowledgments: The work was partially supported by the UGC research fund 2020–2021. Authors are gratefully acknowledging this support. Authors also acknowledge the support of Jashore University of Science and Technology.

Conflicts of Interest: The authors declare there is no conflict of interest.

References

1. Lee, C.Y.; Fu, L.M. Recent advances and applications of micromixers. *Sens. Actuators B Chem.* **2018**, *259*, 677–702. [CrossRef]
2. Park, T.; Lee, S.; Seong, G.H.; Choo, J.; Lee, E.K.; Kim, Y.S.; Ji, W.H.; Hwang, S.Y.; Gweon, D.G.; Lee, S. Highly sensitive signal detection of duplex dye-labelled DNA oligonucleotides in a PDMS microfluidic chip: Confocal surface-enhanced Raman spectroscopic study. *Lab Chip* **2005**, *5*, 437–442. [CrossRef] [PubMed]
3. Rapp, B.E.; Gruhl, F.J.; Länge, K. Biosensors with label-free detection designed for diagnostic applications. *Anal. Bioanal. Chem.* **2010**, *398*, 2403–2412. [CrossRef] [PubMed]
4. Rahman, M.; Rebrov, E. Microreactors for gold nanoparticles synthesis: From faraday to flow. *Processes* **2014**, *2*, 466–493. [CrossRef]
5. Stone, H.A.; Stroock, A.D.; Ajdari, A. Engineering flows in small devices. *Annu. Rev. Fluid Mech.* **2004**, *36*, 381–411. [CrossRef]
6. Squires, T.M.; Quake, S.R. Microfluidics: Fluid physics at the nanoliter scale. *Rev. Mod. Phys.* **2005**, *77*, 977–1026. [CrossRef]
7. Hessel, V.; Löwe, H.; Schönfeld, F. Micromixers—A review on passive and active mixing principles. *Chem. Eng. Sci.* **2005**, *60*, 2479–2501. [CrossRef]
8. Naas, T.T.; Hossain, S.; Aslam, M.; Rahman, A.; Hoque, A.S.M.; Kim, K.-Y.; Islam, S.M.R. Kinematic measurements of novel chaotic micromixers to enhance mixing performances at low Reynolds numbers: Comparative study. *Micromachines* **2021**, *12*, 364. [CrossRef]
9. Raza, W.; Hossain, S.; Kim, K.Y. A Review of Passive Micromixers with a Comparative Analysis. *Micromachines* **2020**, *11*, 455. [CrossRef]
10. Ingham, C.J.; van Hylckama Vlieg, J.E.T. MEMS and the microbe. *Lab Chip* **2008**, *8*, 1604–1616. [CrossRef]
11. Schulte, T.H.; Bardell, R.L.; Weigl, B.H. Microfluidic technologies in clinical diagnostics. *Clin. Chim. Acta* **2002**, *321*, 1–10. [CrossRef]
12. Razzacki, S.Z.; Thwar, P.K.; Yang, M.; Ugaz, V.M.; Burns, M.A. Integrated microsystems for controlled drug delivery. *Adv. Drug Deliv. Rev.* **2004**, *56*, 185–198. [CrossRef] [PubMed]

13. Hossain, S.; Lee, I.; Kim, S.M.; Kim, K.Y. A micromixer with two-layer serpentine crossing channels having excellent mixing performance at low Reynolds numbers. *Chem. Eng. J.* **2017**, *327*, 268–277. [CrossRef]
14. Afzal, A.; Kim, K.Y. Passive split and recombination micromixer with convergent-divergent walls. *Chem. Eng. J.* **2012**, *203*, 182–192. [CrossRef]
15. Xia, H.M.; Wan, S.Y.M.; Shu, C.; Chew, Y.T. Chaotic micromixers using two-layer crossing channels to exhibit fast mixing at low Reynolds numbers. *Lab Chip* **2005**, *5*, 748–755. [CrossRef] [PubMed]
16. Hossain, S.; Kim, K.Y. Parametric investigation on mixing in a micromixer with two-layer crossing channels. *Springerplus* **2016**, *5*, 794. [CrossRef]
17. Hossain, S.; Kim, K.Y. Mixing analysis in a three-dimensional serpentine split-and-recombine micromixer. *Chem. Eng. Res. Des.* **2015**, *100*, 95–103. [CrossRef]
18. Raza, W.; Hossain, S.; Kim, K.Y. Effective mixing in a short serpentine split-and-recombination micromixer. *Sens. Actuators B Chem.* **2018**, *258*, 381–392. [CrossRef]
19. Raza, W.; Kim, K.Y. Unbalanced Split and Recombine Micromixer with Three-Dimensional Steps. *Ind. Eng. Chem. Res.* **2020**, *59*, 3744–3756. [CrossRef]
20. Naas, T.T.; Lasbet, Y.; Benzaoui, A.; Loubar, K. Characterization of Pressure Drops and Heat Transfer of Non-Newtonian Power-Law Fluid Flow Flowing in Chaotic Geometry. *Int. J. Heat Technol.* **2016**, *34*, 251–260. [CrossRef]
21. Naas, T.T.; Lasbet, Y.; Aidaoui, L.; Ahmed, L.B.; Khaled, L. High performance in terms of thermal mixing of non-Newtonian fluids using open chaotic flow: Numerical investigations. *Therm. Sci. Eng. Prog.* **2020**, *16*, 100454. [CrossRef]
22. Naas, T.T.; Kouadri, A.; Khelladi, S.; Laib, L. Thermal mixing performances of shear thinning non-Newtonian fluids inside two-layer crossing channels micromixer using entropy generation method: Comparative study. *Chem. Eng. Process.* **2020**, *156*, 108096.
23. Antar, T.; Kacem, M. Theoretical investigation of laminar flow convective heat transfer in a circular duct for a non-Newtonian nanofluid. *Appl. Therm. Eng.* **2017**, *112*, 1027–1039.
24. Antar, T.; Kacem, M. Analytical solution by Laplace-ritz variational method for non-Newtonian nanofluid inside a circular tube. *Int. J. Mech. Sci.* **2018**, *135*, 596–608.
25. Mehryan, S.A.M.; Mohammad, G.; Mohammad, V.; Seyed, M.H.Z.; Nima, S.; Obai, Y.; Ali, J.C.; Hani, A. Non-Newtonian phase change study of nano-enhanced n-octadecane comprising mesoporous silica in a porous medium. *Appl. Math. Model.* **2021**, *97*, 463–482. [CrossRef]
26. Nadeem, S.; Rashid, M.; Noreen, S.A. Non-orthogonal stagnation point flow of a nano non-Newtonian fluid towards a stretching surface with heat transfer. *Int. J. Heat Mass Transf.* **2013**, *57*, 679–689. [CrossRef]
27. Liang, C.; Nawaz, M.; Hajra, K.; Alaoui, M.K.; Abdellatif, S.; Chuanxi, L.; Hamid, A. Flow and heat transfer analysis of elastoviscoplastic generalized non-Newtonian fluid with hybrid nano structures and dust. *Int. Commun. Heat Mass Transf.* **2021**, *126*, 105275.
28. Kourosh, J.; Habib, K.; Abbas, K. Numerical study of heat transfer enhancement of non-Newtonian nanofluid in porous blocks in a channel partially. *Powder Technol.* **2021**, *383*, 270–279.
29. Kumar, M.S.; Raju, C.S.K.; El-Sayed, M.S.; Ebrahem, A.A.; Bilal, S.; Harri, J. A Comprehensive Physical Insight about Enhancement in Thermo Physical Features of Newtonian Fluid Flow by Suspending of Metallic oxides of Single Wall Carbon Nano Tube Structures. *Surf. Interfaces* **2021**, *23*, 100838. [CrossRef]
30. Amin, S.; Seyed, S.A.; Askari, I.B.; Muhammad, H.A. Numerical investigation of the effect of corrugation profile on the hydrothermal characteristics and entropy generation behavior of laminar forced convection of non-Newtonian water/CMC-CuO nanofluid flow inside a wavy channel. *Int. Commun. Heat Mass Transf.* **2021**, *121*, 105–117.
31. Yang, J.C.; Li, F.C.; Zhou, W.W.; He, Y.R.; Jiang, B.C. Experimental investigation on the thermal conductivity and shear viscosity of viscoelastic-fluid-based nanofluids. *Int. J. Heat Mass Transf.* **2012**, *55*, 3160–3166. [CrossRef]
32. Li, F.C.; Yang, J.C.; Zhou, W.W.; He, Y.R.; Jiang, B.C. Experimental study on the characteristics of thermal conductivity and shear viscosity of viscoelastic-fluid-based nanofluids containing multi-walled carbon nanotubes. *Acta* **2013**, *556*, 47–53.
33. Xuan, Y.; Li, Q. Investigation on convective heat transfer and flow features of nanofluids. *J. Heat Transf.* **2003**, *125*, 151–155. [CrossRef]
34. Esmaeilnejad, A.; Aminfar, H.; Neistanak, M.S. Numerical investigation of forced convection heat transfer throughmicrochannels with non-Newtonian nanofluids. *Int. J. Therm. Sci.* **2014**, *75*, 76–86. [CrossRef]
35. Evangelos, K.; Christos, L.; Theodoros, K.; Ioannis, S. Mixing of Particles in Micromixers under Different Angles and Velocities of the Incoming Water. *Proceeding* **2018**, *2*, 577.
36. Pouya, B. Numerical assessment of heat transfer and mixing quality of a hybrid nanofluid in a microchannel equipped with a dual mixer. *Int. J.* **2021**, *12*, 100111.
37. Mehdi, B.; Nima, M.; Masoud, A. Development of chaotic advection in laminar flow of a non-Newtonian nanofluid: A novel application for efficient use of energy. *Appl. Therm. Eng.* **2017**, *124*, 1213–1223.
38. Santra, A.K.; Sen, S.; Chakraborty, N. Study of heat transfer due to laminar flow of copper—Water nanofluid through two isothermally heated parallel plates. *Int. J. Therm. Sci.* **2009**, *48*, 391–400. [CrossRef]
39. Delplace, F.; Leuliet, J.C. Generalized Reynolds number for the flow of power-law fluids in cylindrical ducts of arbitrary cross-section. *Chem. Eng. J.* **1995**, *56*, 33–37. [CrossRef]

40. Metzne, A.B.; Reed, J.C. Flow of non-Newtonian fluids—Correlation of the laminar transition, and turbulent-flow regions. *AIChE J.* **1955**, *1*, 434–440. [CrossRef]
41. Embarek, D.; Samir, L.; Amar, K.; Naas, T.T.; Sofiane, K.; Abdelylah, B. High hydrodynamic and thermal mixing performances of efficient chaotic micromixers: A comparative study. *Chem. Eng. Process.-Process Intensif.* **2021**, *164*, 108394.
42. ANSYS, Inc. *ANSYS Fluent User's Guide*; Release 15, 2013; ANSYS, Inc.: Canonsburg, PA, USA, 2013; pp. 1–814.
43. Li, S.-N.; Zhang, H.-N.; Li, X.-B.; Li, Q.; Li, F.-C.; Qian, S.; Joo, S.W. Numerical study on the heat transfer performance of non-Newtonian fluid flow in a manifold microchannel heat sink. *Appl. Therm.* **2017**, *115*, 1213–1225. [CrossRef]
44. Jibo, W.; Guojun, L.; Xinbo, L.; Fang, H.; Xiang, M. A Micromixer with Two-Layer Crossing Microchannels Based on PMMA Bonding Process. *Int. J. Chem. React. Eng.* **2019**, *17*, 20180265. [CrossRef]
45. Gidde, R.R. Concave wall-based mixing chambers and convex wall-based constriction channel micromixers. *Int. J. Environ. Anal. Chem.* **2019**, *101*, 561–583. [CrossRef]

MDPI

Article

Mixing Enhancement of Non-Newtonian Shear-Thinning Fluid for a Kenics Micromixer

Abdelkader Mahammedi [1], Naas Toufik Tayeb [2], Kwang-Yong Kim [3,*] and Shakhawat Hossain [4,*]

[1] Department of Mechanical Engineering, University of Djelfa, Djelfa 17000, Algeria; abdelkader.mahammedi@gmail.com
[2] Gas Turbine Joint Research Team, University of Djelfa, Djelfa 17000, Algeria; toufiknaas@gmail.com
[3] Department of Mechanical Engineering, Inha University, 253 Younghyun-dong, Nam-gu, Incheon 402-751, Korea
[4] Department of Industrial and Production Engineering, Jashore University of Science and Technology, Jessore 7408, Bangladesh
* Correspondence: kykim@inha.ac.kr (K.-Y.K.); shakhawat.ipe@just.edu.bd (S.H.); Tel.: +880-130-852-6191 (S.H.)

Abstract: In this work, a numerical investigation was analyzed to exhibit the mixing behaviors of non-Newtonian shear-thinning fluids in Kenics micromixers. The numerical analysis was performed using the computational fluid dynamic (CFD) tool to solve 3D Navier-Stokes equations with the species transport equations. The efficiency of mixing is estimated by the calculation of the mixing index for different cases of Reynolds number. The geometry of micro Kenics collected with a series of six helical elements twisted 180° and arranged alternately to achieve the higher level of chaotic mixing, inside a pipe with a Y-inlet. Under a wide range of Reynolds numbers between 0.1 to 500 and the carboxymethyl cellulose (CMC) solutions with power-law indices among 1 to 0.49, the micro-Kenics proves high mixing Performances at low and high Reynolds number. Moreover the pressure losses of the shear-thinning fluids for different Reynolds numbers was validated and represented.

Keywords: Kenics micromixer; numerical simulation; mixing index; non-Newtonian fluids; CMC solutions; low Reynolds number

Citation: Mahammedi, A.; Tayeb, N.T.; Kim, K.-Y.; Hossain, S. Mixing Enhancement of Non-Newtonian Shear-Thinning Fluid for a Kenics Micromixer. *Micromachines* **2021**, *12*, 1494. https://doi.org/10.3390/mi12121494

Academic Editor: Yangcheng Lu

Received: 8 November 2021
Accepted: 27 November 2021
Published: 30 November 2021

1. Introduction

Different applications of micromixers can be found in biomedical, environmental industries and chemical analysis, where they are essential components in the micro-total analysis systems for such applications requiring the rapid and complete mixing of species for a variety of tasks [1,2]. Various characteristics of micromixers have been developed to produce fast and homogenous mixing; micromixers are usually classified according to their mixing principles as active or passive devices [3–6]. Active micromixers need an external energy supply to mix species. Passive micromixers are preferable due to their simple structures and easy manufacturing and greater robustness and stability [7]. To get enhanced mixing flows, the chaotic advection technique is employed as one of the strongest passive mixing methods for non-Newtonian flows. One of the potential chaotic geometries which can present a good method to advance the performances of the hydrodynamics is called the Kenics mixer. A Kenics mixer is a passive mixer created for conditions of laminar flow; it is generally constituted of a series of helical elements, and each element rotated 90° relatives to the second. The helical elements are designed to divide the flow into two or more flows, turn them and afterward recombine them [8,9]. Kurnia [10] observed that the inserted of a twisted tape in a T-junction micromixer creates a chaotic movement that improves the convection mass transfer at the expense of a higher pressure drop. For dean instability, Fellouah et al. [11] investigated experimentally the flow field of power-law and Bingham fluids inside a curved rectangular duct. Pinho and White law [12] studied the effect of dean number on flow behavior of non-Newtonian laminar fluid. For twisted pipes, Stroock et al. [13] realized a twisting flow microsystem with diagonally oriented ridges on the bottom wall in a microchannel. They attained chaotic mixing by alternating velocity fields. Tsai et al. [14] calculated the mixing fluid of non-Newtonian carboxymethyl cellulose (CMC) solutions in three serpentine micromixers. They summarized that the curvature-induced vortices develop in a long way the mixing efficiency. Bahiri et al. [15], using grooves integrated on the bottom wall of a curved surface, studied numerically the mixing of non-Newtonian shear-thinning fluids. They illustrated that the grooves elevated the chaotic advection and augmented the mixing performance. Naas et al. [16,17] and Kouadri et al. [18] used the two-layer crossing channels micromixer to evaluate the mixing rate hydrodynamics and thermal mixing performances, finding that the mixing rate was nearly 99%, at a very low Reynolds number. Kim et al. [19] experimentally characterized the barrier embedded Kenics micromixer. They showed that the mixing rate decreases as the Reynolds number augments for the suggested BEKM chaotic micromixer. Hossain et al. [20] experimentally and numerically analyzed a model of micromixer with TLCCM that can achieve 99% mixing over a series of Reynolds number values (0.2–110).

There are not many works in the literature for non-Newtonian fluid mixing using Kenics micromixers. Therefore, the idea of the current study is to investigate the performance of a microKenics for mixing shear thinning fluids, trying to attain a high mixing quality and pressure drop. Using CFD code, numerical simulations were carried out at Reynolds numbers ranging from 1 to 500 in order to examine the flow structures and the hydrodynamic mixing performances within the concerned Kenics micromixer. Various CMC concentrations were proposed to investigate the chaotic flow formation and thermal mixing performances within the suggested micromixer. In order to get important homogenization of the fluids' indices and pressure losses will be appraised.

2. Governing Equations and Geometry Discretion

Steady conservation equations of incompressible fluid are solved numerically in a laminar regime by using the ANSYS Fluent^{TM} 16 CFD software (Ansys, Canonsburg, PL, USA) [21], which is fundamentally based on the finite volumes method. We choose the SIMPLEC scheme for velocity coupling and pressure. A second-order upwind scheme was nominated to solve the concentration and momentum equations. The numeric's were ensured and simulated to be converged at 10^{-6} of root mean square residual values.

A non-Newtonian solution of carboxyméthyl cellulose (CMC) is used as working fluid for the simulation of fluid flows. The density of CMC solutions according to Fellouah et al. [11] and Pinho et al. [12], is 1000 kg/m^3. The coherence coefficient and the power law indexes of the CMC solutions are indicated in Table 1, where the diffusion coefficient equals 1×10^{-11} m^2/s.

Table 1. Rheological properties of CMC solutions.

CMC%	N	k (Pa·s^n)
0	1	0.000902
0.1	0.9	0.0075
0.2	0.85	0.0252
0.3	0.73	0.15
0.5	0.6	0.67
0.7	0.49	2.75

The 3D governing equations for incompressible and steady flows are continuity, momentum and species mass fraction convection diffusion equation:

$$\nabla U = 0 \tag{1}$$

$$\rho U \cdot \nabla U = -\nabla P + \mu \nabla^2 U \tag{2}$$

$$U \cdot \nabla C = D_i \nabla^2 C_i \tag{3}$$

where U (m/s) denotes the fluid velocity, ρ (kg/m^3) is the fluid density, P (Pa) is the static pressure, μ (w/m·k) is the viscosity, C_i is the local mass fraction of each species by solving the convection-diffusion equation for the *i*-th species. D_i is the mass diffusion coeffcient of the species "*i*" in the mixing.

For power-law non-Newtonian fluids the apparent viscosity is:

$$\mu_a = k\dot{\gamma}^{n-1} \tag{4}$$

where k (w/m·k) is called the consistency coefficient and n is the power-law index $\dot{\gamma}$ (s^{-1}) is the shear rate.

For a shear thinning fluid (Ostwald model), the generalized Reynolds number (Re_g) is defined as [6]:

$$Re_g = \frac{\rho U^{2-n} D_h{}^n}{\frac{k}{8}\left(\frac{6n+2}{2}\right)^n} \tag{5}$$

where D_h (m) is the hydraulic diameter of the micromixer.

To measure the efficiency of the Kenics micromixers, mixing rate is defined as follows [16]:

$$MI = 1 - \frac{\sigma}{\sigma_0} \tag{6}$$

where σ signifies the standard deviation of mass fraction and characterized as:

$$\sigma^2 = \frac{1}{N}\sum_{i=1}^{N}\left(C_i - \overline{C}\right)^2 \tag{7}$$

N indicates the number of sampling locations inside the transversal division, C_i is the mass fraction at examining point *i*, and $\overline{C}$ is the ideal mixing mass fraction of C_i, and it is equal to 0.5, σ_0 (Pa·s^{-1}) is the standard deviation at the inlet part.

The boundary conditions are a condition of adhesion on the walls where the velocities are considered to be zero, uniform velocities are executed at the inlets, the mass fraction of the fluid at the inlet 1 equal to 1 and that of the inlet 2 equal to 0, an atmospheric pressure condition is considered at the exit. All walls are considered adiabatic.

The configuration (Figure 1) is based on the Kenics KM static mixer. It consists of a tube with a diameter D = 1.2 mm and a length L = 16.5 mm, with six helical parts. Each part has a thickness t = 0.025 mm and length li = 1.5 mm. The final helical blade element is placed at the distance l = 3 mm from the tube outlet. The angle between the two inlet entrances is $\alpha = 35°$.

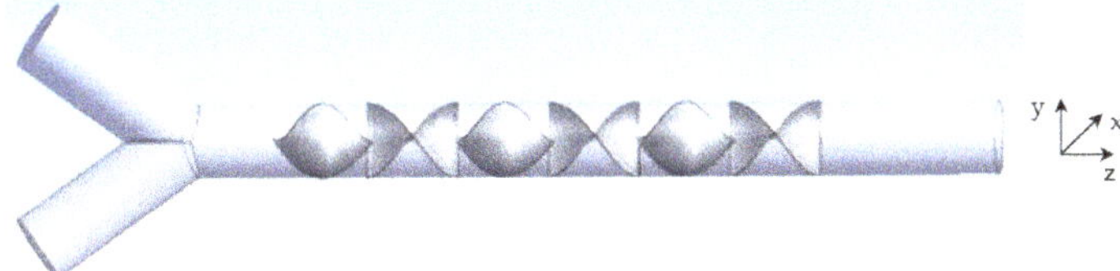

Figure 1. Kenics micromixer.

3. Grid Independence Test

In this work, which investigated fluid mixing in laminar flow in which the convergence is limited by the pressure-velocity coupling, a converged and stable solution was obtained using SIMPLE algorithm. The pressure correction under relaxation factor is given at 0.3, which facilitates the acceleration of convergence for the second order upwind scheme. The convergence of iterative calculations was attained when the specified value of the residual quantities are less than 10^6.

To choose an adequate mesh; an unstructured mesh has been generated with tetrahedral elements; several grids were tested for the present proposed geometry (Table 2). Numerical variations of the mixing index are not important after the marked cell size; therefore this can be studied as the better mesh for the calculation. This mesh size with 338,438 cells will give better results with less time as compared to the finer mesh.

Table 2. Mesh independency test.

Mesh Elements	MI for Re = 80	MI for Re = 10
87,304	0.8924	0.9834
137,592	0.8994	0.9877
241,322	0.8996	0.9890
338,438	0.9017	0.9941
593,476	0.9017	0.9942

4. Numerical Validation

A numerical study of the pressure drop in a T-Junction Passive micromixer to verify the accuracy of the CFD with that of C. Kurnia et al. [10], see Table 3. The comparison illustrated a good agreement where the relative error of the numerical results is less than 1%.

Table 3. Comparison of current computational results for Pressure drop as a function of various Reynolds numbers.

	Re	1	5	10	50	100
Pressure Drop	Kurnia et al. [10]	1.48	7.5	15.22	86.59	205
	Present simulation	1.5	7.6	15.3	86.6	205.5

5. Results and Discussion

To compare the numerical results a quantitative comparison was made for Newtonian fluids in a Kenics micromixer as shown in Figure 2. The mixing performance of the microKenics was compared with other three micromixers [15]: the SHG micromixer (staggered herringbone), a mixer based on patterns of grooves on the floor of the channel and a 3D serpentine micromixer with repeating "L-shape" units and the TLSCC (two-layer serpentine crossing channels) a micromixer which the principle serpentine channels with an angle of 90° regarding the inlets.

The mixing indices were compared using the range of Reynolds number between 0.1 to 120. The Kenics micromixer and TLSCC displayed exceptional mixing performance compared to the other two micromixers for Re < 30, with superiority of the microKenics, as the Kenics device shows almost complete mixing (MI > 0.999) at low Reynolds numbers (Re = 0.1–5).

The mixing index at the exit of Newtonian and non-Newtonian fluid in Kenics was compared with the curved micromixer of Bahiri et al. [15]. For range numbers of Reynolds (0.1–500) and shear thinning index n (1, 0.85, and 0.6), as shown in Figure 3, the Kenics proved a high mixing efficiency.

The effects of the Reynolds number and the behavior index on the chaotic mixing mechanism were qualitatively analyzed by presenting the contours of the mass fraction at the various transverse planes P1–P7 and the exit. Figure 4 shows the improvement of the mass fraction distribution on the y-plane along the micromixer at $Re_g = 25$ and n = 0.73. The twist of element and the sharp change of angle between blades affected the intensity of the fluid particles' movement and the mixing performance.

Figure 5 show the streamlines flow in the microKenics for fluid behavior 0.73 and for Re_g 0.1 and 50. The flow field enhanced the secondary flow along the micromixer for all cases of Re due to the blade conurbation of the Kenics device.

Figures 6 and 7 present the mass fraction contours for different power-law indices (n = 1 and 0.49), with Reynolds numbers ranging between 0.1 and 50, at the different cross-sectional planes. Table 4 gives the distances between different plans to analyze the local flow behavior. For n = 1, the flow behavior presented by the mass fraction contours shows that the fluid layers for P1 to P4 advance in the same mode of molecular diffusion.

When the Reynolds number increases to 50, the quality of the mixing begins to improve, therefore a homogeneous mixture is obtained at the exit plane in the microKenics for all the values of the behavior index.

Figure 8 shows the variation of mixing index versus generalized Reynolds number for different values of power-law index inside the microKenics. It can be seen that for all values of n, the micromixers have nearly the same high mixing index (Mi = 0.99).

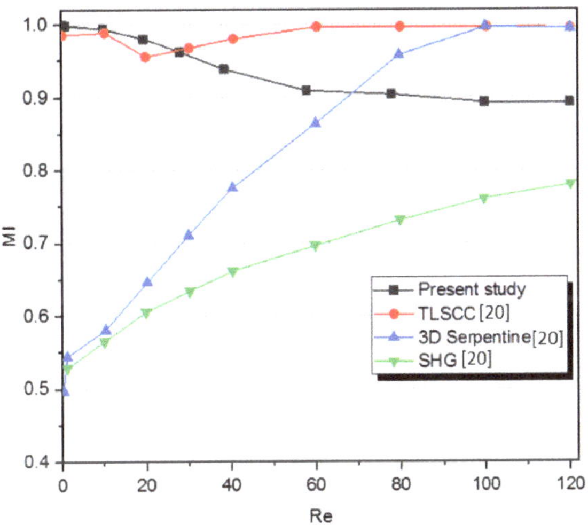

Figure 2. Evolution of mixing index (MI) with Reynolds number for a Newtonian fluid compared with Houssain et al. [20].

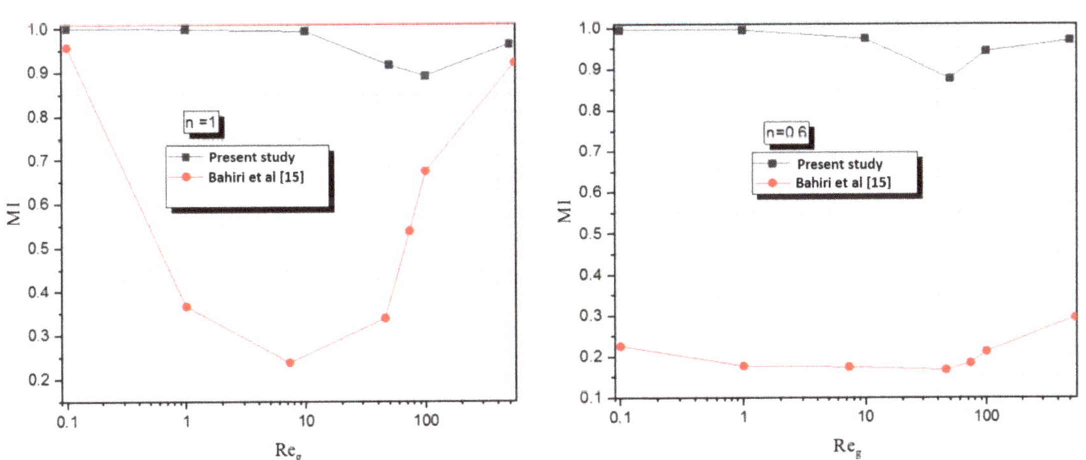

Figure 3. *Cont.*

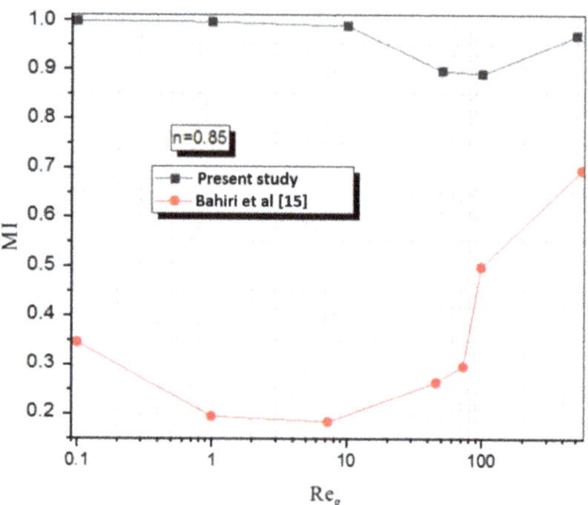

Figure 3. Evolution of mixing index of Reynolds number for a Newtonian and non-Newtonian fluid compared with Bahiri et al. [15].

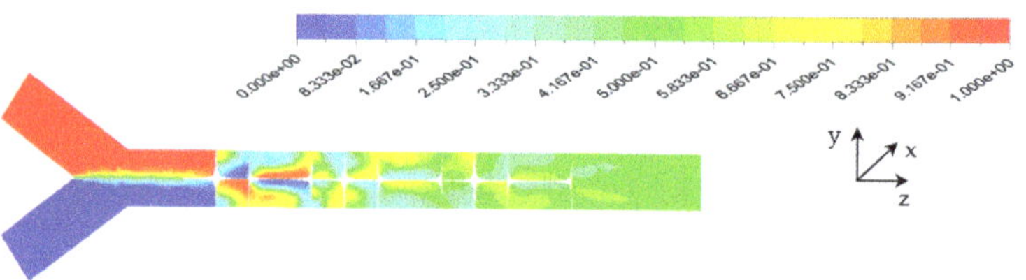

Figure 4. Distribution of mass fraction at $Re_g = 25$, n = 0.73.

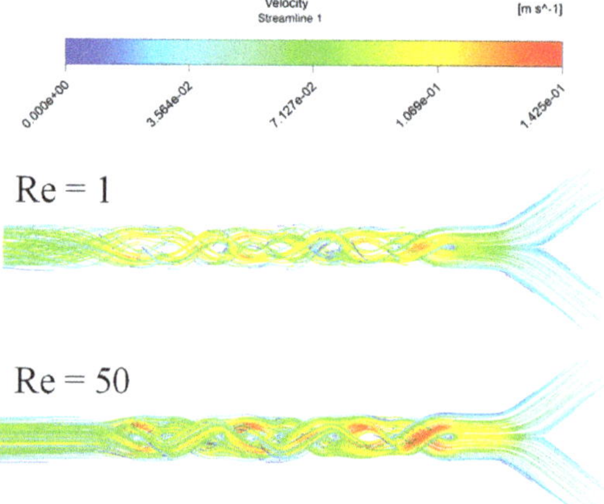

Figure 5. Velocity streamlines of n = 0.73 with different Reynolds number.

For low Reynolds numbers (Re = 0.1–5), when the species have more contact time to achieve a perfect mixing with the Kenics, the mixing index loses a part corresponding to nearly 14% of its value. In addition, by increasing the Reynolds number the fluid homonezation augments due to increases of secondary flows and advection, compared with TLCC micromixers for all cases of the power-law index.

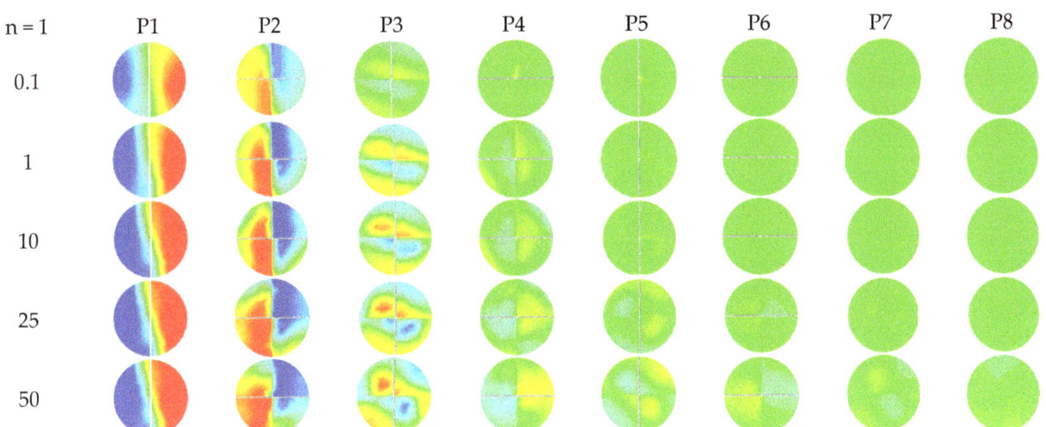

Figure 6. Contour of mass fraction at cross-section P1–P8 for different Reynolds numbers with n = 1.

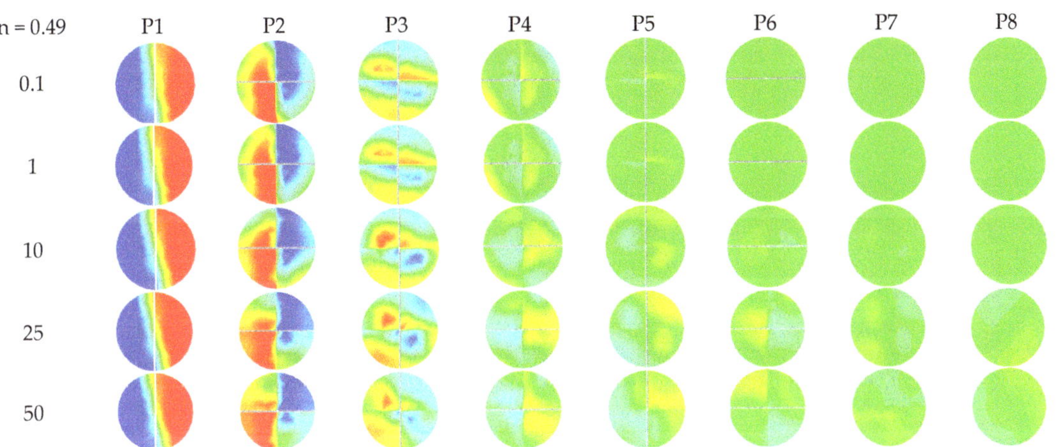

Figure 7. Contours of mass fraction at cross-section P1–P8 for different Reynolds numbers with n = 0.49.

Table 4. Position of the planes in the micromixer.

Plans	Z mm
P1	4.5
P2	6
P3	7.5
P4	9
P5	10.5
P6	12
P7	13.5
P8	16.5

Figure 9 shows the evolution of MI along the micro-Kenics, at different planes, with various values of the behavior index and for Re = 1, 5, 10, 25 and 100. For all cases of n, we can see from this figure that MI grows progressively and reaches high values when approaching the exit plane. Thus, as mentioned before we can see in all figures, the mixing performance increases with the increase in the behavior index, for $Re_g \leq 50$.

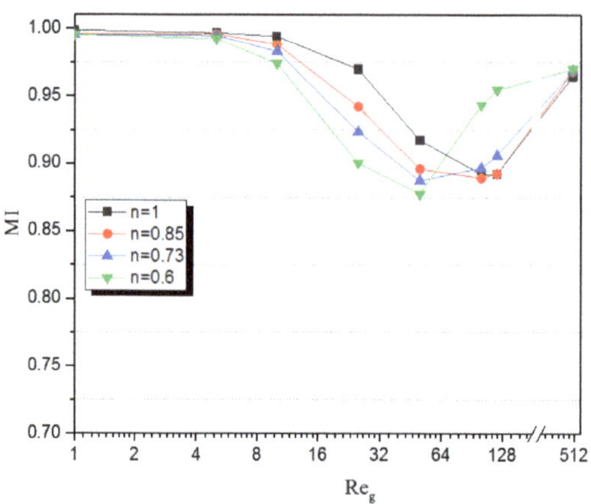

Figure 8. Development of mixing index versus generalized Reynolds number for differences values of power-law index.

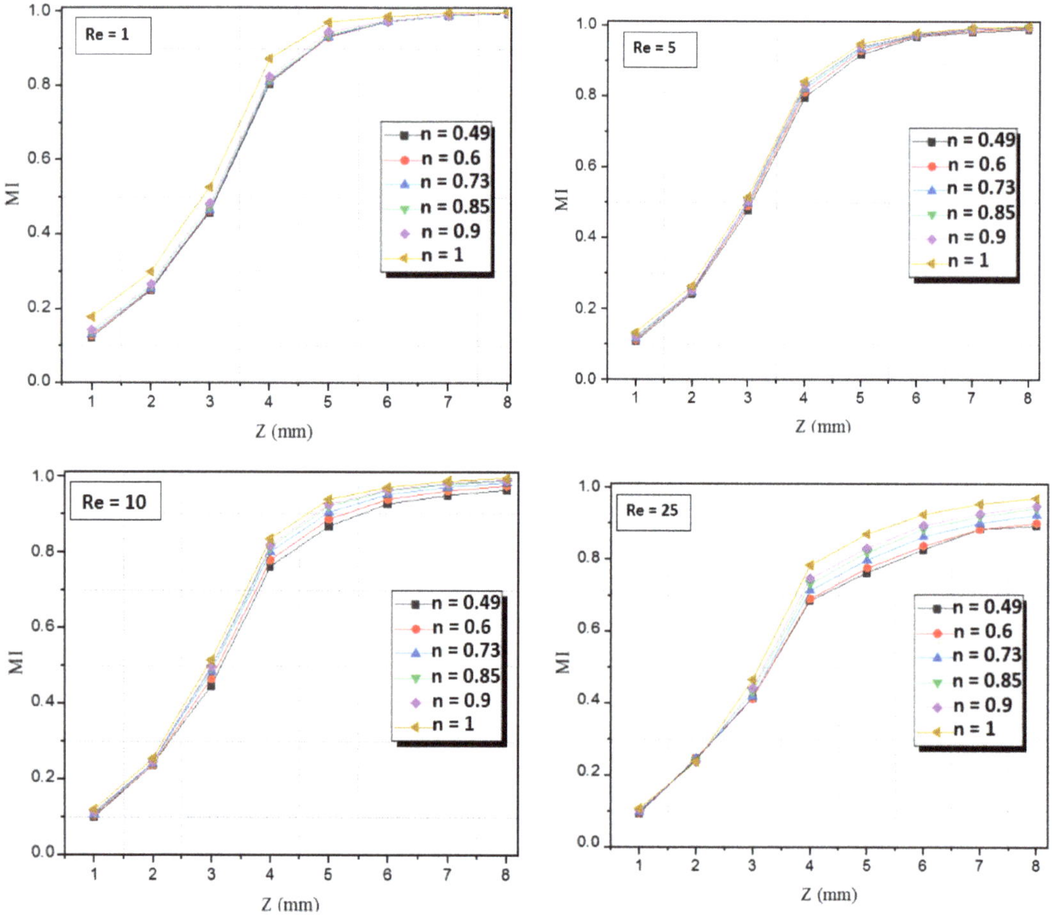

Figure 9. *Cont.*

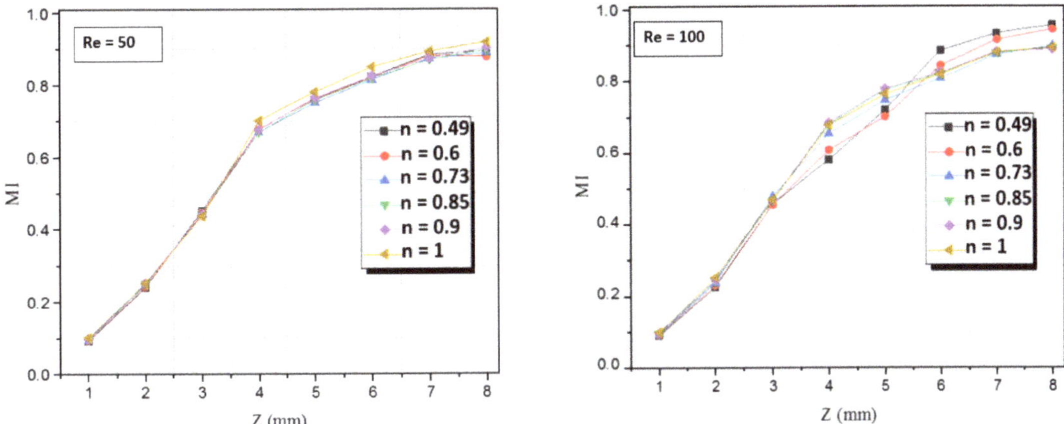

Figure 9. Development of mixing index along the mixing helical for different power-law indices and Reynolds numbers.

The Newtonian fluid with n = 1 is independent of shear rate and maintains a constant value of viscosity among different numbers of Reynolds (Figures 10 and 11). Besides, the decrease of value of n induces the increases of the apparent viscosity of non-Newtonian fluid which also depends on the consistency coefficient of the fluid and the shear rate.

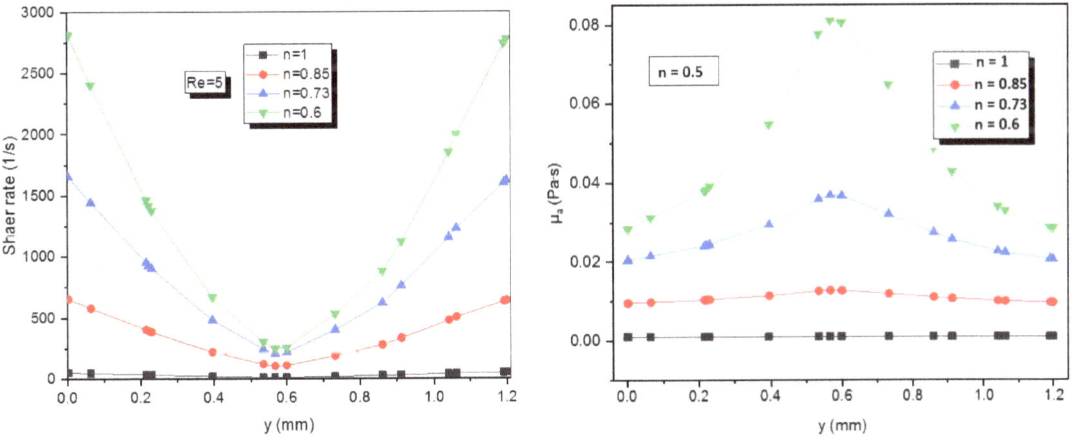

Figure 10. Shear rate profiles and apparent viscosity profiles on line x = 0 at the exit plane for different power-law indices.

Therefore, Figure 10 indicates that for a known shear rate, where the fluid with a lower power-law index has a higher apparent viscosity furthermore the apparent viscosity increases by reducing the power-law index.

Figure 11 shows the apparent viscosity on line x = 0 at the exit of micro Kenics for all power-law indices. It is obvious that the apparent viscosity decreases by rising the Reynolds number.

The pressure drop obtained from CFD simulations was compared with the TLCC micromixer [18], for the cases with the same CMC solutions and flow speed. As remarked in Figure 12, the pressure loss of Kenics is less than that from TLCC; so the best advantage has been obtained by the Kenics.

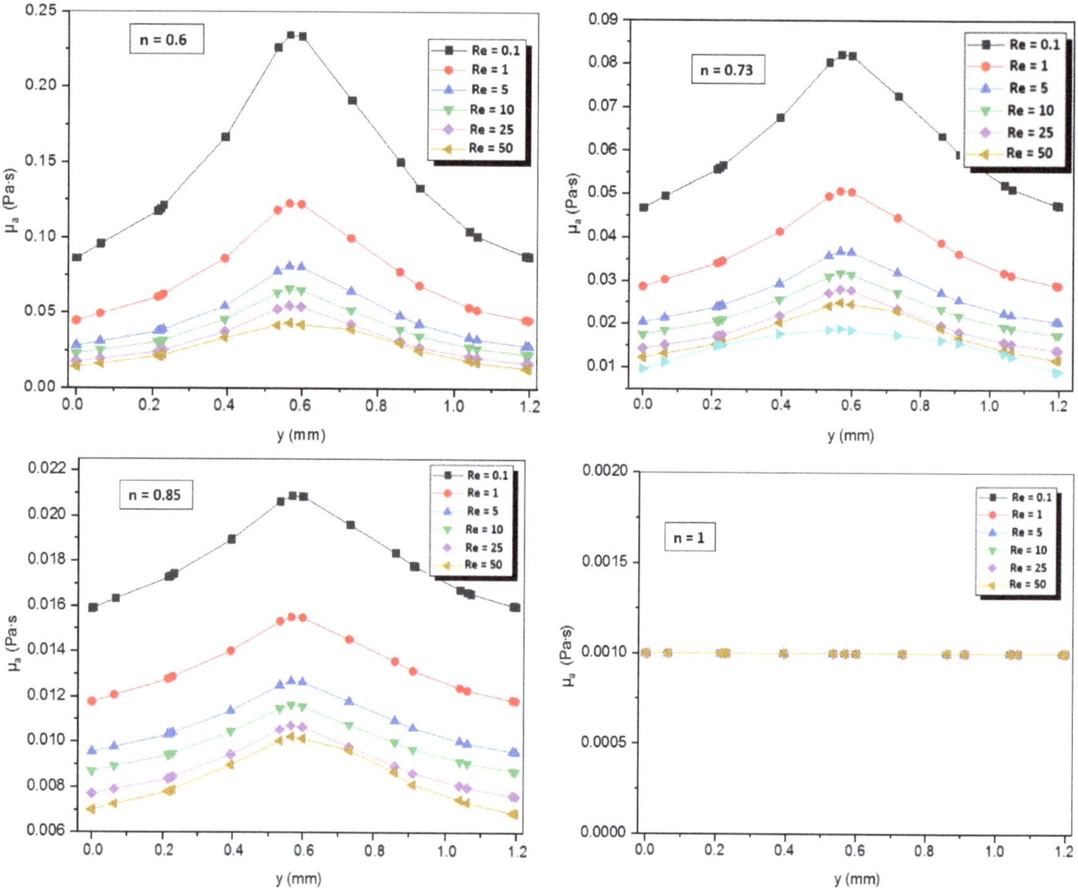

Figure 11. Apparent viscosity profiles on line x = 0 at the exit plane for different Reynolds numbers and different power-law indices.

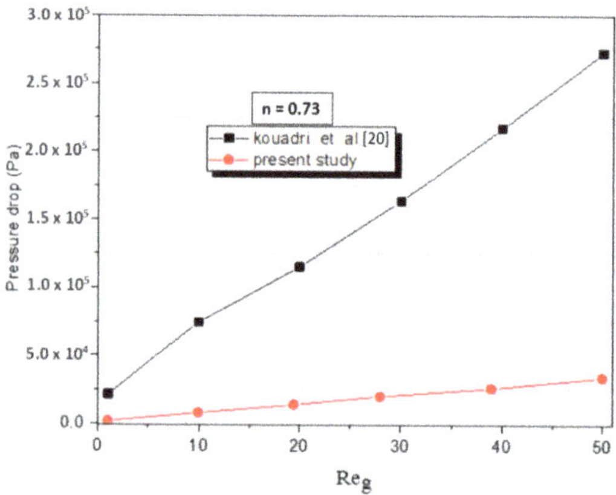

Figure 12. Pressure drop vs. Reynolds numbers compared with TLCC micromixer for n = 0.73.

A high mixing performance of the micromixer is generally associated with a high-pressure loss that involves the required energy input for the mixing process. Figure 13 shows the pressure loss increases with the increases of generalized Reynolds and concentration level. It is evident that a decreasing power-law index leads to an increment of the apparent viscosity and consequently a rising pressure loss.

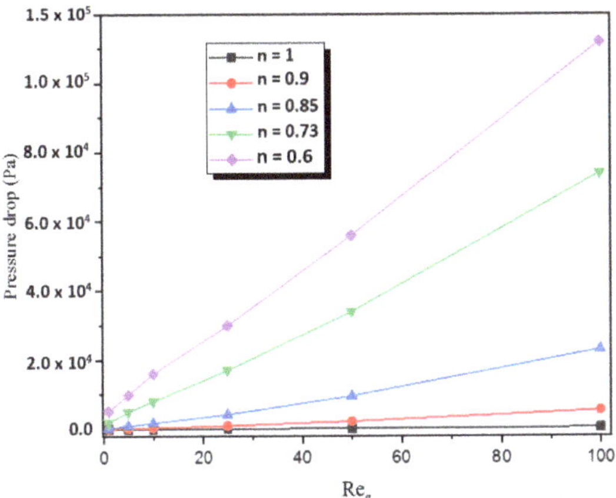

Figure 13. Variations of the pressure drop with generalized Reynolds number for different power-law indices.

6. Conclusions

In this work, mixing of CMC non-Newtonian fluids in a microKenics device was numerically investigated for different regimes (Re = 0.1–500), using CFD code. The analyses showed that the mixing performances of the Kenics micromixers consisting of repeating short twisted helical configurations is better than that of other micromixers at low Reynolds numbers.

It can be achieved that for fluids with all power-law indices studied (n = 0.49, 0.6, 0.73, 0.85, 0.9, and 1) and low Reynolds numbers (less than 8) the micromixer is an excellent one, while for the fluids with Re $>$ 12 MI start decreasing for all power-law indices, but for low power law index (n = 0.6), the MI is reached from numbers of Re $\geq$ 60. At elevated Reynolds numbers (Re $\geq$ 120), the micromixer performance is improved for all values of the power-law indices. The results confirmed that the apparent viscosity of CMC solutions decreases with the increase of the shear rate, while, the pressure drop increases rapidly with increasing Reynolds number and power-law index. Nevertheless it is still the slightest loss compared to other micromixers in the literature with the same mean flow speed and apparent viscosity.

Author Contributions: Conceptualization, A.M. and S.H.; methodology, N.T.T.; software, S.H.; validation, A.M.; formal analysis, N.T.T.; investigation, A.M.; resources, S.H.; writing—original draft preparation, A.M. writing—review and editing, S.H. and A.M.; visualization, N.T.T.; supervision, K. Y.K.; project administration, K. Y.K.; funding acquisition, K. Y.K. All authors have read and agreed to the published version of the manuscript.

Funding: This work was supported by the National Research Foundation of Korea (NRF) grant funded by the Korean government (MSIT) (No. 2019R1A2C1007657).

Data Availability Statement: Data would be available upon reader's request.

Conflicts of Interest: The authors declare no conflict of interest.

References

1. Jeong, G.S.; Chung, S.; Kim, C.B.; Lee, S.H. Applications of Micromixing Technology. *Analyst* **2010**, *135*, 460–473. [CrossRef] [PubMed]
2. Minye, L. Computational study of convective—Diffusive mixing in a microchannel mixer. *Chem. Eng. Sci.* **2011**, *66*, 2211–2223.
3. Nguyen, N.T.; Wu, Z. Micromixers—A review. *J. Micromech. Microeng.* **2005**, *15*, 1. [CrossRef]
4. Hessel, V.; Lowe, H.; Schönfeld, F. Micromixers—A review on passive and active mixing principles. *Chem. Eng. Sci.* **2005**, *60*, 2479–2501. [CrossRef]
5. Okuducu, M.B.; Aral, M.M. Performance Analysis and Numerical Evaluation of Mixing in 3-D T-Shape Passive Micromixers. *Micromachines* **2018**, *9*, 210. [CrossRef]
6. Okuducu, M.B.; Aral, M.M. Computational Evaluation of Mixing Performance in 3-D Swirl-Generating Passive Micromixers. *Processes* **2019**, *7*, 121. [CrossRef]
7. Bakker, A.; LaRoche, R.D.; Marhsall, E.M. Laminar Flow in Static Mixers with Helical Elements. Online CFM Book. 2000. Available online: https://www.researchgate.net/publication/237224585_Laminar_Flow_in_Static_Mixer_with_Helical_Elements (accessed on 26 November 2021).
8. Byrde, O.; Sawley, M.L. Optimization of a Kenics static mixer for non-creeping flow conditions. *Chem. Eng. J.* **1999**, *72*, 163–169. [CrossRef]
9. Mahammedi, A.; Ameur, H.; Ariss, A. Numerical Investigation of the Performance of Kenics Static Mixers for the Agitation of Shear Thinning Fluids. *J. Appl. Fluid Mech.* **2017**, *10*, 989–999. [CrossRef]
10. Kurnia, J.C.; Sasmito, A.P. Performance Evaluation of Liquid Mixing in a T-Junction Passive Micromixer with a Twisted Tape Insert. *Ind. Eng. Chem. Res.* **2020**, *59*, 3904–3915. [CrossRef]
11. Fellouah, H.; Castelain, C.; Ould-El-Moctar, A.; Peerhossaini, H. The Dean instability in power-law and Bingham fluids in a curved rectangular duct. *J. Non-Newton. Fluid Mech.* **2010**, *165*, 163–173. [CrossRef]
12. Pinho, F.T.; Whitelaw, J.H. Flow of non-Newtonian fluids in pipe. *J. Non-Newton. Fluid Mech.* **1990**, *34*, 129–144. [CrossRef]
13. Stroock, A.D.; Dertinger, S.K.; Ajdari, I.A.; Mezić, I.; Stone, H.A.; Whitesides, G.M. Chaotic mixer for microchannels. *Science* **2002**, *295*, 647–651. [CrossRef] [PubMed]
14. Tsai, R.T.; Wu, C.Y.; Chang, C.Y.; Kuo, M.Y. Mixing Behaviors of Shear-Thinning Fluids in Serpentine Channel Micromixers. *Int. J. Mech. Aerosp. Ind. Mechatron. Manuf. Eng.* **2015**, *9*, 1329–1335.
15. Islami, S.B.; Khezerloo, M.; Gharraei, R. The effect of chaotic advection on mixing degree and pressure drop of non-Newtonian fluids flow in curved micromixers. *J. Braz. Soc. Mech. Sci. Eng.* **2017**, *39*, 813–831. [CrossRef]
16. Naas, T.T.; Hossain, S.; Aslam, M.; Rahman, A.; Hoque, A.S.M.; Kim, K.-Y.; Islam, S.M.R. Kinematic measurements of novel chaotic micromixers to enhance mixing performances at low Reynolds numbers: Comparative study. *Micromachines* **2021**, *12*, 364. [CrossRef] [PubMed]
17. Naas, T.T.; Kouadri, A.; Khelladi, S.; Laib, L. Thermal mixing performances of shear thinning non-Newtonian fluids inside two-layer crossing channels micromixer using entropy generation method: Comparative study. *Chem. Eng. Process.* **2020**, *156*, 108096.
18. Kouadri, A.; Embarek, D.; Yahia, L.; Naas, T.T.; Khelladi, S.; Makhlouf, M. Comparative study of mixing behaviors using non-Newtonian fluid flows in passive micromixers. *Int. J. Mech. Sci.* **2021**, *201*, 106472. [CrossRef]
19. Kim, D.S.; Lee, H.; Kwon, T.H.; Cho, D.W. A barrier embedded Kenicsmicromixer. *J. Micromech. Microeng.* **2004**, *14*, 1294–1301. [CrossRef]
20. Hossain, S.; Lee, I.; Kim, S.M.; Kim, K.Y. A micromixer with two-layer serpentine crossing channels having excellent mixing performance at low Reynolds numbers. *Chem. Eng. J.* **2017**, *327*, 268–277. [CrossRef]
21. ANSYS, Inc. *ANSYS Fluent User's Guide, Release 16*; ANSYS, Inc.: Canonsburg, PL, USA, 2016.

micromachines

MDPI

Article

Mixing Performance of the Modified Tesla Micromixer with Tip Clearance

Makhsuda Juraeva and Dong-Jin Kang *

School of Mechanical Engineering, Yeungnam University, Gyoungsan 38541, Korea
* Correspondence: djkang@yu.ac.kr; Tel.: +82-53-810-2463

Abstract: A passive micromixer based on the modified Tesla mixing unit was designed by embedding tip clearance above the wedge-shape divider, and its mixing performance was simulated over a wider range of the Reynolds numbers from 0.1 to 80. The mixing performance was evaluated in terms of the degree of mixing (DOM) at the outlet and the required pressure load between inlet and outlet. The height of tip clearance was varied from 40 μm to 80 μm, corresponding to 25% to 33% of the micromixer depth. The numerical results show that the mixing enhancement by the tip clearance is noticeable over a wide range of the Reynolds numbers Re < 50. The height of tip clearance is optimized in terms of the DOM, and the optimum value is roughly h = 60 μm. It corresponds to 33% of the present micromixer depth. The mixing enhancement in the molecular diffusion regime of mixing, Re $\leq$ 1, is obtained by drag and connection of the interface in the two sub-streams of each Tesla mixing unit. It appears as a wider interface in the tip clearance zone. In the intermediate range of the Reynolds number, 1 < Re $\leq$ 50, the mixing enhancement is attributed to the interaction of the flow through the tip clearance and the secondary flow in the vortex zone of each Tesla mixing unit. When the Reynolds number is larger than about 50, vortices are formed at various locations and drive the mixing in the modified Tesla micromixer. For the Reynolds number of Re = 80, a pair of vortices is formed around the inlet and outlet of each Tesla mixing unit, and it plays a role as a governing mechanism in the convection-dominant regime of mixing. This vortex pattern is little affected as long as the tip clearance remains smaller than about h = 70 μm. The DOM at the outlet is little enhanced by the presence of tip clearance for the Reynolds numbers Re $\geq$ 50. The tip clearance contributes to reducing the required pressure load for the same value of the DOM.

Keywords: degree of mixing (DOM); modified Tesla micromixer; tip clearance; symmetric counter-rotating vortices; drag and connection of interface

Citation: Juraeva, M.; Kang, D.-J. Mixing Performance of the Modified Tesla Micromixer with Tip Clearance. *Micromachines* **2022**, *13*, 1375. https://doi.org/10.3390/mi13091375

Academic Editor: Kwang-Yong Kim

Received: 18 July 2022
Accepted: 18 August 2022
Published: 23 August 2022

1. Introduction

Micromixers are widely used in many microfluidic systems for biochemistry analysis, chemical synthesis, biomedical diagnostics, and drug delivery [1–3]. As the microfluidic systems aims to achieve several characteristics such as reduced consumption of reagent, fast processing, low cost, and portability [4], they require rapid and complete mixing. Micromixing is therefore one of the fundamental technologies utilized in microfluidic applications.

The mixing in most microfluidics systems is governed by molecular diffusion, slow fluid velocity and microscale geometry. The associated flow corresponds to very low Reynolds number regime, and mixing is inevitably slow and inefficient. Therefore, it is critical to develop a more efficient micromixer for the progress of the microfluidic industry. Mixing enhancement is still a crucial design goal, even though various technologies have been proposed to enhance the efficiency of microfluidic mixing [2].

A variety of micromixers have been proposed to enhance the mixing in microfluidic systems, and they are usually categorized as either active or passive. An active micromixer utilizes an external energy source to improve mixing efficiency. The external energy source is mostly used to generate flow disturbance, and contributes to the enhancement of mixing. Typical energy sources are acoustic [5], magnetic [6], electric [7], thermal [8], and pressure [9]. As each active micromixer employs an external energy source, the resulting structure of an active micromixer is more complicated and expensive, compared with passive micromixers. This characteristic limits the usage of active technologies in microfluidic systems. On the contrary, passive micromixers rely on the modification of geometric structures to generate a chaotic flow field and have no moving parts. Therefore, they are much simpler to integrate into a microfluidic system. Various geometric modifications have been shown to generate a chaotic flow field. Some of them include a staggered herringbone [10], channel wall twisting [11], repeated surface groove and baffles [12,13], block in the junction [14], split-and-recombine (SAR) [15,16], Tesla structure [17], stacking of mixing units in the cross-flow direction [18], optimization of lateral structure [19] and submergence of planar structures [20].

There are several approaches to enhance the degree of mixing (DOM): complex three-dimensional structures, modification of planar geometry, and manipulation of flow conditions. Using a pulsatile inlet flow is an example to control flow condition. For example, McDonough et al. [21] showed that the micromixing time decreased with an increased velocity ratio of oscillatory velocity to net velocity; baffle designs were used. However, this kind of approach requires an extra device to generate the pulsatile flow. In this paper, a simpler geometric approach is studied to enhance the mixing performance, based on a geometric modification.

Generally, a complex three-dimensional (3D) micromixer may result in a better mixing performance than that of a two-dimensional (2D) micromixer of similar size [22]. However, the entire fabrication process of a 3D micromixer is much more complicated, and costs more compared with a planar design. In addition, some 2D planar micromixers were shown to generate effective 3D flow characteristics such as multidirectional vortex and Dean vortex. For example, Hong et al. [23] proposed a modified Tesla micromixer, which is based on the Coanda effect. The Coanda effect allows the fluid to follow the angles surface, and the Tesla structure is placed serially in the opposite direction to enhance the transverse dispersion of fluid. Hossain et al. [17] optimized this modified Tesla micromixer and showed that it generates a couple of symmetric counter-rotating vortices in the cross section for the Reynolds numbers $Re \geq 2$. Raza et al. [22] recommended this modified Tesla structure in the intermediate ($1 < Re \leq 40$) and high Reynolds number ranges ($Re > 40$) and recommended 3D micromixers in the low Reynolds range ($Re \leq 1$). On the other hand, Makhsuda et al. [20] showed that a submergence of planar structure enhances the mixing performance in the Reynolds number range of $Re \geq 5$; a vortex burst of the two Dean vortices promotes mixing performance. Chung et al. [24] showed that short planar baffles with a gap promotes vortex creation due to the sudden expansion around the baffles and enhance the mixing performance in the diffusion-dominant flow rate ($Re < 1$) and the convection-dominant flow rate ($Re > 40$). Bazaz et al. [25] studied a hybrid micromixer combining six planar mixing units such as modified Tesla, ellipse-like, nozzle, pillar, teardrop and obstruction, and optimized the combination: one nozzle, one pillar, three obstacles in a curved channel, and two modified Tesla units. They obtained a mixing performance improvement for a wide range of Reynolds number ($Re \leq 1$ and $22 \leq Re \leq 45$).

In this paper, tip clearance was embedded into the modified 2D Tesla micromixer to enhance the mixing performance by combining the Coanda effect and flow disturbance due to tip clearance of planar structures. The present micromixer consists of several modified Tesla units, and tip clearance is present above the wedge-shape divider geometry of each modified Tesla unit. As the structure of the present micromixer is slightly modified from that of a planar micromixer, microfabrication techniques such as Xurography [26] can be easily applied. The Xurography technique uses thin, pressure-sensitive double-sided adhesive flexible films so that the tip clearance zone is simply tailored using a cutter plotter; it cuts off a film along the perimeter of the modified Tesla micromixer. The tailored film and the planar structure can be simply assembled to complete the present micromixer. The number of the modified Tesla units was varied from three to five. The tip clearance between the divider structure and the micromixer wall is expected to play a key role in generating flow disturbance and is varied in the range from 40 μm to 80 μm. It is to 20~40% of the present micromixer depth. The mixing performance was simulated in terms of the degree of mixing (DOM) at the outlet and the required pressure load between the inlets and outlet and compared with those of the modified planar Tesla micromixer without tip clearance.

A numerical approach has several benefits, such as easy visualization of the mixing process and the associated flow patterns. Accordingly, it is widely used in studying the mixing performance of a micromixer. For a numerical study, a commercial software is commonly used. For example, Makhsuda et al. [19] used the commercial software ANSYS® Fluent [27] to study the mixing performance. Rhoades et al. [28] used the commercial software COMSOL Multiphysics 5.1 (COMSOL, Inc., Burlington, MA, USA) to simulate the mixing performance of a grooved serpentine micro-channel. Volpe et al. [29] used the lattice Boltzmann method (LBM) to study the flow dynamics of a continuous size-based sorter microfluidic device. In this paper, the mixing performance of the present micromixer was simulated using the commercial software ANSYS® Fluent 2021 R2 [27].

Most micromixers for biological and chemical applications operate in the range of millisecond mixing time, and the corresponding Reynolds number is less than about 100 [30–32]. In this range of the Reynolds number, the micromixing is governed by two distinct mechanisms such as the molecular diffusion and convection [18,22]. The Reynolds number is usually categorized into three regimes according to the dominant mixing mechanism: molecular dominance, transition, and convection dominance. The present numerical study was carried out to cover all of the three mixing regimes. Therefore, the Reynolds number was varied from 0.1 to 80, and the corresponding volume flow rate ranged from 1.3 μL/min to 964.6 μL/min.

2. Modified Tesla Micromixer with Tip Clearance

Figure 1 shows a modified Tesla mixing unit. It is placed serially in the present passive micromixer. The wedge-shape divider splits the fluid stream into two sub-streams which recombine downstream: sub-stream 1 and sub-stream 2. Therefore, the mixing performance of each modified Tesla mixing unit shows some dependence on the geometry of the divider. The detailed geometry of the modified Tesla mixing unit is the same as one optimized by Hossain et al. [17]. In Figure 1b, h is the height of tip clearance, and it is varied from 40 μm to 80 μm. According to previous research [17,23], multiple vortices form at the vortex zone as the Reynolds number increases; the outlet of each Tesla mixing unit is named as the vortex zone.

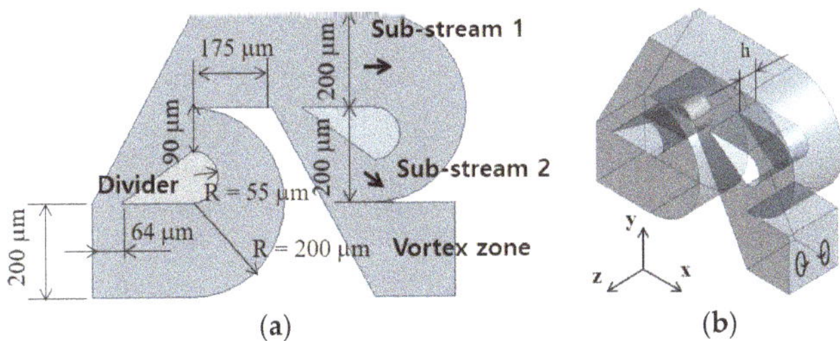

Figure 1. Schematic diagram of the modified Tesla mixing unit (non-proportional): (**a**) Front view and (**b**) Three-dimensional view.

Figure 2 shows a schematic diagram of the present passive micromixer. The cross section of the inlet and outlet branches is rectangular: 200 μm wide and 200 μm deep. Both inlets 1 and 2 are 1000 μm long while the outlet branch is 500 μm long. Even Figure 2 shows four modified Tesla mixing units, and the actual number of the modified Tesla mixing units is varied from three to five. The Ss in Figure 2 indicate the cross section at the outlet of each modified Tesla mixing unit. For example, S4 is the cross section after the fourth Tesla mixing unit. As the two inlets are facing opposite to each other, micromixing takes place mainly in the modified Tesla mixing units.

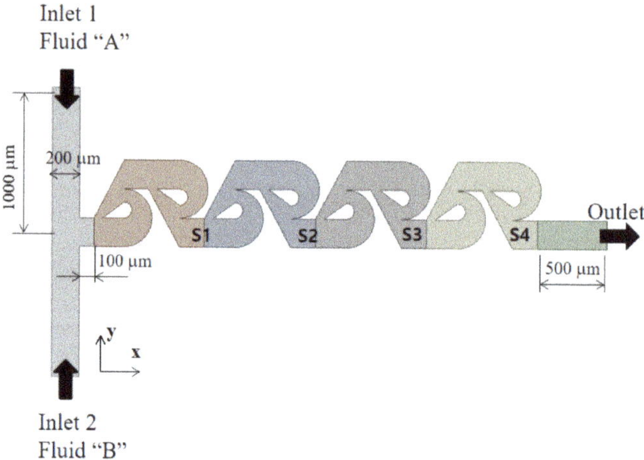

Figure 2. Schematic diagram of present micromixer (non-proportional).

3. Governing Equations and Computational Procedure

As the fluid was assumed Newtonian and incompressible, the following continuity and Navier–Stokes equations are the governing equations:

$$\left(\vec{u}\cdot\nabla\right)\vec{u} = -\frac{1}{\rho}\nabla p + \nu\nabla^2\vec{u} \tag{1}$$

$$\nabla\cdot\vec{u} = 0 \tag{2}$$

where $\vec{u}$, p, and ν are the velocity vector, pressure, and kinematic viscosity, respectively. The evolution of mixing was simulated by solving an advection-diffusion equation:

$$\left(\vec{u}\cdot\nabla\right)\varphi = D\nabla^2\varphi \tag{3}$$

where D and φ are the mass diffusivity and mass fraction of fluid A, respectively.

ANSYS® FLUENT 2021 R2, Canonsburg, PA, USA [25] was used to solve the governing Equations (1)–(3). It is based on the finite volume method. The QUICK scheme (quadratic upstream interpolation for convective kinematics) was used to discretize the convective terms in Equations (1) and (3), and its theoretical accuracy is third order. The velocity distribution at the two inlets was assumed as uniform, and the outflow condition was used at the outlet. The no-slip boundary condition was specified along the all walls were treated as a no-slip boundary. The mass fraction of fluid A is $\varphi = 1$ at inlet 1 and $\varphi = 0$ at inlet 2.

The mixing performance of a combined micromixer was evaluated using the degree of mixing (DOM) and mixing energy cost (MEC). The DOM is defined in the following form:

$$DOM = 1 - \frac{1}{\xi}\sqrt{\sum_{i=1}^{n}\frac{(\varphi_i - \xi)^2}{n}}, \tag{4}$$

where φ_i and n are the mass fraction of fluid A in the ith cell and the total number of cells, respectively; $\xi = 0.5$, which means equal mixing of the two fluids. The MEC is widely used to evaluate the effectiveness of the present micromixer and is defined by combining the pressure load and DOM in the following form [33,34]:

$$MEC = \frac{\Delta p / \rho u_{mean}^2}{DOM \times 100}, \tag{5}$$

where u_{mean} is the average velocity at the outlet, and Δp is the pressure load between the inlet and the outlet.

The aqueous fluids flowing into the two inlets were assumed to have the same properties, the same as the physical properties of the water. Therefore, the density, diffusion constant, and viscosity of the fluid are ρ = 997 kg/m^3, D = 1.0 × 10^{-10} m^2 s^{-1}, and ν = 0.89 × 10^{-6} m^2 s^{-1}, respectively. The corresponding Schmidt (Sc) number is approximately 10^4 (the ratio of the kinetic viscosity and the mass diffusivity of the fluid). The Reynolds number was defined as $\mathrm{Re} = \frac{\rho U_{mean} d_h}{\mu}$, where ρ, U_{mean}, d_h, and μ mean the density, the mean velocity at the outlet, the hydraulic diameter of the outlet channel, and the dynamic viscosity of the fluid, respectively.

4. Validation of the Numerical Study

Accurate numerical simulation is still a challenging problem to study the mixing in micromixers, especially for high Sc numbers. Many research papers do not deal with this computational issue. In general, the numerical diffusion can deteriorate the accuracy of the simulated results for high Sc number simulations. To obtain a quantitatively more rigorous numerical solution, we could use either a particle-based simulation method such as Monte Carlo method [35] or decrease the cell Peclet number for a grid-based method. Here, the cell Peclet number is defined as $Pe = \frac{U_{cell} l_{cell}}{D}$, where U_{cell} and l_{cell} are the local flow velocity and cell size, respectively. However, these methods are computationally expensive to adopt in a study such as this paper. As a practical remedy, most numerical studies prefer a detailed study of grid independence by comparing with experimental data [15,36].

The present numerical approach was validated by simulating the micromixer examined by Chung et al. [24]. Figure 3 shows a schematic diagram of the micromixer, and it consists of three mixing units. Each mixing unit contains three rectangular baffles and the associated gaps; the thickness of the baffles is 80 μm. As each baffle is shorter than the micromixer width, a gap is created. The first two baffles form a gap in the center while the third baffle makes two gaps around its edges as shown in Figure 3. The width of the inlet 1 and two side inlets, inlet 2 and inlet 3, are 400 μm and 200 μm, respectively. The depth of the micromixer is 130 μm.

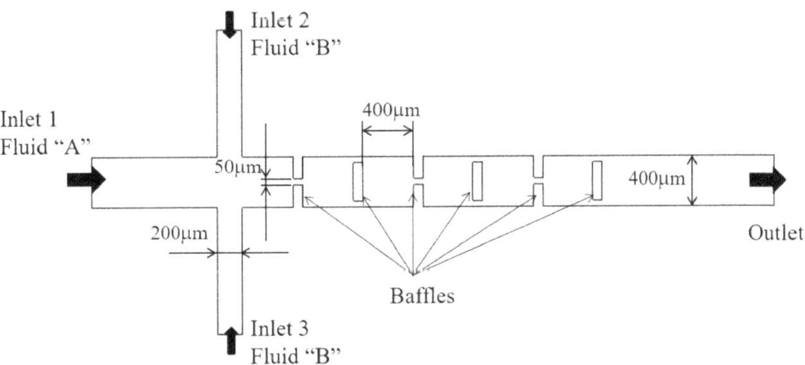

Figure 3. Diagram of the micromixer experimented by Chung et al. [24].

The density, diffusion constant and viscosity of the fluid are ρ = 997 kg/m^3, D = 3.6 × 10^{-10} m^2 s^{-1}, and ν = 0.89 × 10^{-6} m^2 s^{-1}, respectively. Therefore, the Schmidt (Sc) number is approximately 40,000. The simulation was carried out and compared with the corresponding experimental data for Reynolds numbers Re = 60. Here, the Reynolds number is defined as $\mathrm{Re} = \frac{\rho U_{mean} d_h}{\mu}$, where ρ, U_{mean}, d_h, and μ indicate the density, the mean velocity at the outlet, the hydraulic diameter of the outlet channel ($d_h = 196.2$ μm), and the dynamic viscosity of the fluid, respectively. Structured hexahedral cells were used to mesh the computational domain; the total number of cells is about 3.75 million.

Figure 4 compares the present simulation images with the corresponding experimental data reported by Chung et al. [24]. The mixing images at the two different depth show that the mixing process is quite depth-dependent: a strong mixing in the cross-flow direction. The comparison confirms that the present numerical simulation captures all the important mixing features such as the formation of vortices around short baffles.

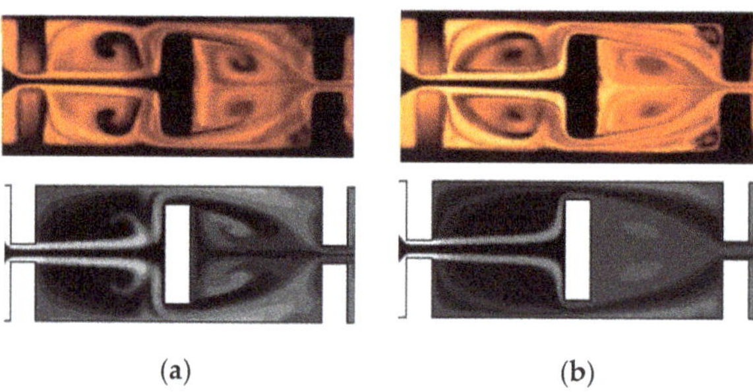

(**a**) (**b**)

Figure 4. Comparison of the mixing images of the first two chambers for Re = 60: (**a**) At z = 32.5 μm and (**b**) At z = 65 μm.

Prior to the present numerical study, an additional set of preliminary simulations was carried out to determine an appropriate cell size for the present micromixer. For this study, the edge size of cells was varied from 4.5 μm to 6 μm for three modified Tesla units. The corresponding number of mesh varies from 1.8×10^6 to 3.8×10^6. The simulation was carried out for Re = 0.5. Figure 5 shows the dependence of the calculated DOM on the edge size. The deviation of 5 μm solution from that of 4.5 μm is about 1%. Therefore, 5 μm is small enough to obtain grid independent solutions.

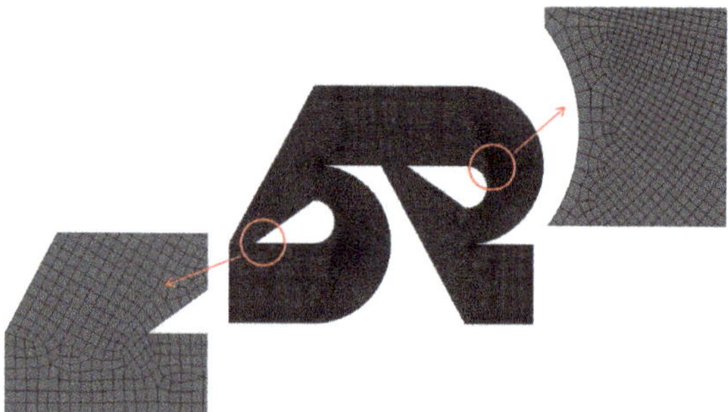

Figure 5. Grid for a modified Tesla mixing unit.

Using the numerical solutions, the grid convergence index (GCI) was also calculated to quantify the uncertainty of grid convergence [37,38]. According to the Richardson extrapolation methodology, the GCI is calculated as follows:

$$GCI = F_s \frac{|\varepsilon|}{r^p - 1}, \tag{6}$$

$$\varepsilon = \frac{f_{coarse} - f_{fine}}{f_{fine}}, \tag{7}$$

where F_s, r, and p are the safety factor of the method, grid refinement ratio, and the order of accuracy of the numerical method, respectively. f_{coarse} and f_{fine} are the numerical results obtained with a coarse grid and fine grid, respectively. F_s was specified at 1.25 as suggested by Roache [37]. For the edge size of 4.5 μm, 5 μm, and 6 μm, the corresponding number of nodes are 3.8×10^6, 2.98×10^6, and 1.8×10^6 for three Tesla mixing units, respectively; and 5.9×10^6, 4.4×10^6, and 2.6×10^6 for five mixing units, respectively. As a result, the GCI of the computed DOM is reduced from 5.7% to 1.1%. Therefore, the edge size of 5 μm was chosen to obtain the present numerical solutions.

According to Okuducu et al. [39], the accuracy of numerical solutions is also dependent on the type of cells. Structured hexahedral cells show the most reliable numerical solution, in comparison with tetrahedral and prism cells. In this paper, most cells were generated to be hexahedral as can be seen in Figure 5. The number of prism cells was minimized; refer to the red circle in Figure 5.

5. Results and Discussion

The present micromixer with tip clearance was simulated to assess its mixing performance by comparing with that of the Tesla micromixer without any tip clearance for Reynolds numbers from 0.1 to 80. The velocity at the two inlets was specified as uniform in the range from 0.2512 mm/s to 200.96 mm/s. Therefore, the corresponding volume flow rates range from 1.2 μL/min to 964.6 μL/min. The mixing performance was evaluated in terms of the DOM at outlet and the corresponding MEC.

Figure 6 shows the DOM of the modified Tesla micromixer with tip clearance h = 60 μm against that of no tip clearance; N indicates the number of the modified Tesla mixing units. A noticeable enhancement of the DOM is observed in the range of the Reynolds numbers Re < 50, and the amount of improvement increases with the number of the modified Tesla mixing units. For the Reynolds number of Re = 20 and N = 4, the DOM with tip clearance h = 60 μm is 24% higher than that with no tip clearance. When the Reynolds number is larger than about 50, the tip clearance allows a more efficient operation in terms of the required pressure load. For example, the DOM of tip clearance h = 40 μm for Re = 80 and N = 5 shows almost the value with the case of no tip clearance while it reduces the pressure load by about 8%.

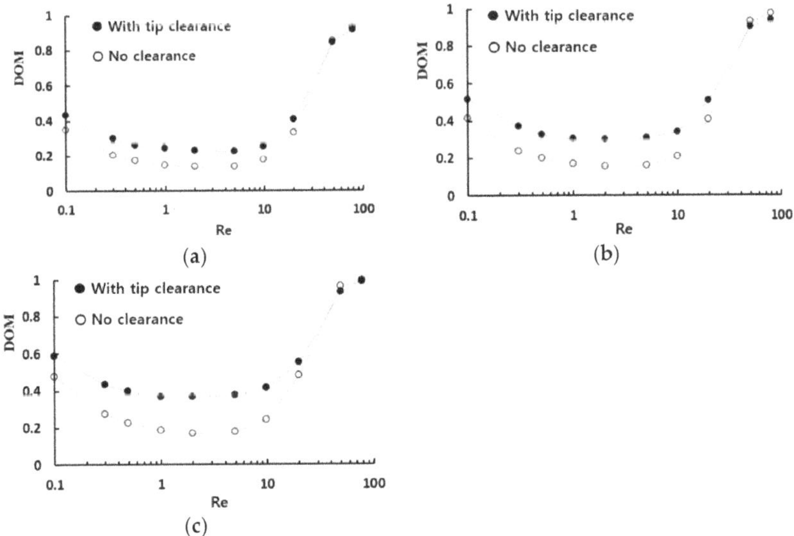

Figure 6. Enhancement of the DOM by tip clearance: (**a**) N = 3, (**b**) N = 4, and (**c**) N = 5.

Figure 7 shows the mixing performance map in terms of the DOM versus the required pressure load for four Reynolds numbers, Re = 0.1, 1, 10, and 50. The dotted line in the figure indicates the variation of the DOM for the case with no tip clearance. The DOM shows a linear relationship with the number of the mixing units. It means that the DOM can be improved linearly at the expense of the pressure load between the inlets and outlet, increasing the number of mixing units. On the other hand, for the case of tip clearance, all the DOM except for Re = 50 show an additional increment of the DOM from that of the modified Tesla micromixer. For example, the DOM of tip clearance h = 60 μm is 94% higher than that of the case of no tip clearance for Re = 5 and N = 4. At the same time, the required pressure load is 12% reduced. It is also noteworthy that the height of tip clearance is optimized in terms of the DOM. The optimum value of h is roughly h = 60 μm for most cases, and it is about 33% of the present micromixer depth. As the Reynolds number is increased to larger than about 50, the effects of tip clearance become less significant. It is associated with twis iso symmetric counter-rotating vortices formed at the outlet branch of each modified Tesla unit; refer to the vortex zone in Figure 1a. This flow characteristic is also reported by Hossain et al. [17] and is described in detail later.

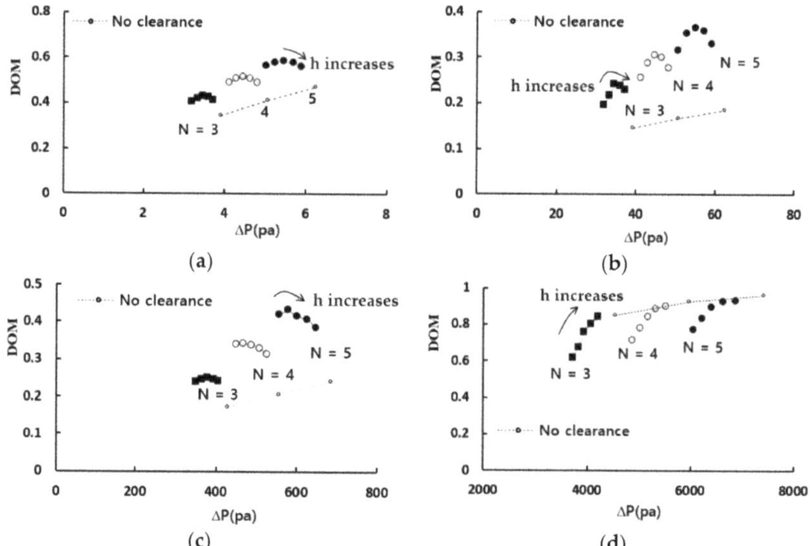

Figure 7. Mixing performance map: (**a**) Re = 0.1, (**b**) Re = 1, (**c**) Re = 10, and (**d**) Re = 50.

In the low Reynolds number regime of Re $\leq$ 1, the molecular diffusion dominates mixing process so that a straight channel is a good reference to compare with. Figure 8 compares the mixing evolution of the present micromixer with that of a straight channel; the DOM was obtained just after each Tesla mixing unit. The micromixers based on the modified Tesla mixing units show a noticeable mixing enhancement from the straight channel, and the enhancement increases with the number of mixing units. For the Reynolds number of Re = 0.1, the mixing enhancement of the modified Tesla mixing unit after one mixing unit is 63% and increases to 75% after five mixing units. The tip clearance results in an additional enhancement. It is 13% after one mixing unit and increases to 25% after five mixing units. A similar enhancement of the DOM is observed for the Reynolds number of Re = 1.

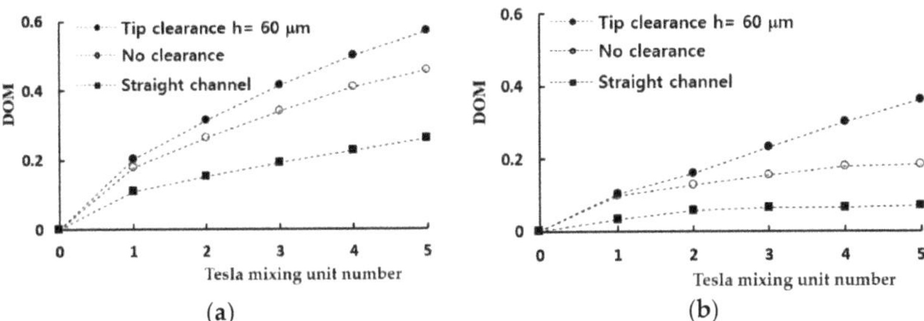

Figure 8. Comparison of mixing evolution in the axial direction: (**a**) Re = 0.1 and (**b**) Re = 1.

Figure 9 shows the mixing effectiveness of the present micromixer in comparison with that of the modified Tesla micromixer. In the figure, a smaller value of the MEC means more effective and requires a lesser pressure load to obtain the same degree of mixing. Therefore, the tip clearance of the present micromixer is found to reduce significantly the required pressure load for all Reynolds numbers, even though the effect is quite limited for the Reynolds number of Re = 50. Another interesting thing is that the tip clearance is optimized to minimize the MEC. The optimum value is about 60 μm, and close to the value for maximizing the DOM shown in Figure 6. This suggests that the size of tip clearance can be determined to enhance the DOM as well as minimize the required pressure load.

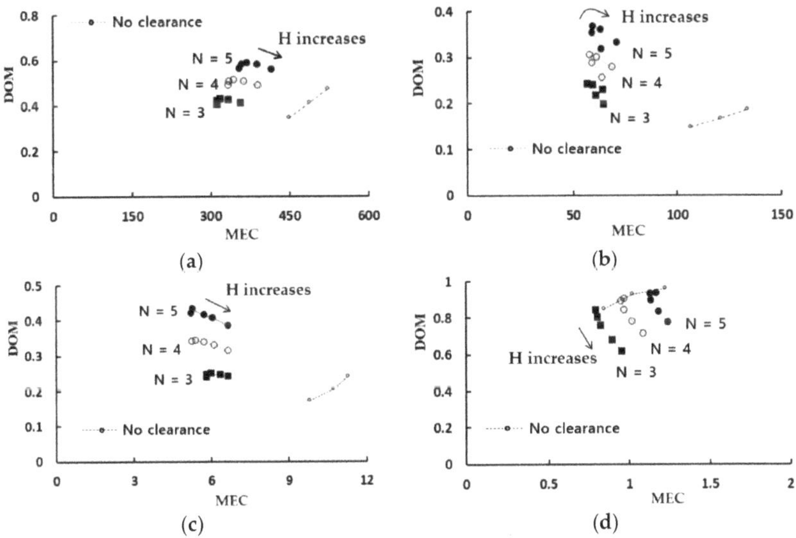

Figure 9. Mixing effectiveness map: (**a**) Re = 0.1, (**b**) Re = 1, (**c**) Re = 10, and (**d**) Re = 50.

Figure 10 shows the increment of the DOM obtained by embedding tip clearance into the modified Tesla mixing units. The vertical axis indicates the increment of the DOM obtained with tip clearance from the DOM with tip clearance. Therefore, a negative value means that the tip clearance affects the DOM in a negative way. For Re = 0.1, the effects of tip clearance is significant throughout the whole mixing unit; in the molecular dominance regime of mixing. On the contrary, the tip clearance in the first mixing unit takes little or negative effects on mixing for Re = 1, 5 and 10; in the intermediate range of the Reynolds number. The increment of the DOM increases as it goes downstream. This suggests that the flow characteristics associated with the mixing enhancement in the two ranges are different from each other.

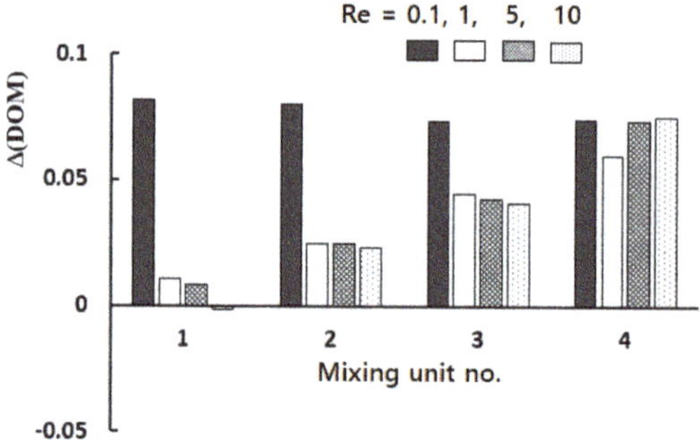

Figure 10. Increment of the DOM by tip clearance.

The mixing enhancement mechanism is analyzed further in the three different regimes of mixing: molecular diffusion dominance, transition, and convection dominance. Figure 11 shows the concentration contours on two planes for the Reynolds number of Re = 0.1: at z = 30 μm and 100 μm. Here, the plane at z = 100 μm corresponds to the mid-depth plane of the present micromixer while the plane at z = 30 μm is in the middle of the tip clearance. The case of tip clearance h = 60 μm shows a wider interface between the two fluids on the plane at z = 30 μm; this means that the mixing along the interface is more active. Figure 12 plots the concentration and the velocity vector on the yz plane for Re = 0.1 and h = 60 μm. The flow through tip clearance drags the interface of the sub-stream 1 and connects it to the interface of the sub-stream 2: drag and connection of the interface by tip clearance flow. This kind of flow characteristic is seen both on section 1 and 2 while the two interfaces for the case of no tip clearance is separated by the structure, as seen in Figure 11b. This explains why the increment of the DOM for Re = 1 is significant throughout the whole mixing unit. Therefore, the drag and connection of interface is the main flow mechanism for the mixing enhancement caused by the tip clearance in the molecular diffusion regime of mixing.

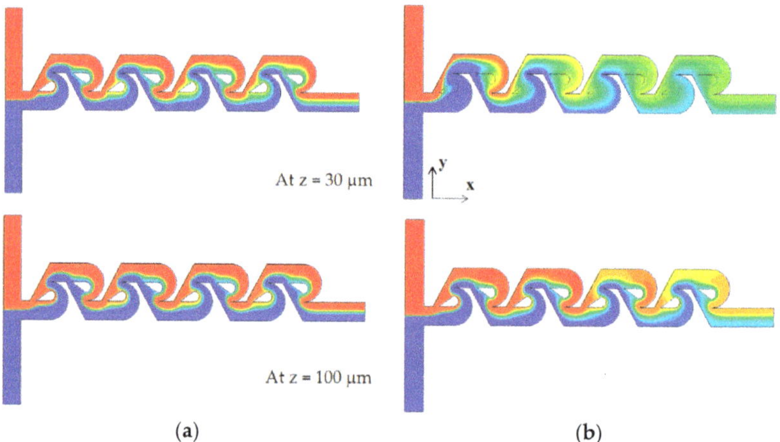

Figure 11. Comparison of concentration contours on the xy planes for Re = 0.1: (**a**) no clearance and (**b**) with tip clearance h = 60 μm.

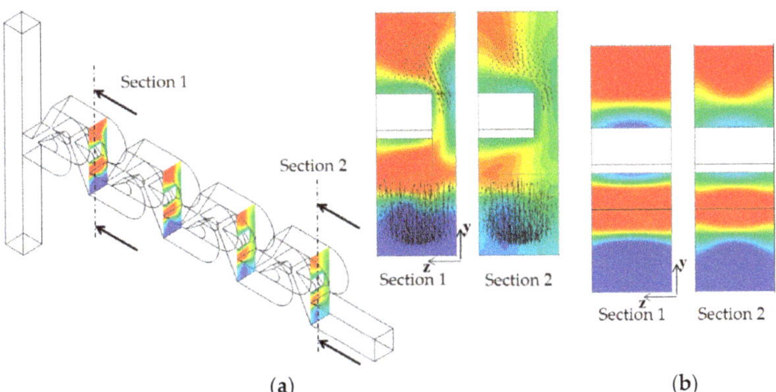

Figure 12. Evolution of mixing and velocity vector on the yz planes for Re = 0.1: (**a**) tip clearance h = 60 μm and (**b**) no tip clearance.

As the Reynolds number increases, the convective mixing becomes significant, and the mixing enhancement by tip clearance is obtained in a different way. Figure 13 compares the concentration contours on the xy planes for the Reynolds number of Re = 5. For the case of tip clearance, the interface between the two fluids appears wavier and multiple times and is a result of the mixing enhancement. The difference caused by tip clearance is more obvious on the plane at z = 100 μm. Figure 14 compares the concentration contours with the corresponding velocity vector on the two yz planes for Re = 5: section 1 and 2. For the case of no tip clearance, a vortex forms on the cross-section 1, and develops into a pair of two counter-rotating vortices on the cross-section 2.

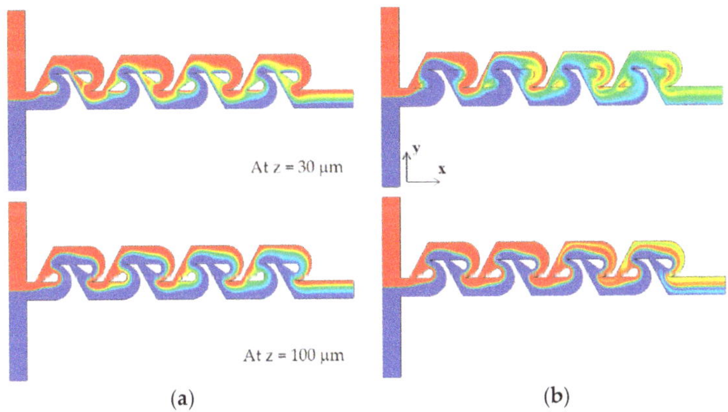

Figure 13. Comparison of concentration contours on the xy planes for Re = 5: (**a**) no clearance and (**b**) with tip clearance h = 60 μm.

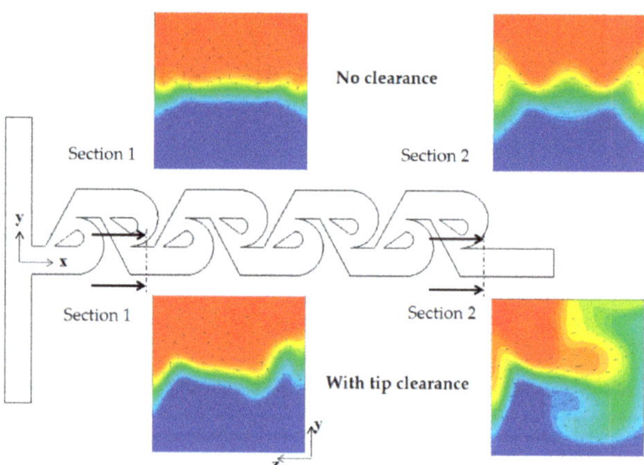

Figure 14. Concentration contours with velocity vector on the yz planes for Re =5.

Contrarily, the vortex flow on section 1 is agitated and developed in a different way on section 2 for the case of tip clearance. This different flow evolution suggests that the flow passing through tip clearance interacts with the secondary flow generated on section 1 and develops into an agitated vortex flow on section 2. The agitated vortex flow plays a significant role in the mixing enhancement for the Reynolds numbers $1 \leq Re < 50$. Figure 15 compares the concentration contours with the corresponding velocity vector on the two yz planes for Re = 10: section 1 and 2. For the case of tip clearance, the flow through tip clearance agitates the secondary flow generated in the vortex zone and confirms the agitated vortex flow to cause the mixing enhancement.

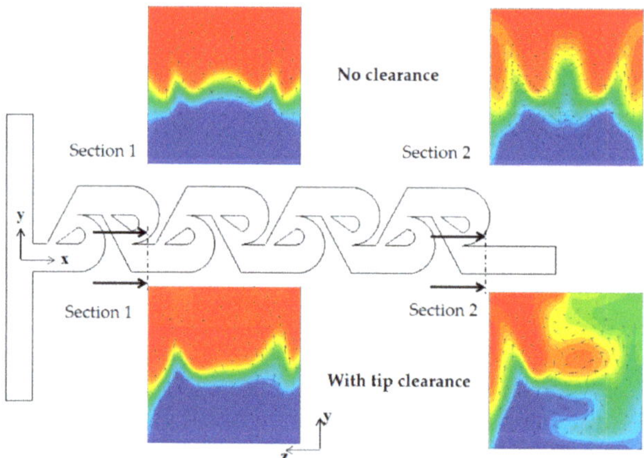

Figure 15. Concentration contours with velocity vector on the yz planes for Re = 10.

As the Reynolds number increases further, the vortex flow formed in the vortex zone of each Tesla mixing unit develops into a pair of strong counter-rotating vortices. They are separated distinctly and become stronger. Figure 16 compares the concentration contours on the two xy planes for the Reynolds number of Re = 50. Comparing the concentration contours on the plane at z = 100 μm, the mixing seems to be processed in a similar way, even though there is a little difference locally. On the contrary, the concentration contours on the plane at z = 30 μm show a more vivid difference between them. The case of tip clearance shows a more complicated pattern of the interface. This is caused by the asymmetric geometry due to tip clearance. This flow pattern is observed on both cross sections of section 1 and 2. This difference of interface pattern suggests that the influence of tip clearance is localized, and there is another significant mechanism of mixing. Figure 17 compares the concentration contours with velocity vector on the two yz planes. It shows that there are two distinct counter-rotating vortices on both planes of section 1 and 2, irrespective of tip clearance. They seem almost symmetric even for the case of tip clearance. This suggests that the mixing is mainly governed by the two counter-rotating vortices for the Reynolds numbers Re $\geq$ 50; this pair of counter-rotating vortices was also reported in the previous study [17,21].

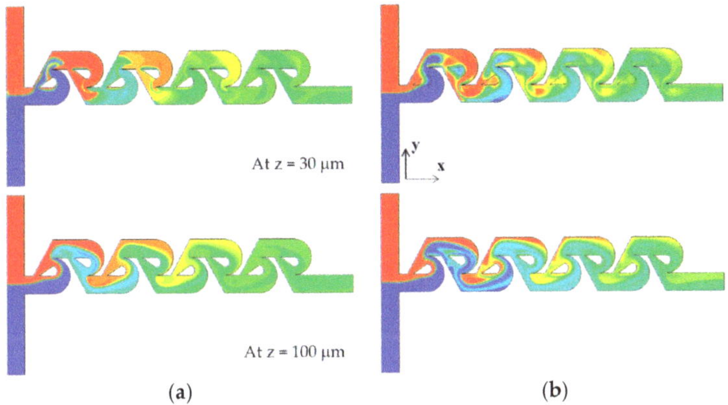

Figure 16. Comparison of concentration contours on the xy planes for Re = 50: (**a**) no clearance and (**b**) with tip clearance h = 60 μm.

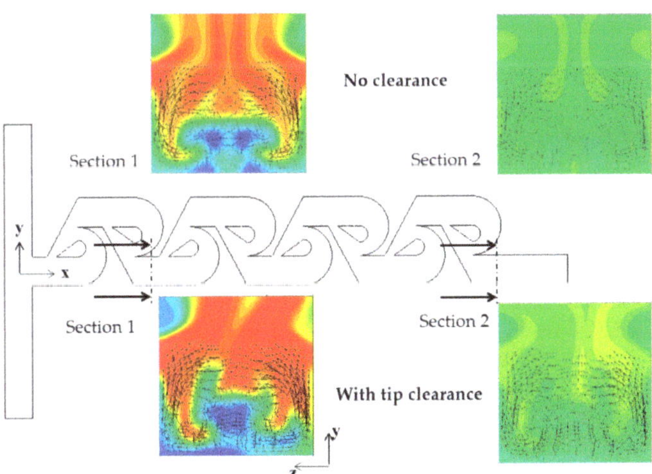

Figure 17. Concentration contours and velocity vector on the yz planes for Re = 50.

Figure 18 shows how the presence of tip clearance affects the vortex patterns and mixing performance in the present micromixer for the Reynolds of Re = 80. At the cross section 1 and 2, a pair of vortices seems formed, and this is generated as the flow follows the circular passage of the micromixer, at the first and last Tesla mixing unit. It implies that the centrifugal force plays a significant role. Another interesting thing is that the vortex pattern is little affected as long as the tip clearance remains about h $\leq$ 60 μm. On the other hand, the vortex close to the tip clearance zone (lower vortex) was observed to have noticeably shrunk for the tip clearance h = 70 μm, as can be seen in Figure 18a. This suggests that the convection flow is strong enough to localize the effects of tip clearance as long as the tip clearance is smaller than about h = 70 μm. Another pair of vortices are seen at section 3 and 4, which are located around the outlet of the first and last Tesla mixing units.

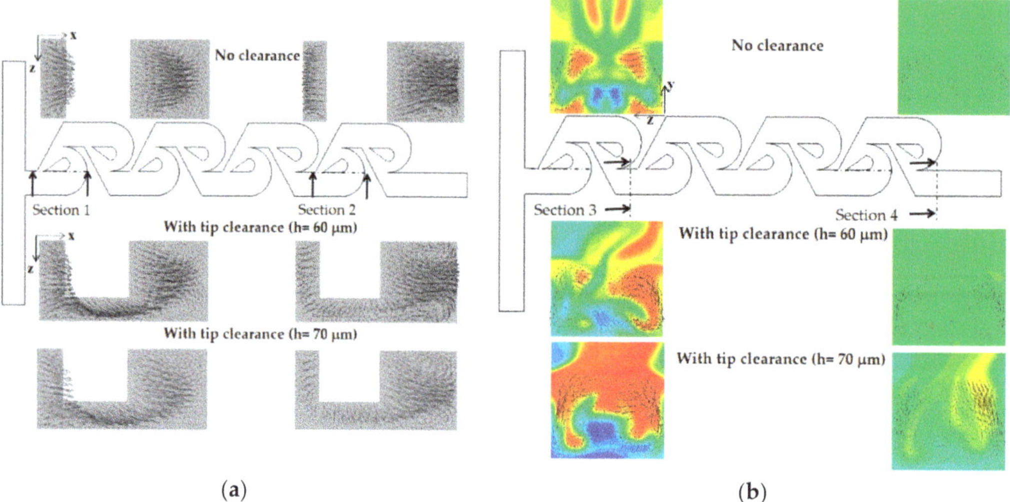

Figure 18. Concentration contours and velocity vector on the several planes for Re = 80: (**a**) velocity vector (**b**) concentration contours.

Accordingly, the flow characteristics such as a pair of vortices play a role as a governing mechanism in the convection-dominant regime of mixing. As a result, the concentration on plane 4 for no tip clearance and h = 60 μm is almost identical, as can be seen in Figure 18b. Therefore, the presence of tip clearance contributes a little to the mixing enhancement for the Reynolds numbers Re $\geq$ 50 as long as the tip clearance remains smaller than about h = 70 μm.

Figure 19 shows how the mixing evolves throughout the mixing units for Re = 50. Unlike the case of no tip clearance in Figure 19a, the case of tip clearance in Figure 19b shows asymmetric concentration contours on the upper section of yz planes of section 1 and 2; the upper section corresponds to the sub-stream 1. However, symmetry seems to be recovered mostly on the lower section of the yz planes; refer to the box of red dotted lines in the figure. The lower section corresponds to the yz plane in sub-stream 2. This recovery is attributed to the two symmetric counter-rotating vortices depicted in Figure 15 and suggests that the effects of tip clearance are localized for Re = 50. The two counter-rotating vortices generated in the vortex zone are strong enough to recover the asymmetric mixing pattern in the tip clearance zone. No significant increment of the DOM is achieved by the tip clearance for Re $\geq$ 50, as shown in Figure 6.

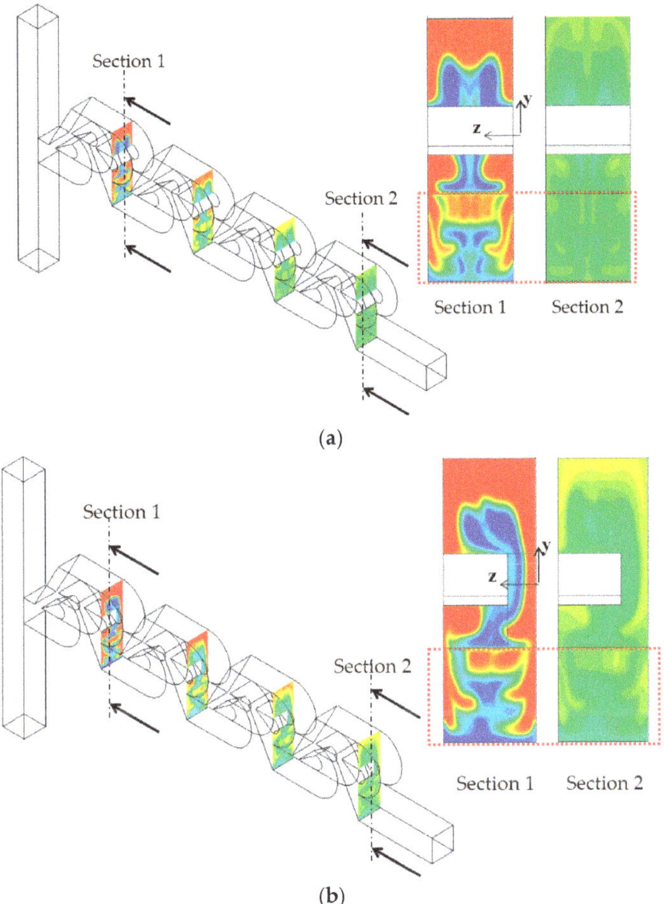

Figure 19. Evolution of mixing on the yz planes for Re = 50: (**a**) no tip clearance and (**b**) tip clearance h = 60 μm.

6. Conclusions

This paper studied numerically the effects of tip clearance on the mixing performance of the modified Tesla micromixer. The present micromixer consists of several modified Tesla mixing units, and each mixing unit has tip clearance above the wedge-shape divider. The numerical simulation was carried out for the Reynolds numbers $0.1 \leq Re \leq 80$ and three different numbers of mixing units: 3, 4 and 5. The mixing performance was assessed in terms of the DOM at the outlet and the required pressure load between the inlets and outlet. The mixing performance was simulated using the commercial software ANSYS® Fluent 2021 R2.

The effects of tip clearance were found noticeably over a wide range of the Reynolds numbers, $Re < 50$. For example, the DOM of tip clearance h = 60 μm is 94% higher than that with no tip clearance for Re = 5 and N = 4, and in addition, the required pressure load is 12% reduced. The height of tip clearance is optimized in terms of the DOM, and the optimum value is roughly h = 60 μm for most cases. It corresponds to 33% of the present micromixer depth. The tip clearance is also optimized to minimize the MEC. The optimum value is close to that for maximizing the DOM. The size of tip clearance can be determined to enhance the DOM as well as minimize the required pressure load.

The mixing enhancement due to tip clearance was obtained by different mixing mechanisms in accordance with the Reynolds number. In the molecular diffusion regime of mixing, Re $\leq$ 1, the mixing enhancement is obtained mainly by connection of the two interfaces in sub-stream 1 and sub-stream 2. The flow through the tip clearance drags the interface in sub-stream 1 and connects it to the interface in sub-stream 2. This flow characteristic causes the mixing to happen actively and the interface to become wider in the tip clearance zone.

The mixing enhancement in the intermediate range of the Reynolds number, 1 < Re $\leq$ 50, is attributed to the interaction of the flow through tip clearance and the secondary flow in the vortex zone of each Tesla mixing unit. Unlike for the case of no tip clearance, the flow through tip clearance agitates the secondary flow formed in the vortex zone of each Tesla mixing unit, and it leads to an increment in the DOM.

When the Reynolds number is larger than about 50, vortices are formed at various locations and drive the mixing in the modified Tesla micromixer. For the Reynolds number of Re = 80, a pair of vortices is formed around the inlet and outlet of each Tesla mixing unit. This vortex pattern is little affected by the presence of tip clearance as long as the tip clearance remains smaller than about h = 70 μm. It plays a role as a governing mechanism for the present micromixer in the convection-dominant regime of mixing. As a result, the DOM at the outlet is little enhanced by the presence of tip clearance. The tip clearance contributes only to reduce the required pressure load for the same value of the DOM.

The tip clearance embedded into the modified Tesla micromixer was shown to improve the mixing performance over a wide range of the Reynolds numbers. The improvement of mixing performance is achieved in terms of the DOM enhancement as well as the reduction of the corresponding pressure load. The mixing enhancement mechanism is dependent on the magnitude of the Reynolds number. The tip clearance is easily realized by lowering the height of the wedge-shape divider of the modified Tesla micromixer.

Author Contributions: Simulation, M.J.; formal analysis, D.-J.K.; visualization, M.J.; writing—original draft, D.-J.K. All authors have read and agreed to the published version of the manuscript.

Funding: This work was supported by a BOKUK© 2022 Research Grant.

Conflicts of Interest: The authors declare no conflict of interest.

References

1. Sackmann, E.K.; Fulton, A.I.; Beebe, D.J. The present and future role of microfluidics in biomedical research. *Nature* **2014**, *507*, 181–189. [CrossRef] [PubMed]
2. Lee, C.Y.; Chang, C.L.; Wang, Y.N.; Fu, L.M. A review on micromixers. *Micromachines* **2017**, *8*, 274.
3. Liu, L.; Cao, W.; Wu, J.; Wen, W.; Chang, D.C.; Shang, P. Design and integration of an all in one biomicrofluidic chip. *Biomicrofluidics* **2008**, *2*, 034103. [CrossRef]
4. Zhang, J.; Yan, S.; Yuan, D.; Alici, G.; Nguyen, N.T.; Wakiani, M.E.; Li, W. Fundamentals and applications of inertial microfluidics: A review. *Lab Chip* **2016**, *16*, 10–34. [CrossRef]
5. Huang, P.H.; Xie, Y.; Ahmed, D.; Rufo, J.; Nama, N.; Chen, Y.; Chan, C.Y.; Huang, T.J. An acoustofluidic micro-mixer based on oscillating sidewall sharp-edges. *Lab Chip* **2013**, *13*, 3847–3852. [CrossRef]
6. Mao, L.; Koser, H. Overcoming the Diffusion Barrier: Ultrafast microscale mixing via ferrofluids. In Proceedings of the 2007 International Solid-State Sensors, Actuators and Microsystems Conference (TRANSDUCERS 2007), Lyon, France, 10–14 June 2007; pp. 829–1832.
7. Wu, Y.; Ren, Y.; Tao, Y.; Hu, Q.; Jiang, H. A novel micromixer based on the alternating current flow field effect transistor. *Lab Chip* **2016**, *17*, 186–197. [CrossRef]
8. Huang, K.R.; Chang, J.S.; Chao, S.D.; Wung, T.S.; Wu, K.C. Study of active micro-mixer driven by electrothermal force. *Jpn. J. Appl. Phys.* **2012**, *51*, 047002. [CrossRef]
9. Wang, X.; Ma, X.; An, L.; Kong, X.; Xu, Z.; Wang, J. A pneumatic micromixer facilitating fluid mixing at a wide range of flow rate for the preparation of quantum dots. *Sci. China Chem.* **2012**, *56*, 799–805. [CrossRef]
10. Stroock, A.D.; Dertinger, S.K.; Ajdari, A.; Mezic, I.; Stone, H.A.; Whitesides, G.M. Patterning flows using grooved surfaces. *Science* **2002**, *295*, 647–651. [CrossRef]
11. Kang, D.J. Effects of channel wall twisting on the mixing in a T-shaped microchannel. *Micromachines* **2019**, *11*, 26. [CrossRef]
12. Somashekar, V.; Olse, M.; Stremler, M.A. Flow structure in a wide microchannel with surface grooves. *Mech. Res. Comm.* **2009**, *36*, 125–129. [CrossRef]

13. Kang, D.J. Effects of baffle configuration on mixing in a T-shaped microchannel. *Micromachines* **2015**, *6*, 765–777. [CrossRef]
14. Kang, D.J.; Song, C.H.; Song, D.J. Junction contraction for a T-shaped microchannel to enhance mixing. *Mech. Res. Comm.* **2012**, *33*, 739–746.
15. Raza, W.; Kim, K.Y. Asymmetric split-and-recombine micromixer with baffles. *Micromachines* **2019**, *10*, 844. [CrossRef] [PubMed]
16. Makhsuda, J.; Kang, D.J. Optimal combination of mixing units using the design of experiments method, a Cross-Channel Split-and-Recombine Micro-Mixer Combined with Mixing Cell. *Micromachines* **2021**, *12*, 985.
17. Hossain, S.; Ansari, M.A.; Husain, A.; Kim, K.-Y. Analysis and optimization of a micromixer with a modified Tesla structure. *Chem. Eng. J.* **2010**, *158*, 305–314. [CrossRef]
18. Makhsuda, J.; Kang, D.J. Mixing performance of a passive micromixer with mixing units stacked in the cross flow direction. *Micromachines* **2021**, *12*, 1350.
19. Zhou, T.; Xu, Y.; Liu, Z.; Joo, S.W. An enhanced one-layer passive microfluidic mixer with an optimized lateral structure with the Dean effect. *J. Fluids Eng.* **2015**, *137*, 091102. [CrossRef]
20. Makhsuda, J.; Kang, D.J. Mixing enhancement of a passive micromixer with submerged structures. *Micromachines* **2022**, *13*, 1050.
21. McDonough, J.R.; Oates, M.F.; Law, R.; Harvey, A.P. Micromixing in oscillatory baffles flows. *Chem. Eng. J.* **2019**, *361*, 508–518. [CrossRef]
22. Raza, W.; Hossain, S.; Kim, K.Y. A review of passive micromixers with comparative analysis. *Micromachines* **2020**, *11*, 455. [CrossRef] [PubMed]
23. Hong, C.-C.; Choi, J.-W.; Ahn, C.H. A novel in plane passive microfluidic mixer with modified Tesla structures. *Lab Chip* **2004**, *4*, 109–113. [CrossRef]
24. Chung, C.K.; Shiha, T.R.; Changa, C.K.; Lai, C.W.; Wu, B.H. Design and experiments of a short-mixing-length baffled microreactor and its application to microfluidic synthesis of nanoparticles. *Chem. Eng. J.* **2011**, *168*, 790–798. [CrossRef]
25. Bazaz, S.R.; Mehrizi, A.A.; Ghorbani, S.; Vasilescu, S.; Asadnia, M.; Warkiani, M.E. A hybrid micromixer with planar mixing units. *RSC Adv.* **2018**, *8*, 33101–33120. [CrossRef] [PubMed]
26. Bartholomeusz, D.A.; Boutte, R.W.; Andrade, J.D. Xurography: Rapid prototyping of microstructures using a cutting plotter. *J. Microelectromech. Syst.* **2005**, *14*, 1364–1374. [CrossRef]
27. *ANSYS Fluent Tutorial Guide 2020 R2*; ANSYS Inc.: Canonsburg, PA, USA, 2020.
28. Rhoades, T.; Kothapalli, C.R.; Fodor, P. Mixing optimization in grooved serpentine microchannels. *Micromachines* **2020**, *11*, 61. [CrossRef] [PubMed]
29. Volpe, A.; Paie, P.; Ancona, A.; Osellame, R.; Lugara, P.M. A computational approach to the characterization of a microfluidic device for continuous size-based inertial sorting. *J. Phys. D Appl. Phys.* **2017**, *50*, 255601. [CrossRef]
30. Hessel, V.; Löwe, H.; Schönfeld, F. Micromixers—A review on passive and active mixing principles. *Chem. Eng. Sci.* **2006**, *60*, 2479–2501. [CrossRef]
31. Liao, Y.; Mechulam, Y.; Lassale-Kaiser, B. A millisecond passive micromixer with low flow rate, low sample consumption and easy fabrication. *Sci. Rep.* **2021**, *11*, 20119. [CrossRef]
32. Cook, K.J.; Fan, Y.F.; Hassan, I. Mixing evaluation of a passive scaled-up serpentine micromixer with slanted grooves. *ASME J. Fluids Eng.* **2013**, *135*, 081102. [CrossRef]
33. Ortega-Casanova, J. Enhancing mixing at a very low Reynolds number by a heaving square cylinder. *J. Fluids Struct.* **2016**, *65*, 1–20. [CrossRef]
34. Valentin, K.; Ekaterina, S.B.; Wladimir, R. Numerical and experimental investigations of a micromixer with chicane mixing geometry. *Appl. Sci.* **2018**, *8*, 2458.
35. Vikhansky, A. Quantification of reactive mixing in laminar micro-flows. *Phys. Fluids* **2004**, *16*, 4738–4741. [CrossRef]
36. Sarkar, S.; Singh, K.K.; Shankar, V.; Shenoy, K.T. Numerical simulation of mixing at 1-1 and 1-2 microfludic junctions. *Chem. Eng. Process.* **2014**, *85*, 227–240. [CrossRef]
37. Roache, P. Perspective: A method for uniform reporting of grid refinement studies. *ASCE J. Fluids Eng.* **1994**, *116*, 307–357. [CrossRef]
38. Roache, P. *Verification and Validation in Computational Science and Engineering*; Hermosa: Albuquerque, NM, USA, 1998.
39. Okuducu, B.; Aral, M.M. Performance analysis and numerical evaluation of mixing in 3-D T-shape passive micromixers. *Micromachines* **2018**, *9*, 210. [CrossRef] [PubMed]